Collins

Cambridge Lower Secondary

Science

STAGE 9: TEACHER'S GUIDE

Mark Levesley, Aidan Gill, Gemma Young, Sheila Tarpey, Beverly Rickwood, Nigel Saunders

William Collins' dream of knowledge for all began with the publication of his first book in 1819.
A self-educated mill worker, he not only enriched millions of lives, but also founded a flourishing publishing house.
Today, staying true to this spirit, Collins books are packed with inspiration, innovation and practical expertise. They place you at the centre of a world of possibility and give you exactly what you need to explore it.

Collins. Freedom to teach.
Published by Collins
An imprint of HarperCollins*Publishers*
The News Building
1 London Bridge Street
London
SE1 9GF

HarperCollins*Publishers*
1st Floor
Watermarque Building
Ringsend Road
Dublin 4
Ireland

10 9 8 7 6 5 4 3 2 1

ISBN 978-0-00-836436-6

British Library Cataloguing-in-Publication Data
A catalogue record for this publication is available from the British Library.

Updating and contributing authors: Mark Levesley, Aidan Gill, Gemma Young, Sheila Tarpey, Beverly Rickwood, Nigel Saunders
Contributing authors: Lucy Hawkins, Louise Smiles, Chris Meunier, Fran Eardley
Development editors: Amanda Harman, Lynette Woodward, Gillian Lindsey, Tony Wayte, Fiona McDonald, Sarah Binns
Project manager: Amanda Harman
Product manager: Joanna Ramsay
Content editor: Tina Pietron
Copyeditors: Debbie Oliver
Proofreader: Jan Schubert
Safety checker: Joe Jefferies
Illustrator: Jouve India Private Limited
Cover designer: Gordon MacGilp
Cover artwork: Maria Herbert-Liew
Internal designer: Jouve India Private Limited
Typesetter: Jouve India Private Limited
Production controller: Lyndsey Rogers
Printed and bound in the UK using 100% Renewable Electricity at CPI Group (UK) Ltd

The publishers gratefully acknowledge the permission granted to reproduce the copyright material in this book. Every effort has been made to trace copyright holders and to obtain their permission for the use of copyright material. The publishers will gladly receive any information enabling them to rectify any error or omission at the first opportunity.

Acknowledgements
The publishers wish to thank the following for permission to reproduce photographs. Every effort has been made to trace copyright holders and to obtain their permission for the use of copyright materials. The publishers will gladly receive any information enabling them to rectify any error or omission at the first opportunity.
(t = top, c = centre, b = bottom, r = right, l = left)

Print book:
p 176 Webspark/Shutterstock.

Practical notes:
11.2b(2) Fouad A. Saad/Shutterstock.

Worksheets:
2.1b La Gorda/Shutterstock, 3.5a t giedrius_b/Shutterstock, 3.5a bl Melinda Fawver/Shutterstock, 3.5a br Yasar Ugurlu/Shutterstock, 4.1a tl Cat Downie/Shutterstock, 4.1a tr Zhiltsov Alexandr/Shutterstock, 4.1a bl skywing/Shutterstock, 4.1a br David Carbo/Shutterstock, 4.1c(1) l FotoRequest/Shutterstock, 4.1c(1) r Warren Metcalf/Shutterstock, 12.1b t zeber/Shutterstock, 12.1b b Filip Bjorkman/Shutterstock, 13.1e(1) Graph by Robert Simmon, using CO2 data from the NOAA Earth System Research Laboratory and temperature data from the Goddard Institute for Space Studies, 14.2c VectorMine/Shutterstock.

Worksheets answers:
12.1b tinkivinki/Shutterstock,
14.2c Activity 14.5 VectorMine/Shutterstock.

Student's Book answers:
Activity 14.2 Nasky/Shutterstock,
Activity 14.5 VectorMine/Shutterstock.

FSC www.fsc.org
MIX
Paper from responsible sources
FSC™ C007454

This book is produced from independently certified FSC™ paper to ensure responsible forest management.

For more information visit:
www.harpercollins.co.uk/green

Contents

Downloads of editable learning episodes, worksheets, practical notes and answers are available at www.collins.co.uk/CambridgeInternationalDownloads. These are all supplied in Word format as well as pdf, so that teachers are able to edit the resources to suit their own teaching methods and lesson plans.

Introduction

Welcome to the Collins Cambridge Stage 9 Lower Secondary Science course, which has been written by a team of highly experienced authors and teachers to help you deliver an effective and successful Cambridge Lower Secondary Science Stage 9 curriculum framework.

Student's Book

The Student's Book provides full coverage of the Cambridge Lower Secondary Science curriculum framework and includes the following features:

'You will learn', a list of key concepts that will be covered in each topic.

Starting points that can be used as part of a quick baseline assessment. Previous ideas, terminology and skills are outlined to aid recall and help students to build links between topics.

Questions throughout, including **challenge questions**, for formative assessment (aiding continuous feedback and monitoring of progress).

Key terms and **glossary** that provide language support for students and help them to develop their use of scientific vocabulary.

Activities that encourage students to apply their knowledge to practical and research-based contexts. These can then be used as a basis for formatively assessing depth of knowledge and understanding of both topic material and of scientific enquiry skills.

Key facts and **Check your skills progress** that enable students to take ownership of their own learning (and identify specific areas of knowledge and understanding that still need attention).

End of chapter and End of stage reviews

These can be used for summative assessment and may also be used formatively. The mark schemes can be found as digital downloads accompanying this Teacher's Guide.

The End of chapter reviews start with quick questions on knowledge, and then move on to questions that test deeper understanding. These are designed to be used after a chapter has been completed, and can be used for formative assessment or as summative assessment (with marks being given, although note that each End of chapter review is not worth the same number of marks).

The End of stage review is designed to be used after all chapters in a stage have been completed. The End of stage review can be used in a number of ways:

- Ask students to answer the questions under test conditions (give students 45 minutes to complete questions worth 50 marks). You could give students the End of stage review to complete as a practice for a mock test.
- Alternatively, ask students to go through the questions, or a selection of them, and prepare mark schemes. Students then work together and amend their mark schemes as necessary, before comparing their mark schemes with the one given in the Teacher's Guide. Ask students which part of their mark schemes was most difficult to write, and use those topics for focused revision.
- Check the subject matter for one or two of the questions, and ask students to revise these topics for homework. In class, ask students to complete the questions. Checking how well students have done will provide feedback on lingering misconceptions and the quality of students' revision.
- Or, ask students to look at the mark scheme for a few of the questions (in the Teacher's Guide) and challenge them to write questions to match the mark scheme.

Teacher's Guide

This Teacher's Guide comprises two components – this printed book and downloadable material.

Printed book

The printed book mirrors the Student's Book topics and contains:

- **Chapter overviews** summarising all You will learn, learning objectives, and the match to the Cambridge Lower Secondary curriculum framework, along with notes on common misconceptions, glossary words and starting point activities as an informal baseline assessment to find out what students already know and to get students thinking in the right direction for what they are about to learn.
- **Activity plans** with resource lists, key concepts that will be covered, and detailed guidance including ideas for differentiation and extension.
- Teaching notes on **checking students' progress** for each chapter.

Downloadable material

Printable, editable resources in MS Word format are available as digital downloads at www.collins.co.uk/CambridgeInternationalDownloads so that you can tailor your lessons and activities. The content has been arranged in folders by section and then by topic, with all the materials you need so that you can quickly and easily prepare for your lessons:

- **Worksheets** for students – some are designed to support teaching and learning, others contain guidance for practical work.
- **Practical activities** including teacher demonstrations and student experiments.
- **Practical notes** – technician's or teacher's notes with lists of equipment, resources and set-up information are included with the relevant lesson topic. All the notes for each section are also collated in one handy document.
- A **planning guide** that maps out all the learning episodes, learning and thinking and working scientifically focus and lesson resources in a convenient chart.
- **Answers** to all the questions in the Stage 9 Student's Book, Workbook and worksheets.

Learning episodes

The Teacher's Guide has been matched topic by topic to the Student's Book to allow easy cross-referencing between the books. Each topic is introduced with a brief introduction and an overview of the activities in an easy-to-navigate table. Activity plans give guidance on delivering learning episodes that can be combined for longer lessons to give you complete flexibility and control.

Activity plans cover the following:

- **Learning episodes** aligned with the 'You will learn' lists given for each topic in the Student's Book
- **Common misconceptions** that students hold, which may need addressing
- **Resources** you will need for activities
- A detailed pick-up-and-teach **approach** for varied tasks
- Suggestions for **differentiation** and **extension** tasks
- **Homework suggestions**, including those from the associated Workbook

Activity plans are supported by **worksheets, practical notes** and all **answers**. The questions in the workbook and some of the worksheets give students an opportunity to continue to learn and test their knowledge and recall at home.

Worksheets

Worksheets provide engaging activities for students. Those that deal with practical work give step-by-step guidance for students to ensure that they carry out their investigations safely and successfully.

All worksheets are available in editable MS Word format as well as in PDF format.

Practical notes

Detailed **teacher's** or **technician's notes** are provided for lessons with lists of resources needed for practical activities and/or demonstrations, set-up instructions and safety notes.

For Today's Safety Information, which you will need to check before setting up any practical activity or demonstration, please refer to your employer's guidelines or visit: science.cleapss.org.uk (a subscription is necessary to access all CLEAPSS resources).

Practical notes for technicians or teachers appear as separate printable files.

It is always best for the technician or teacher to try out a practical first and determine the best concentrations to use; it is always a good idea to get a set of specimen results.

Answers

Answers to all questions in the Student's Book, Workbook and worksheets are available in MS Word and PDF format and are arranged by chapter.

Planning guide

The detailed **planning guide** provides a comprehensive overview of the Cambridge Lower Secondary Science curriculum framework, matching each learning episode to its learning and thinking and working scientifically focus and showing at a glance where the practical activities and demonstrations occur so that you can plan ahead.

Deeper learning at the heart of the course

This course has been designed with the idea of deep learning in mind. Deeper learning is best enhanced through a student's own engagement and exploration of a topic, reinforced with strong practice, consolidation and reflection. Deep learning is encouraged through our features and activities, Workbook practice and assessment features. The Student's Book identifies prior knowledge; indicating to students where content has been covered previously. To further this, the Teacher's Guide includes questions for discussion, encouraging students to make links between topics.

Activities throughout all the resources in this course encourage students to work in different ways; individually, in pairs and in groups, all of which contribute to the development of skills such as communication, teamwork, analysis and evaluation. The importance of science in a global context, and its importance in encouraging students to gain a deeper understanding of the subject, is highlighted throughout this course.

Cambridge Global Perspectives™

Cambridge Global Perspectives is a unique programme that helps learners develop outstanding transferable skills, including critical thinking, research and collaboration. Cambridge Global Perspectives is available for learners aged 5–19, from Cambridge Primary through to Cambridge Advanced. For Cambridge Primary and Lower Secondary learners, the programme is made up of a series of Challenges covering a wide range of topics, using a personal, local and global perspective. These curriculum frameworks are available to Cambridge schools but participation is voluntary.

However, whether or not your school is involved with Cambridge Global Perspectives, the six skills it focuses on are relevant to all students in the modern world. These skills are: research, evaluation, analysis, communication, collaboration and reflection.

Collins supports Cambridge Global Perspectives by including activities, tasks and projects in our Cambridge Primary and Lower Secondary courses, in order to develop and apply these skills. Activities that link to the skills are listed at the back of this book on page 228.

Registered Cambridge International Schools benefit from high-quality programmes, assessments and a wide range of support so that teachers can effectively deliver Cambridge Lower Secondary.

Visit www.cambridgeinternational.org/lowersecondary to find out more.

Mapping document

	Student's Book	Workbook	Teacher's Guide
	Topic	Topic	Topic
Thinking and Working Scientifically			
Models and representations			
9TWSm.01 Understand that models and analogies reflect current scientific evidence and understanding, and can change.	3.2, 3.5, 5.1, 5.2, 8.2, 12.1, 14.1, 14.2	5.1, 14.1, 14.2	3.2, 3.5, 5.1, 5.2, 8.2, 12.1, 14.1, 14.2
9TWSm.02 Describe some important models, including analogies, and discuss their strengths and limitations.	2.1, 3.3, 5.1, 6.1, 6.2, 7.4, 7.5, 8.1, 8.3, 10.1, 10.2, 11.2, 12.1, 12.2, 14.1, 14.2	2.1, 3.2, 5.1, 6.1, 12.1, 12.2	2.1, 3.3, 5.1, 6.1, 6.2, 7.4, 7.5, 8.1, 8.3, 10.1, 10.2, 11.2, 12.1, 12.2, 14.1, 14.2
9TWSm.03 Use symbols and formulae to represent scientific ideas.	5.1, 6.1, 7.2, 7.3, 7.4, 10.1, 10.2	5.1, 5.2, 7.2, 9.1, 10.1, 10.2	5.1, 6.1, 7.2, 7.3, 7.4, 10.1, 10.2
Scientific enquiry: purpose and planning			
9TWSp.01 Suggest a testable hypothesis based on scientific understanding.	1.1, 1.2, 8.3, 9.1, 11.1, 14.2	1.1, 1.2	1.1, 1.2, 8.3, 9.1, 11.1, 14.2
9TWSp.02 Describe examples where scientists' unexpected results from enquiries have led to improved scientific understanding.	2.1, 12.1		2.1, 12.1
9TWSp.03 Make predictions of likely outcomes for a scientific enquiry based on scientific knowledge and understanding	1.1, 1.2, 6.2, 7.1, 7.4, 7.5, 8.1, 8.2, 8.3, 9.1, 10.2, 11.1, 11.2, 12.1, 13.2, 14.1	7.1, 7.5, 9.1, 10.2, 12.1, 14.2	1.1, 1.2, 6.2, 7.1, 7.4, 7.5, 8.1, 8.2, 8.3, 9.1, 10.2, 11.1, 11.2, 12.1, 13.2, 14.1
9TWSp.04 Plan a range of investigations of different types to obtain appropriate evidence when testing hypotheses.	1.1, 1.2, 7.5, 8.3, 9.1, 11.1, 14.2	1.1, 6.3, 7.4, 7.5, 9.1, 10.2	1.1, 1.2, 6.3, 7.5, 8.3, 9.1, 11.1, 14.2
9TWSp.05 Make risk assessments for practical work to identify and control risks.	1.1, 2.1, 7.3, 7.5, 8.2, 8.3, 13.2, 14.2	1.1, 2.1, 7.3, 8.3, 12.2	1.1, 2.1, 7.3, 7.5, 8.2, 8.3, 13.2, 14.2

	Student's Book	Workbook	Teacher's Guide
	Topic	Topic	Topic
Carrying out scientific enquiry			
9TWSc.01 Sort, group and classify phenomena, objects, materials and organisms through testing, observation, using secondary information, and making and using keys.	5.2, 6.2, 9.1, 14.1, 14.2	6.2, 6.3, 14.1, 14.2	5.2, 6.2, 9.1, 14.1, 14.2
9TWSc.02 Decide what equipment is required to carry out an investigation or experiment and use it appropriately.	1.1, 6.3, 7.3, 13.2	7.3, 10.1	1.1, 6.3, 7.3
9TWSc.03 Decide when to increase the range of observations and measurements, and increase the extent of repetition, to give sufficiently reliable data.	1.1, 7.5		1.1, 7.5
9TWSc.04 Take appropriately accurate and precise measurements, explaining why accuracy and precision are important.	1.2, 7.1, 8.3, 10.1, 10.2		1.2, 7.1, 8.3, 10.1, 10.2
9TWSc.05 Carry out practical work safely, supported by risk assessments where appropriate.	1.1, 2.1, 7.1, 7.3, 7.4, 7.5, 8.2, 8.3, 10.1, 10.2, 12.1, 13.2, 14.1		1.1, 2.1, 7.1, 7.3, 7.4, 7.5, 8.2, 8.3, 10.1, 10.2, 12.1, 13.2, 14.1
9TWSc.06 Make an informed decision whether to use evidence from first-hand experience or secondary sources.	1.2, 3.3, 8.1		1.2, 3.3, 8.1
9TWSc.07 Collect, record and summarise sufficient observations and measurements, in an appropriate form.	1.1, 1.2, 3.1, 3.5, 6.3, 7.3, 7.4, 7.5, 8.2, 8.3, 9.1, 10.1, 10.2, 11.1		1.1, 1.2, 3.1, 3.5, 6.3, 7.3, 7.4, 7.5, 8.2, 8.3, 9.1, 10.1, 10.2, 11.1
Scientific enquiry: analysis, evaluation and conclusions			
9TWSa.01 Evaluate the strength of the evidence collected and how it supports, or refutes, the prediction.	1.1, 11.1, 12.1, 14.1, 14.2	6.3, 7.5, 10.1, 10.2, 14.1, 14.2	1.1, 11.1, 12.1, 14.1, 14.2
9TWSa.02 Describe trends and patterns in results, identifying any anomalous results and suggesting why results are anomalous.	1.1, 5.2, 7.4, 7.5, 11.1, 12.1, 13.2	1.1, 10.2, 11.1, 13.1	1.1, 5.2, 7.4, 7.5, 11.1, 12.1, 13.2

	Student's Book	Workbook	Teacher's Guide
	Topic	Topic	Topic
9TWSa.03 Make conclusions by interpreting results, explain the limitations of the conclusions and describe how the conclusions can be further investigated.	1.2, 2.1, 3.1, 8.3, 9.1, 10.1, 11.1, 14.1	1.1, 5.2, 9.1, 14.1	1.2, 2.1, 3.1, 8.3, 9.1, 10.1, 11.1, 14.1
9TWSa.04 Evaluate experiments and investigations, including those by others, and suggest improvements, explaining any proposed changes.	1.2, 4.1,14.1	2.1, 7.1, 7.3, 13.1	1.2, 4.1, 14.1
9TWSa.05 Present and interpret results, and predict results between the data points collected.	1.1, 3.1, 3.5, 4.1, 5.2	4.1, 5.2, 7.4, 7.5	1.1, 3.1, 3.5, 4.1, 5.2

	Student's Book	Workbook	Teacher's Guide
	Topic	**Topic**	**Topic**
Biology			
Structure and function			
9Bs.01 Describe the pathway of water and mineral salts from the roots to the leaves in flowering plants, including absorption in root hair cells, transport through xylem and transpiration from the surface of leaves.	1.2	1.2	1.2
9Bs.02 Describe the structure of the human excretory (renal) system and its function (limited to kidneys filtering blood to remove urea, which is excreted in urine).	2.1	2.1	2.1
9Bs.03 Know that chromosomes contain genes, made of DNA, and that genes contribute to the determination of an organism's characteristics.	3.2	3.2	3.2
Life processes			
9Bp.01 Describe the fusion of gametes to produce a fertilised egg with a new combination of DNA.	3.3	3.3	3.3
9Bp.02 Describe the inheritance of sex in humans in terms of XX and XY chromosomes.	3.3	3.3	3.3
9Bp.03 Describe variation within a species and relate this to genetic differences between individuals.	3.1	3.1, 3.2, 3.5	3.1, 3.2, 4.1
9Bp.04 Describe the scientific theory of natural selection and how it relates to genetic changes over time.	3.5	3.5	3.5
9Bp.05 Know that plants require minerals to maintain healthy growth and life processes (limited to magnesium to make chlorophyll and nitrates to make protein).	1.2	1.2	1.2

	Student's Book	Workbook	Teacher's Guide
	Topic	**Topic**	**Topic**
9Bp.06 Know that photosynthesis occurs in chloroplasts and is the process by which plants make carbohydrates, using the energy from light.	1.1	1.1	1.1
9Bp.07 Know and use the summary word equation for photosynthesis (carbon dioxide + water → glucose + oxygen, in the presence of light and chlorophyll).	1.1	1.1	1.1
9Bp.08 Discuss how fetal development is affected by the health of the mother, including the effect of diet, smoking and drugs.	3.4	3.4	3.4
Ecosystems			
9Be.01 Describe what could happen to the population of a species, including extinction, when there is an environmental change.	4.1	4.1	4.1

	Student's Book	Workbook	Teacher's Guide
	Topic	Topic	Topic
Chemistry			
Materials and their structure			
9Cm.01 Understand that the structure of the Periodic Table is related to the atomic structure of the elements and the Periodic Table can be used to predict an element's structure and properties.	5.1	5.1	5.1
9Cm.02 Understand that a molecule is formed when two or more atoms join together chemically, through a covalent bond.	6.1	6.1	6.1
9Cm.03 Describe a covalent bond as a bond made when a pair of electrons is shared by two atoms (limited to single bonds).	6.1	6.1	6.1
9Cm.04 Describe an ion as an atom which has gained at least one electron to be negatively charged or lost at least one electron to be positively charged.	6.1	6.1	6.1
9Cm.05 Describe an ionic bond as an attraction between a positively charged ion and a negatively charged ion.	6.1	6.1	6.1
Properties of materials			
9Cp.01 Understand that the groups within the Periodic Table have trends in physical and chemical properties, using group 1 as an example.	5.2	5.2	5.2
9Cp.02 Describe how the density of a substance relates to its mass in a defined volume.	6.3	6.3	6.3
9Cp.03 Calculate and compare densities of solids, liquids and gases.	6.3	6.3	6.3
9Cp.04 Know that elements and compounds exist in structures (simple or giant), and this influences their physical properties.	6.2	6.3	6.2

	Student's Book	Workbook	Teacher's Guide
	Topic	**Topic**	**Topic**
Changes to materials			
9Cc.01 Use word equations and symbol equations to describe reactions (balancing symbol equations is not required).	7.2	7.2	7.2, 7.3
9Cc.02 Identify examples of displacement reactions and predict products (limited to reactions involving calcium, magnesium, zinc, iron, copper, gold and silver salts).	7.4	7.4	7.4
9Cc.03 Describe how to prepare some common salts by the reactions of metals with acids, and metal carbonates with acids, and purify them, using filtration, evaporation and crystallisation.	7.3	7.3	7.3
9Cc.04 Describe the effects of concentration, surface area and temperature on the rate of reaction, and explain them using the particle model.	7.5	7.5	7.5
9Cc.05 Understand that in chemical reactions mass and energy are conserved.	7.1	7.1	7.1

	Student's Book	Workbook	Teacher's Guide
	Topic	**Topic**	**Topic**
Physics			
Forces and energy			
9Pf.01 Use density to explain why objects float or sink in water.	9.1	9.1	9.1
9Pf.02 Describe the difference between heat and temperature.	8.2	8.2	8.2
9Pf.03 Know that energy is conserved, meaning it cannot be created or destroyed.	8.1	8.1	8.1
9Pf.04 Know that thermal energy will always transfer from hotter regions or objects to colder ones, and this is known as heat dissipation.	8.2	8.2	8.2, 8.3
9Pf.05 Describe thermal transfer by the processes of conduction, convection and radiation.	8.2, 8.3	8.2	8.3
9Pf.06 Explain cooling by evaporation.	8.3	8.3	8.3
Light and sound			
9Ps.01 Draw and interpret waveforms, and recognise the link between loudness and amplitude, pitch and frequency.	11.1	11.1	11.1
9Ps.02 Use waveforms to show how sound waves interact to reinforce or cancel each other.	11.2	11.2	11.2
Electricity and magnetism			
9Pe.01 Describe how current divides in parallel circuits.	10.2	10.2	10.2
9Pe.02 Know how to measure current and voltage in series and parallel circuits, and describe the effect of adding cells and lamps.	10.1, 10.2	10.1, 10.2	10.1

	Student's Book	Workbook	Teacher's Guide
	Topic	**Topic**	**Topic**
9Pe.03 Calculate resistance (resistance = voltage / current) and describe how resistance affects current.	10.1	10.1, 10.2	10.1
9Pe.04 Use diagrams and conventional symbols to represent, make and compare circuits that include cells, switches, resistors (fixed and variable), ammeters, voltmeters, lamps and buzzers.	10.1, 10.2	10.1, 10.2	10.1, 10.2

	Student's Book	Workbook	Teacher's Guide
	Topic	**Topic**	**Topic**
Earth and Space			
Planet Earth			
9ESp.01 Explain the movement of tectonic plates in terms of convection currents.	12.2	12.2	12.2
9ESp.02 Explain why the jigsaw appearance of continental coasts, location of volcanoes and earthquakes, fossil record and alignment of magnetic materials in the Earth's crust are all evidence for tectonic plates.	12.1	12.1	12.1
Cycles on Earth			
9ESc.01 Describe the carbon cycle (limited to photosynthesis, respiration, feeding, decomposition and combustion).	13.1	13.1	13.1
9ESc.02 Describe the historical and predicted future impacts of climate change, including sea level change, flooding, drought and extreme weather events.	13.2	13.2	13.2
Earth in space			
9ESs.01 Describe the consequences of asteroid collision with the Earth, including climate change and mass extinctions.	14.1	14.1	14.1
9ESs.02 Describe the evidence for the collision theory for the formation of the Moon.	14.1	14.1	14.1
9ESs.03 Know that nebulae are clouds of dust and gas, and can act as stellar nurseries.	14.2	14.2	14.2

	Student's Book	Workbook	Teacher's Guide
	Topic	**Topic**	**Topic**
Science in Context			
9SIC.01 Discuss how scientific knowledge is developed through collective understanding and scrutiny over time.	3.2, 3.5, 5.2	14.2	3.2, 3.5, 5.2
9SIC.02 Describe how science is applied across societies and industries, and in research.	1.2, 3.1, 3.4, 7.2, 7.3, 7.4, 8.1, 11.1, 11.2, 13.2	13.2, 14.2	1.2, 3.1, 3.4, 7.2, 7.3, 7.4, 8.1, 11.1, 11.2, 13.2
9SIC.03 Evaluate issues which involve and/or require scientific understanding.	1.1, 4.1, 8.1, 11.2, 12.1, 13.2	14.1	1.1, 4.1, 8.1, 11.2, 12.1, 13.2
9SIC.04 Describe how people develop and use scientific understanding as individuals and through collaboration, e.g. through peer-review.	1.1, 3.2, 3.5, 5.2, 7.2		1.1, 3.2, 3.5, 5.2, 7.2
9SIC.05 Discuss how the uses of science can have a global environmental impact.	1.1, 1.2, 3.1, 4.1, 7.1, 7.3, 8.1, 12.1, 13.2	13.2	1.1, 1.2, 3.1, 4.1, 7.1, 7.3, 8.1, 12.1, 13.2

Learning episodes

Chapter 1 Photosynthesis and plant growth

Chapter overview

In this chapter students will learn about photosynthesis, what plants require to be able to do this, where they get these resources from, where photosynthesis happens and what the products are. They will also consider the role of water and mineral salts and how these affect plant growth.

Topic title	Number of 40-minute periods	Learning Objectives	Thinking and Working Scientifically Learning Objectives
1.1 Photosynthesis	3–4	9Bp.06 Know that photosynthesis occurs in chloroplasts and is the process by which plants make carbohydrates, using the energy from light 9Bp.07 Know and use the summary word equation for photosynthesis (carbon dioxide + water → glucose + oxygen, in the presence of light and chlorophyll)	9TWSp.01 Suggest a testable hypothesis based on scientific understanding 9TWSp.03 Make predictions of likely outcomes for a scientific enquiry based on scientific knowledge and understanding 9TWSp.04 Plan a range of investigations of different types to obtain appropriate evidence when testing hypotheses 9TWSp.05 Make risk assessments for practical work to identify and control risks 9TWSc.02 Decide what equipment is required to carry out an investigation or experiment and use it appropriately 9TWSc.03 Decide when to increase the range of observations and measurements, and increase the extent of repetition, to give sufficiently reliable data 9TWSc.05 Carry out practical work safely, supported by risk assessments where appropriate 9TWSc.07 Collect, record and summarise sufficient observations and measurements, in an appropriate form 9TWSa.01 Evaluate the strength of the evidence collected and how it supports, or refutes, the prediction 9TWSa.02 Describe trends and patterns in results, identifying any anomalous results and suggesting why results are anomalous 9TWSa.05 Present and interpret results, and predict results between the data points collected
1.2 Transport of water and mineral salts in plants	3	9Bs.01 Describe the pathway of water and mineral salts from the roots to the leaves in flowering plants, including absorption in root hair cells, transport through xylem and transpiration from the surface of leaves 9Bp.05 Know that plants require minerals to maintain healthy growth and life processes (limited to magnesium to make chlorophyll and nitrates to make protein)	9TWSp.01 Suggest a testable hypothesis based on scientific understanding 9TWSp.03 Make predictions of likely outcomes for a scientific enquiry based on scientific knowledge and understanding 9TWSp.04 Plan a range of investigations of different types to obtain appropriate evidence when testing hypotheses 9TWSc.04 Take appropriately accurate and precise measurements, explaining why accuracy and precision are important 9TWSc.06 Make an informed decision whether to use evidence from first-hand experience or secondary sources

			9TWSc.07 Collect, record and summarise sufficient observations and measurements, in an appropriate form 9TSa.03 Make conclusions by interpreting results, explain the limitations of the conclusions and describe how the conclusions can be further investigated 9TWSa.04 Evaluate experiments and investigations, including those by others, and suggest improvements, explaining any proposed changes
End of chapter questions	1	Check students' progress	

Background science

Plants come in all shapes and sizes and are imperative to life on Earth. In order to survive and grow, green plants need to photosynthesise.

Many reactions occur during the process of photosynthesis. The resultant changes that occur during theses reactions can be summarised using the word equation:

carbon dioxide + water → oxygen + glucose

Only the substances that are used in the chemical reactions are shown in the equation. Both light energy and chlorophyll are required for the reaction to take place and therefore these are sometimes shown above and below the arrow.

Photosynthesis involves a series of chemical reactions that happen inside the chloroplasts in the cells of green plants. Chloroplasts contain chlorophyll, which is the photosynthetic pigment that absorbs energy from light. This energy allows the reactions to occur during the process of photosynthesis. This reaction is endothermic, meaning that the products contain a greater store of energy than the reactants. This energy has come ultimately from the Sun.

Plants absorb the water that is needed for photosynthesis through their roots. This is a passive process (which occurs via osmosis) and does not require energy. Plant roots contain specialised cells, which have a large surface area to facilitate water absorption. They also contain many mitochondria. Mitochondria are found in almost all eukaryotic cells and they are the structures in which aerobic respiration takes place. Aerobic respiration is a chemical reaction which releases energy. Some of the energy released from glucose (in the mitochondria) in the root hair cells will be used in active transport to allow the plants to absorb mineral salts from the soil against the concentration gradient. Plants need various elements from mineral salts for many different purposes, including respiration, to make chlorophyll and for healthy growth.

Gas exchange happens in the leaves. Gases diffuse in and out of the leaves through the stomata (singular: stoma). Stomata are tiny holes, found mainly on the lower surface of leaves, as shown in the diagram below.

Carbon dioxide gas moves into the leaves by diffusion and oxygen (which is produced by photosynthesis) diffuses out. Diffusion of gasses happens down the concentration gradient – there is an overall movement of gas particles from where they are in higher concentration to where they are in a lower concentration.

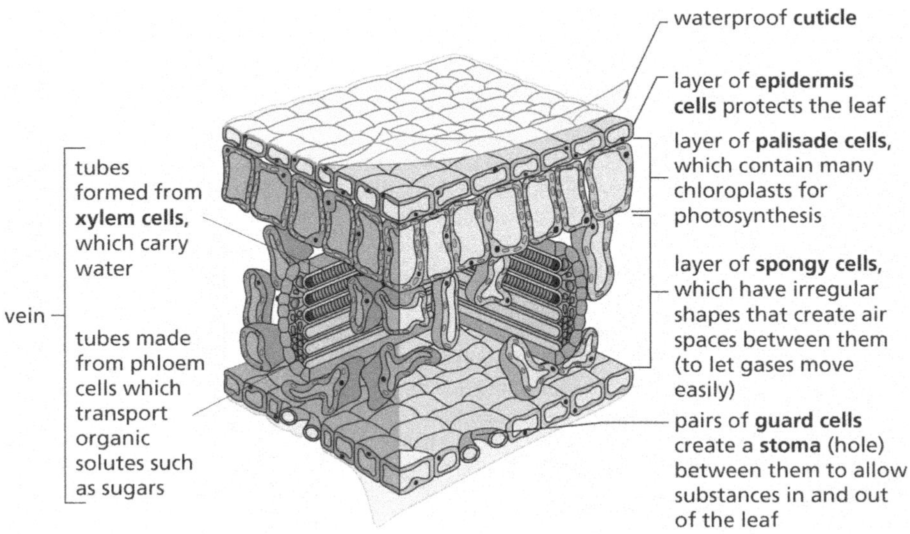

Plants transport water and mineral salts in dead, hollow tubes called xylem vessels, formed from xylem cells. Xylem vessels transport water from the roots to the tips of the plants (including leaves) where it is needed. Xylem vessels only carry water (and dissolved mineral salts) in one direction. Any organic solutes such as sugars are carried in phloem vessels. Phloem vessels carry solutes to all parts of a plant. The solutes may be used for respiration (glucose), building cell parts or be converted into storage materials. (Note: phloem and sugar movement is beyond Stage 9 and is taught at IGCSE level.)

Note that not all plants photosynthesise. Some parasitic plants, such as *Rafflesia* gain all their nutrients from another living organism. Others, including 'carnivorous plants', such as the Venus flytrap (*Dionaea muscipula*), photosynthesise, but also catch insects and digest them to gain the extra vital nitrogen that it requires (carnivorous plants and parasites are introduced to students at Cambridge IGCSE level).

Within this chapter the term 'rate' is used. The 'rate' of something is given by the amount of change occuring over a given unit of time. For example, if a plant is doing more photosynthesis per unit time when there is more sunlight then you would describe this as 'increasing the rate of photosynthesis'.

The image on the first page of Chapter 1 in the Student's Book is of giant *Sequoia* trees found in the Sierra Nevada, California, USA. These are the largest trees on our planet, growing up to 80 m tall. The tallest of these trees is thought to be over 2000 years old demonstrating how successful plants as organisms are in terms of survival.

1.1 Photosynthesis

Topic overview

		Student's Book	Workbook	Downloadable material
1.1a	**Starting point** This short discussion activity provides an opportunity to establish the extent of your students' prior knowledge from Stage 7 on plants	Starting point table		
1.1b	**Leaf structure and photosynthesis** This activity introduces the process of photosynthesis	Questions 1–17		Worksheets 1.1b
1.1c	**Investigating the formation of oxygen** This activity focuses on the enquiry skills of making observations and discussing explanations for results using scientific knowledge; students will observe that a gas is produced by plants and that this gas is oxygen. They then use scientific knowledge to explain the formation of the oxygen	Activity 1.1	Question 4	Practical notes 1.1c Worksheet 1.1c
1.1d	**Investigating the need for chlorophyll or light** This activity focuses on the enquiry skills of making predictions and planning investigations; students will be shown how to test a leaf for starch and then they will use this to plan to show that light and/or chlorophyll is needed for photosynthesis	Questions 18–26 Activity 1.2 Activity 1.3		Practical notes 1.1d Worksheet 1.1d
1.1e	**Investigating the rate of photosynthesis** This content goes beyond the curriculum framework expectations for Stage 9 and is covered in Cambridge IGCSE Biology. It is worth completing this activity but with a focus on investigative skills rather than the learning content.	Activity 1.4	Questions 1–3, 5–8	Practical notes 1.1e

Learning Objectives

- 9Bp.06: Know that photosynthesis occurs in chloroplasts and is the process by which plants make carbohydrates, using the energy from light
- 9Bp.07: Know and use the summary word equation for photosynthesis (carbon dioxide + water → glucose + oxygen, in the presence of light and chlorophyll)
- 9TWSp.01: Suggest a testable hypothesis based on scientific understanding
- 9TWSc.02: Decide what equipment is required to carry out an investigation or experiment and use it appropriately
- 9TWSc.03: Decide when to increase the range of observations and measurements, and increase the extent of repetition, to give sufficiently reliable data
- 9TWSc.05: Carry out practical work safely, supported by risk assessments where appropriate
- 9TWSp.05 Make risk assessments for practical work to identify and control risks
- 9SIC.03 Evaluate issues which involve and/or require scientific understanding
- 9TWSc.07: Collect, record and summarise sufficient observations and measurements, in an appropriate form
- 9TWSa.01: Evaluate the strength of the evidence collected and how it supports, or refutes, the prediction
- 9TWSa.02: Describe trends and patterns in results, identifying any anomalous results and suggesting why results are anomalous
- 9TWSa.05: Present and interpret results, and predict results between the data points collected
- 9TWSp.03 Make predictions of likely outcomes for a scientific enquiry based on scientific knowledge and understanding
- 9TWSp.04 Plan a range of investigations of different types to obtain appropriate evidence when testing hypotheses
- 9SIC.04 Describe how people develop and use scientific understanding as individuals and through collaboration, e.g. through peer-review
- 9SIC.05 Discuss how the uses of science can have a global environmental impact

Links to other topics

- Topics 1.3 and 1.4 in Stage 7 provide essential prior understanding for this topic. In those topics students will have learnt about plant cells, together with their parts and functions. They will have also been introduced to the idea of cells being specialised and the use of microscopes.
- Activity 1.1d revises work from Topic 2.1 in Stage 8 (food test for starch).
- The idea of compounds, met in Stage 7 Chemistry, is important in this chapter. Students may also have studied molecules in Stage 9 Chemistry at this point. If they have, draw attention to the fact that water, carbon dioxide and glucose are all molecules.
- Topic 13.1 Photosynthesis
- There are many opportunities in Stage 8 for students to write risk assessments for practical work. Students will need these skills in this topic.

Using models

- Students should appreciate that the diagrams of the plant cell and the leaf in the Student's Book are models.
- There is an opportunity for students to plan (and perhaps create) their own physical models of the internal structure of a leaf. Physical models like this can help us to understand how certain processes occur in a leaf (for example, gas exchange and the movement of gases).
- There is an opportunity here to address learning objective 9TWSm.02 where the students are asked to describe their model and discuss the strengths and limitations of the models produced.

Science in context

Questions 1 and 2 in the Science in context: Artificial photosynthesis box in Topic 1.1 of the Student's Book cover the following learning objectives:

- 9SIC.03 Evaluate issues which involve and/or requires scientific understanding
- 9SIC.04 Describe how people develop and use scientific understanding as individuals and through collaboration.
- 9SIC.05 Discuss how the uses of science can have a global environmental impact.

Teaching ideas

Most students should be able to make the link between energy being trapped by the plant and then released again when the plants are burned. They should also see that the plants take carbon dioxide out of the atmosphere when grown, and release it again when burnt. If there were no need for fuels to farm these crops, such crops would be 'carbon neutral'.

Common misunderstandings

- Students can confuse the products and reactants of photosynthesis, especially when they also learn about respiration. Using diagrams of the leaf to show that carbon dioxide enters and oxygen exits through the stomata. During the night when the light levels are low respiration occurs instead of photosynthesis, and the carbon dioxide exits the plant through the stomata.
- Students may think that water enters the leaves through stomata rather than this being the point of exit. Teachers should emphasise the role roots and stems play in taking in and transporting water from the soil to the leaves.

1.1a Starting point

Show students a plant with green leaves. Remind students that, like all living things, the plant requires nutrition. Ask students to discuss how the plant gets its food; they may recall from primary science that the plants need light and water in order to survive. They may also recall the role of chloroplasts in making food for plants from Stage 7 work on plant cells, and that plants provide the energy storage materials for all the organisms in a food chain.

1.1b Leaf structure and photosynthesis

Students will learn:

- To describe how plants make their own food, using photosynthesis
- To model photosynthesis using a word equation
- To collect and record observations and measurements appropriately
- To choose the appropriate equipment and use it correctly

Glossary words

chemical reaction, chlorophyll, chloroplast, cuticle, diffusion, epidermis cell, guard cell, glucose, mineral or mineral salt, palisade cell, photosynthesis, plant vein, product, raw material, reactant, spongy cell, stoma, xylem cell

Resource preparation

Image showing the inside of a leaf (for example, from the internet)

For microscope activity: microscopes, pre-prepared slides of leaf sections, light source(s)

Worksheet 1.1b

Practical notes 1.1b

Teaching and learning ideas

- As a whole class, look at the list of resources that plants need at the start of the Student's Book Topic 1.1. Ask students what process plants need light, carbon dioxide and water for. Establish the idea that these are all needed for the plant to make its own food using a chemical reaction called photosynthesis. Ask students to answer questions 1 and 2.
- Discuss the idea of leaves being the food factories of plants. Explain that the food they make is called glucose and that at the same time the plant makes glucose it makes oxygen. Read the section entitled Photosynthesis in the Student's Book and encourage students to answer questions 3–8.
- Note that Activity 1.1 is covered in the next section.
- Look at figure 1.3 of the plant cell. Ask students to work in groups to list the function of each part of the cell (revising material from Stage 7 Biology). You could ask students to present their work as labelled diagrams or ask a spokesperson from each group to give a function that you then write on the board. Correct any misconceptions and reinforce the idea that the chloroplasts are where the plant makes its food, and this is where photosynthesis occurs.
- Ask students to look at figure 1.6 in the Student's Book. Point out the veins and explain that they are connected to the roots. Ask students to suggest why they are needed for photosynthesis. Ensure they understand that water from the soil travels from the roots through the veins to the leaf cells so that it can be used in photosynthesis. Explain that there are two types of 'tubes' in the veins and that water is carried in one (called the xylem). Some students may realise that the glucose made is then transported from the leaf through the veins. Explain that this is the case, but that glucose has to be turned into another sugar before it can be transported; this sugar is sucrose. The tubes used to transport the sucrose are called the phloem, a term not needed at this stage but some students may wish to know.
- Point out that palisade cells contain more chloroplasts than spongy cells. This is because palisade cells receive more light and so more photosynthesis occurs in palisade tissue than in spongy tissue.
- Identify the stomata on the leaf structure and discuss their role in gas exchange; carbon dioxide enters through the stomata while oxygen leaves. Explain that stomata is the plural and stoma the singular.
- Remind students of their work on microscopes in Stage 7 Chapter 1. Give students copies of Worksheet 1.1b and use a microscope to demonstrate its operation (following the steps on the worksheet). Then provide students with pre-prepared slides of leaf sections to look at. Ask students to make labelled drawings of the different cells that they find (on the second side of the sheet). Practical details are given in Practical notes 1.1b.
- Read the section entitled Adaptations of a leaf in the Student's Book and encourage students to answer questions 9–17.
- Challenge students to explain why plants growing in shady places may have leaves with a larger surface area than those growing in full sunlight.

Differentiation

For less confident students, substitute using a microscope for showing a variety of images of sections from different plant leaves. These can be downloaded from the internet. Point out the different types of cell in each.

Ask more confident students to look at slides of leaf sections from more than one type of plant, and to compare them.

Extension ideas

Ask students to research why not all leaves are green.

1.1c The formation of oxygen

Students will learn:

- To describe how plants make their own food, using photosynthesis
- To collect and record observations and measurements appropriately

Resource preparation

Image or video of pondweed in which bubbles of gas can be seen; search images on the internet for 'pondweed oxygen'

For Activity 1.1: beakers, pondweed (*Cabomba* or *Elodea*), wooden splints (thin wooden stick used for lighting Bunsen burners), filter funnels, test tubes, eye protection

Practical notes 1.1c

Worksheet 1.1c

Teaching and learning ideas

- Engage students by showing them some pondweed in a beaker of water. Explain that it is a green plant that lives in water but it photosynthesises in the same way as other green plants. Show students a close-up image (or video) of pondweed demonstrating that it is releasing bubbles of gas. Ask students to predict which gas is being released from the pondweed; do not give them the answer, but have a vote to see how many think it is oxygen and how many think it is carbon dioxide.
- Activity 1.1 *Investigating photosynthesis in pondweed* in the Student's Book. Demonstrate how to collect the gas being released by placing a filter funnel and test tube full of water over the pondweed (see Practical notes 1.1c and the image on Worksheet 1.1c). Then show students the same apparatus, but explain that it has been left for a few days with plenty of light. Explain that if you remove the test tube of gas and place a glowing splint inside the tube, the splint will do one of two things: relight (or it may just glow more brightly) if there is oxygen, or go out completely if there is carbon dioxide. Students should then use Worksheet 1.1c to predict what will happen to the splint. Once their predictions are made, remove the test tube and place a glowing splint inside.
- As a whole class, discuss what they found out and then ask students to complete Worksheet 1.1c.
- Challenge students to plan how they could adapt the demonstration to show that light is needed for photosynthesis, including a prediction of what they expect the results to be.

Differentiation

With more confident students, ask them to predict what will happen before you show them the set of apparatus that has been left for a few days. They may be able to predict that the gas released is oxygen and that it will collect at the top of the test tube. Less confident students will need to be shown that the gas has collected before they can predict which gas it is.

Homework suggestions

Students answer question 4 on Topic 1 Photosynthesis in the Workbook.

Alternatively, challenge students to plan an experiment (based on the one in Activity 1.1) to investigate how light intensity affects photosynthesis.

1.1d Investigating the need for light or chlorophyll

Students will learn:

- To describe how plants make their own food, using photosynthesis
- To use scientific knowledge to make predictions
- To identify hazards and identify ways to minimise risks
- To choose the appropriate equipment for an investigation and use it correctly
- To interpret results and form conclusions using scientific knowledge and understanding
- To evaluate how the prediction is supported by the evidence collected
- To carry out practical work in a safe manner to minimise risks
- To describe how industries and society use science
- To use scientific understanding to evaluate issues
- To discuss the global environmental impacts of the uses of science

Glossary words

aerobic respiration, carbohydrate, compound, element, hazard, iodine solution, risk, starch, sugar

Resource preparation

For Activity 1.2 and Activity 1.3: plants, one with all the leaves uncovered and others with leaves partially covered in foil (these should be left out of sight until needed), plants with variegated leaves (green and white areas), iodine solution, ethanol, white tile, tongs, spatula, pipette, boiling tube, beaker, heating equipment (Bunsen burner, tripod, gauze, heat-resistant mat) or kettle, beaker labelled 'waste ethanol', eye protection

Practical notes 1.1d

Worksheet 1.1d

Teaching and learning ideas

- Remind students that during photosynthesis plants make glucose. Explain that glucose is an example of a type of carbohydrate called a sugar. Tell students that carbohydrates are compounds only made of carbon, hydrogen and oxygen atoms. Carbohydrates that are small and soluble are called sugars. There are many different types of sugars, and the one they are most familiar with is sucrose (table sugar used to sweeten drinks and food); glucose is just another example.
- Ask students to suggest what the plant needs this glucose for. Elicit the ideas that glucose is used for energy, or as a storage material or as a raw material to make new materials for the plant (e.g. for growth). Explain that as glucose is made, it is converted into a larger, insoluble carbohydrate called starch. The chloroplasts store the starch while the plant is still photosynthesising. When it is dark, the starch is changed into other sugars to be transported around the plant (and used for energy, storage or as a raw material).
- Ask students to read the Plant carbohydrates section in the Student's Book and encourage them to answer questions 18–23.
- Demonstrate to students how to test a leaf for starch, using ethanol and iodine solution, to show that it has photosynthesised (Worksheet 1.1d). Explain that the starch is only present in areas where photosynthesis has occurred. Students may have met the test for starch in Stage 8 Chapter 2.
- Ask students to read the Hazards and risks section in the Student's Book and encourage them to answer questions 24–26.
- Activity 1.2 *Investigating photosynthesis in leaves 1* in the Student's Book. As a whole class, plan how you could show that light is needed for photosynthesis by testing a leaf for starch. Students may suggest putting one plant in the dark and another in the light, but discuss why this may not be a fair test (because the plants are not exactly same). As a hint, show students some foil and ask them how they could use that to make a fairer test. Eventually they should come up with the idea of covering some of the leaves on a plant. Ask them to suggest what might happen if only part of the leaf was covered. Ask students to work through the questions in the activity box on their own or in groups. Remind less confident students of their work on writing risk assessments from Year 8. Go through the answers.
- Activity 1.3 *Investigating photosynthesis in leaves 2* in the Student's Book. Show students a variegated leaf and explain that there are no chloroplasts in the white area of the leaf. Ask them to predict how this would affect where the starch would be found. Ask students to work through the questions in the activity box on their own or in groups. These questions are very similar to those in Activity 1.2 and will help to identify any lingering misconceptions. Go through the answers.
- Provide each group with leaves to test. For less confident students include only leaves that have been in the dark (i.e. covered with foil) or the light for 48 hours. For more confident students include variegated leaves, and consider including variegated leaves that have been in the dark (i.e. covered with foil) and in the light for 48 hours. Students should then follow the instructions on Worksheet 1.1d (see Practical notes 1.1d).
- Students should then analyse their results using Worksheet 1.1d. Encourage each group to collect results from other groups (including those who have tested leaves that they have not tested) and to compare the results.
- Challenge students to explain why more starch is made on warm sunny days compared with cold cloudy days.

Less confident students can be paired up with more confident students during the investigation. When answering the challenge question on Worksheet 1.1d, less confident students may need a hint. Suggest that they think about covering the leaf with something other than foil.

Differentiation

More confident students should be able to test both light and dark grown leaves and variegated leaves at the same time.

Extension ideas

Ask students to find out ways of comparing the amount of photosynthesis happening in different plants.

Homework suggestions

Ask students to prepare a 10-mark quiz on photosynthesis, together with a mark scheme. Alternatively, ask students to plan an investigation on how light can affect the rate (speed) of photosynthesis. This is Activity 1.4 in the Student's Book.

1.1e Measuring the rate of photosynthesis

Students will learn:

- To use scientific knowledge and understanding to suggest a testable hypothesis
- To decide how to obtain sufficiently reliable data from observations and measurements
- To carry out practical work in a safe manner to minimise risks
- To collect and record measurements and observations appropriately
- To evaluate how the prediction is supported by the evidence collected

Glossary word

anomalous, reliable, repeatable

Resource preparation

Activity 1.4: Pondweed (e.g. *Cabomba*, *Elodea*), large beaker, glass funnel, test tube, metre rule, scissors, electric lamp, thermometer, timer, potassium hydrogencarbonate powder. See Practical notes 1.1e for full equipment list

Practical notes 1.1e

Teaching and learning ideas

- Ask students to think back on their work in this chapter and to come up with some ideas of how to get plants to photosynthesise more quickly. Encourage them to develop a hypothesis in pairs (such as 'the rate/speed/amount of photosynthesis depends on the light intensity/carbon dioxide concentration').
- Remind students of some important words that are used when planning and carrying out investigations. Write the following up on the board, and ask students to submit definitions. Correct misconceptions as they arise and write up agreed definitions on the board for: anomalous, dependent variable, independent variable, reliable, repeatable, variable.
- Activity 1.4 *Investigating how light can affect the rate (speed) of photosynthesis.* Reminder: this content goes beyond the curriculum framework expectations for Stage 9 and is covered in Cambridge IGCSE Biology. Students should work through the activity questions A1–A3 in the Student's Book, using figure 1.10 to give them ideas about how to set up the equipment. It may be helpful to students to see the equipment set up as it is in figure 1.10 before beginning their own trial runs. See Practical notes 1.1e for support and an appropriate method.
- If there is not time for students to carry out the investigations themselves, ask them to use the results below to answer the remaining activity questions.

Distance of lamp (cm)	Number of bubbles per minute			Mean number of bubbles per minute
	1st try	2nd try	3rd try	
10	54	50	52	
20	35	30	31	
30	23	12	25	
40	17	17	18	
50	5	4	5	

- Ask students to share their conclusions with the class, and agree an overall conclusion (the greater the light intensity, the greater the rate of photosynthesis) on the board.
- Finish by asking students to work in groups to come up with five key facts about photosynthesis. Ask a spokesperson from randomly selected groups to share ideas, and write these on the board. Compare these ideas with those in the Student's Book.

Differentiation

Some students will need extra support in planning trial runs and setting up the equipment. Work with them, using the Student's Book diagram to help. Challenge more confident students to plan their own experiment by simply being shown the apparatus that they are able to use. Students could also be asked to research and explain why increasing light intensity increases the rate of photosynthesis. Ask them to consider the energy that the plant is being given and how this alters as the light moves closer to the plant.

Extension ideas

Some students could be encouraged to consider different factors that could alter the rate of photosynthesis and how they could investigate these, for example, they could use potassium hydrogencarbonate to add more carbon dioxide to the water – gradually adding more and more.

Students could also evaluate their investigations, considering the potential problems/flaws with the set up and data that they are collecting. Students will need to be told that an increase in temperature increases the rate of photosynthesis and this should then allow them to consider the potential problem that the lamp will provide heat as well as light. Students should be encouraged to design better ways to control the temperature of the water that the pondweed is in – this could include using low temperature lamps or beakers of water between the lamp and the pondweed beaker to act as a heat sink.

Homework suggestions

Students answer questions 1–3, 5–8 on Topic 1.1 Photosynthesis in the Workbook. Alternatively, ask students draw a concept map to cover what they have learnt about photosynthesis in this chapter. The word at the centre should be 'photosynthesis'.

1.2 Transport of water and mineral salts in plants

Topic overview

		Student's Book	Workbook	Downloadable materials
1.2a	**Starting point** This short activity provides an opportunity to establish the extent of your students' prior knowledge of root hair cells from Stage 7	Starting point table		
1.2b	**Absorbing water and mineral salts** Students revise how root hair cells are adapted for their function, and learn about the importance of water and mineral salts for plants and how they are absorbed by a plant.	Questions 1–6		Practical notes 1.2b(1), 1.2b(2) Worksheets 1.1b, 1.2b
1.2c	**Transport of substances in plants** Students learn how water is carried up plants in xylem tissue and how it is subsequently lost from the leaves. There is also a brief look at the transport of sugars	Questions 7–9 Activity 1.5		Practical notes 1.2c(1), 1.2c(2) Worksheets 1.1b, 1.2c(1), 1.2c(2)
1.2d	**The importance of mineral salts** Students learn about the importance of specific mineral salts for the growth of plants.	Questions 10–16 Activity 1.6	Questions 1–10	Practical notes 1.2d Worksheet 1.2d

Learning Objectives

- 9Bs.01: Describe the pathway of water and mineral salts from the roots to the leaves in flowering plants, including absorption in root hair cells, transport through xylem and transpiration from the surface of leaves
- 9Bp.05: Know that plants require minerals to maintain healthy growth and life processes (limited to magnesium to make chlorophyll and nitrates to make protein)
- 9TWSp.01: Suggest a testable hypothesis based on scientific understanding
- 9TWSp.03: Make predictions of likely outcomes for a scientific enquiry based on scientific knowledge and understanding
- 9TWSp.04: Plan a range of investigations of different types to obtain appropriate evidence when testing hypotheses
- 9TWSa.03: Make conclusions by interpreting results, explain the limitations of the conclusions and describe how the conclusions can be further investigated
- 9TWSa.04: Evaluate experiments and investigations, including those by others, and suggest improvements, explaining any proposed changes
- 9TWSc.04: Take appropriately accurate and precise measurements, explaining why accuracy and precision are important
- 9TWSc.06: Make an informed decision whether to use evidence from first-hand experience or secondary sources
- 9TWSc.07 Collect, record and summarise sufficient observations and measurements, in an appropriate form
- 9SIC.02 Describe how science is applied across societies and industries, and in research
- 9SIC.05 Discuss how the uses of science can have a global environmental impact

Links to other topics

- Chapter 1 in Stage 7 is essential prior understanding for this topic. In this chapter students learnt about cells, and the idea of specialised cells (including root hair cells) was introduced in Topic 1.4.
- The idea of mineral salts being important in the human diet was met in Stage 8 Chapter 2. Students should understand that mineral salts are important for both plants and humans, although they need different amounts of the elements from those salts.
- The idea of salts as chemical compounds was first met in Stage 7 Chemistry (as part of work on neutralisation). There is further coverage of salts in both Stage 8 and Stage 9 Chemistry.

Using models

- Students should appreciate that diagrams showing the internal structure of plants and transpiration are models.
- In Activity 1.2b a model is used to explain how water is used to maintain the structure of a plant cell. A balloon and cardboard box is used in this model. The students could be invited to discuss the strengths and limitations of this model.

Science in context

Questions 17 and 18 below the Science in context: Mineral salts from natural fertiliser box in Topic 1.2 in the Student's Book cover the following learning objectives:

- 9SIC.02 Describe how science is applied across societies and industries, and in research.
- 9SIC.05 Discuss how the uses of science can have a global environmental impact.

Teaching ideas

Students should be aware that there are artificial and natural fertilisers but that each have their advantages and disadvantages.

Common misunderstandings

- The terms 'minerals' and 'mineral salts' essentially mean the same in the context of plant and animal nutrition. Minerals are naturally occurring compounds that are formed by physical (non-living) processes. Some minerals contain chemical elements that organisms need in very small amounts. Important chemical elements for all organisms include calcium, magnesium, phosphorus and potassium (note that phosphorus and potassium are not required knowledge at this stage). Some minerals are in a class of chemical compound called 'salts', and it is from these that organisms obtain the elements that they need. People often refer to each element as being a 'mineral'; it is important to remember that a mineral is a compound that contains a certain element that plants require. These elements are sometimes referred to as 'mineral elements'.
- Some students think that the increase in mass of a plant as it grows is partly accounted for by taking in substances from the soil. Ensure that students understand that, while substances from the soil are important, the mass of those soil substances is incredibly small compared with the mass of the plant.
- Take care with the verb 'to absorb' (spelt with a 'b') and its noun 'absorption' (spelt with a 'p').

1.2a Starting point

Draw a diagram of a root hair cell on the board. Ask students what it is. Establish that it is a root hair cell. Ask students to name its parts (nucleus, cytoplasm, cell membrane, cell wall, large vacuole, mitochondria). Then ask students what it does and how it is adapted for that function. Remind students that its function is to absorb water (and mineral salts) from the ground and it has a large surface area to be able to do this efficiently.

1.2b Absorbing water and mineral salts

Students will learn:

- To describe the importance of elements found in mineral salts (magnesium and nitrates) for the growth and development of plants
- To collect and record observations and measurements appropriately

Glossary words

absorb, root hair cell, wilting, xylem cell

Resource preparation

For demonstration 1: a wilted plant

For demonstration 2: images of 'flaccid' and 'turgid' plant cells downloaded from the internet, optional balloons, cardboard boxes and masses

For demonstration 3: seedling (mustard seeds (e.g. *Brassica nigra*), filter paper, pot made from vermiculite (a natural mineral often added to soils to encourage seed germination and growth) or terracotta, Petri dish/clear plastic bag; optional incubator) with root hairs, hand lenses/magnifying glasses

For activity: pre-prepared slide of root hair tissue, microscope and light source

Practical notes 1.2b(1), 1.2b(2)

Worksheets 1.1b, 1.2b

Teaching and learning ideas

- Ask students what happens if a plant does not get enough water. Remind students that it wilts (droops). Demonstration 1: You could show students a wilted plant, if one is available. Some wilted plants (e.g. tomatoes) respond very quickly after being given water. You could demonstrate this to students.
- Together with students, look at the first section in the Student's Book, before the section on Absorbing water. Encourage students to answer questions 1 and 2. If students have already studied Topic 1.1, remind them that minerals and mineral salts refer to the same thing – important chemical compounds from which organisms obtain chemical elements that they need.
- Demonstration 2: Show students some images of 'flaccid' and 'turgid' plant cells downloaded from the internet (search for 'flaccid turgid plant cells'). Students do not need to know the terms 'flaccid' and 'turgid' but should be able to see how the cell membrane pulls away from the cell wall if there is too little water inside it. Explain to students that the pushing of the cytoplasm against the cell wall helps to support the cell wall. Ask them to think about a balloon being blown up inside a cardboard box and how an inflated balloon helps to support the sides of the box. If you have a small carboard box and a balloon available you could demonstrate this. Start by inflating the balloon so that it fits snugly into the cardboard box. Then show students that pressing on the sides of the box (or adding masses on top of it) will not crumple the box. Then let air out of the balloon (or use another partially inflated balloon) and demonstrate again. Show that the box is no longer so well supported and its sides will crumple.
- Then look at the section on Absorbing water in the Student's Book. Encourage students to answer questions 3 and 4.
- Remind students of their work in Topic 1.4 in Stage 7, looking back at the images and diagrams of root hairs in the Student's Book in that topic.
- Demonstration 3: Root hairs. Show students the root hairs on a freshly grown seedling. Details for preparing this are given in Practical notes 1.2b(1). It may help if you can supply students with hand lenses or magnifying glasses with which to examine the root hairs in more detail. This demonstration could be followed by microscope work.
- Activity: Looking at root hairs. Provide students with pre-prepared slides of root hair tissue to examine using microscopes. Full details are given in Practical notes 1.2b(2). Students can remind themselves of how to use a microscope using Worksheet 1.1b. Encourage students to draw one or two of the root hair cells that they find, and to label the parts that they can see.
- Finish by asking students to complete Worksheet 1.2b.

Differentiation

Help less confident students by reminding them of their work on cells from Stage 7. For some students, it may be worth going through the relevant material about root hair cells in the Stage 7 Student's Book first, before they start on the work here.

Challenge more confident students to plan a way to observe flaccid and turgid plant cells using a microscope. Encourage them to learn the meanings of these terms.

Extension ideas

You could show students how to prepare their own microscope slides to show root hair cells. Provide students with the sections of root to use, and do not allow students to use mounted needles.

Homework suggestions

Ask students to write a paragraph that explains how and why plants absorb water and mineral salts. Alternatively they could present this work as a concept map, with a drawn image of a root hair cell at its centre.

1.2c Transport of substances in plants

Students will learn:

- To describe the importance of elements found in mineral salts (magnesium and nitrates) for the growth and development of plants
- To suggest a testable hypothesis and make predictions of likely outcomes
- To plan methods used to test hypotheses
- To make predictions based on scientific understanding
- To evaluate experimental methods, explaining any improvements suggested
- To collect and record observations and measurements appropriately
- To interpret and present results appropriately
- To interpret results and form conclusions using scientific knowledge and understanding

Glossary words

transpiration, xylem vessel

Resource preparation

For Activity 1.5: plant stems (e.g. celery), beaker or jar of water containing ink or food colouring, knife and board (or hard cutting surface), hand lens/magnifying glass, ruler

For activity: pre-prepared slide of plant stem cross-sections, microscope and light source

Practical notes 1.2c(1), 1.2c(2)

Worksheets 1.2c(1), 1.2c(2)

Teaching and learning ideas

- Together with students, look at figure 1.6 in the Student's Book. For students who have covered Topic 1.1 this will be revision. Ensure that students understand that they are looking at a cross-section through a leaf. Point out the different sorts of cells that are found in a leaf, and then point out the stomata on the bottom of the leaf. Explain to students that this is where gases enter and leave a leaf. Ask students to suggest some gases that might do this. Correct answers include any gas in the air. Establish with students that one of these gases is water vapour. Explain that water evaporates from the surfaces of the cells inside the leaf and the water vapour then diffuses out of the leaf through the stomata. Ensure students realise 'stoma' is the singular and 'stomata' is the plural.
- Together with students, look at the sections on Transporting water and Transpiration in the Student's Book. Encourage students to answer the questions.
- Activity 1.5 *Investigating water transport in stems* in the Student's Book. Read through the activity box in the Student's Book and allow students to come up with some suggestions as to which variable to change. Ideas could include the temperature or the number of leaves. Students may have done a similar experiment in Stage 7. The instructions on Practical notes 1.2c(1) and Worksheet 1.2c ask students to investigate changing the number of leaves, but many students should be able to work independently and plan their own investigations to carry out. Check all students' plans before allowing them to start practical work.
- Finish by asking students to work together to write a step-by-step description of how water is taken into a plant and lost again. Ask a spokesperson from one group to suggest the first step, and then ask spokespeople from other groups to suggest subsequent steps. Write all the steps on the board and then go through each, one at a time, making sure that the whole class agrees that they are correct, and suggesting additional information that could be added.

Differentiation

Help less confident students to choose a variable to investigate in the activity on water transport.

Challenge more confident students to investigate more than one variable in the activity on water transport.

Extension ideas

If the class have used Worksheet 1.2c(1), challenge some students to collect the class data and to present the class results as a scatter graph, showing the correlation between the number of leaves and the height that the dyed water reached. If Worksheet 1.2c(1) has not been used supply some sample results for number of leaves and height that the dyed water reached.

Homework suggestions

Worksheet 1.2c(2) is suitable for homework.

Alternatively ask students to draw a flow chart to describe the movement of water in plants. Challenge students to add additional notes to explain the different parts of their flow charts.

1.2d The importance of mineral salts

Students will learn:

- To describe the importance of elements found in mineral salts (magnesium and nitrates) for the growth and development of plants
- To use scientific knowledge and understanding to suggest a testable hypothesis
- To decide whether evidence from first-hand experience or secondary sources should be used
- To evaluate experimental methods, explaining any improvements suggested
- To collect accurate and precise measurements
- To interpret results and form conclusions using scientific knowledge and understanding

Resource preparation

For demonstration or activity: empty packets of fertiliser or their labels (which could be downloaded from the internet). Ensure that any packets of fertiliser do not contain any remaining traces of fertiliser and do not allow students to touch packets.

Activity 1.6: This activity can be made into an optional practical investigation, which will require duckweed plants *(Lemna)*, glass jar/beakers, commercially bought nutrient solutions, forceps, ruler, vegetable oil

Access to secondary sources of information about mineral salts needed for plant growth (e.g. internet, books)

Practical notes 1.2d

Worksheet 1.2d

Teaching and learning ideas

- Ask students where the increase in mass comes from when a plant grows. Ensure that students understand that most of the mass comes from glucose and the products made from it, which in turn come from water and carbon dioxide. Then ask students what else plants need for growth, apart from water and carbon dioxide. Establish that mineral salts are also needed but in incredibly small amounts. Explain to students that mineral salts are compounds that contain certain elements that plants need.
- Ask students to complete a five-minute 'find-out' research activity using secondary sources (these can include leaflets about plant growth, the internet and the Student's Book). The question is 'What elements do plants need for healthy growth?' Clearly display a timer for all students to see and at the end of the five minutes all students should return to their desks with the information they have found out. Ask students to share and discuss their findings with others.
- Together with students, look at the section in the Student's Book on Mineral salts and healthy plant growth (as far as the activity box). Link with ideas that students came up with in 'find-out' activity and make it very clear that mineral salts dissolve in water and are absorbed by plant roots, where they are taken into the xylem vessels. Ask students to answer questions 10–12 in this section.
- Read through the Activity 1.6 *Investigating minerals and plant* growth in the Student's Book with students. Ask students to work through the activity questions either individually or as pairs. Review with students.
- If time and resources allow, students could then set up their own experiment to investigate how restricting access to different minerals can affect plant growth. See Practical notes 1.2d for guidance for this.
- There are some excellent video clips on the internet about plant mineral salts, how plants take mineral salts in, why they need them and what happens if they do not have them.
- Ask students to complete Worksheet 1.2d. Challenge students to do independent research to add further rows to the table.
- How do farmers make sure that their crops get all the nutrients they need? Discuss this question with students. Look at the image in the Student's Book, which shows the label from some 'plant food' (a fertiliser). If available, show students an empty fertiliser packet or give students copies of labels from various plant foods/fertilisers to look at and ask them to look for which minerals the fertiliser contains and what each would be used for in a plant.
- If time allows, ask students to work in groups to create their own brand of fertiliser and design the label or box. Students should first work together to decide the roles of the different people within the group (e.g. researching minerals using secondary sources, design, wording, drawings, box design/shape, advertising). Students should include information about the main elements that are found in the fertiliser and why plants need these. Challenge students to include hazard labels on their packaging – students may need to research other fertilisers and packaging to support this.
- Look at the pink Science in context feature in the Student's Book. Discuss with students why it might be better to use a natural fertiliser on crops. Ask students to come up with the pros and cons of using manure as a natural fertiliser rather than a synthetic one. Students should consider things such as costs, whether you could be sure of the mineral content of manure, other organisms and the farmers' time.
- To finish, ask students to complete the remaining questions in the Student's Book for this section. Review their answers.

Differentiation

Give less confident students access to the Student's Book to complete Worksheet 1.2d.

Challenge more confident students to do independent research to find out more about the different elements that plants need from mineral salts. They should present their work as a table showing the element, an example of the mineral salt it is found in, why the element is needed and what happens if a plant lacks the element. Or they could add additional rows to the table on Worksheet 1.2d.

Extension ideas

Ask students to research the effects of artificial fertilisers on the environment.

Homework suggestions

Students complete questions 1–10 from the Workbook.

Chapter 1 Checking students' progress

End of chapter reflection

- Review the learning points of the topic
- Test understanding through answering questions

Resources

- Student's Book [End of chapter review questions]
- Workbook [Self-assessment grid, End-of-chapter questions]
- Worksheets 1.1b, 1.2b, 1.2c(2), 1.2d [all of which may have already been used]

Approach

Ask students to complete the Self-assessment section at the end of Chapter 1 in the Workbook. Encourage students to work together in small groups to discuss which aspects of the material are the most difficult to understand, and then have a class vote.

Use the Student's Book pages to go through aspects of the work that students have identified as difficult, using the Worksheets listed above as necessary.

Ask students to complete some or all of the End of chapter review questions in the Student's Book. These can be gone through orally in class, or students could be given specific questions to answer in their books.

Ask students to work together to draw a concept map to show the material in this chapter. They should start by writing the title 'Photosynthesis and plant growth' in the centre of a sheet of paper. Lines then link this to key points from the chapter. Encourage students to try to do this work without access to the Student's Book at first. Once they have finished their concept maps, allow them access to the Student's Book so that they can make corrections and additions to their maps as necessary. Then ask students to say what they had forgotten to include in their concept maps and/or what they needed to correct. Revise these topics.

Use the Context worksheet (Worksheet 1.3 Researching crop farming) as a framework for students to conduct their own research enquiries about crop farming, and the different ways that are used by farmers to increase their yields (the amount of useful crop that they produce). Students then go on to explain how each method works.

Finish by asking students to complete the End-of-chapter questions at the end of Chapter 1 in the Workbook. Provide feedback on students' answers, and work with students to identify areas that still need more work.

Chapter 2 The excretory system

Chapter overview

In this chapter students will learn about the excretory (renal) system, its components and their functions. This chapter builds on ideas about organs and organ systems, met in Stage 7 Chapter 1.

Topic title	Number of 40-minute periods	Learning Objectives	Thinking and Working Scientifically Learning Objectives
2.1 Producing urine	2	9Bs.02 Describe the structure of the human excretory (renal) system and its function (limited to kidneys filtering blood to remove urea, which is excreted in urine)	9TWSm.02 Describe some important models, including analogies, and discuss their strengths and limitations. 9TWSc.05 Carry out practical work safely, supported by risk assessments where appropriate 9TWSp.02: Describe examples where scientists' unexpected results from enquiries have led to improved scientific understanding 9TWSp.05 Make risk assessments for practical work to identify and control risks 9TWSa.03 Make conclusions by interpreting results, explain the limitations of the conclusions and describe how the conclusions can be further investigated
End of chapter questions	1	Check students' progress	

Background science

The renal system is composed of the kidneys, ureters, bladder and urethra. In this course it is mainly referred to as the excretory system, to link it with its primary function - the excretion of wastes. It is also called the urinary system.

Be aware that other substances are excreted by the body (such as carbon dioxide by the lungs). The liver also excretes wastes (such as bilirubin, which is formed from the breakdown of haemoglobin) into the gall bladder. These wastes are then removed as a minor part of faeces. It should be remembered, though, that undigested food forms by far the greatest part of the faeces and that this is not excretory material; excretion refers to the removal of wastes that are made by chemical reactions inside cells.

The main function of the excretory (renal) system is to filter the blood to remove waste products and excess water, and then excrete them from the body in urine.

The human body has two kidneys, which are found in the abdomen – one on the right and one on the left. They are shaped like beans and are both located towards the back of the body above the waist and below the ribs.

The kidneys filter the blood to remove urea and excess water. This filtration process occurs in the nephrons of the kidneys and produces a fluid called urine. The urine is passed out of each kidney into the muscular tube known as the ureter. The urine is passed along the ureter through peristalsis (contraction of the ureter's muscular walls) to the bladder, where it is stored.

Once the bladder is approximately half full, a nerve impulse is triggered and sends a message to the bladder's sphincter muscle, instructing it to relax and allow the urine to flow out of the bladder into the urethra. The urethra is a single muscular tube along which urine passes out of the body.

(At Stage 9 the students do not need to know how this filtration works or the detailed structure of the kidney).

2.1 Producing urine

Topic overview

		Student's Book	Workbook	Downloadable materials
2.1a	**Starting point** This short activity allows students to demonstrate their understanding of excretion, organs and organ systems	Starting point table Questions 1–3	Questions 1–3	
2.1b	**The excretory system** Students learn about the structure of the excretory system and how it removes urea from the body	Questions 4–8	Questions 4–6	Practical notes 2.1b Worksheet 2.1b
2.1c	**Kidney problems** Students have opportunities to explore Thinking and Working Scientifically skills and Science in Context, using the context of the renal system. Note that this is optional and is intended as a context in which to apply knowledge and skills; it is not required learning	Questions 9–12	Questions 7–9	Practical notes 2.1c Worksheet 2.1c

Learning Objectives

- 9Bs.02: Describe the structure of the human excretory (renal) system and its function (limited to kidneys filtering blood to remove urea, which is excreted in urine)
- 9TWSm.02: Describe some important models, including analogies, and discuss their strengths and limitations
- 9TWSp.02: Describe examples where scientists' unexpected results from enquiries have led to improved scientific understanding
- 9TWSp.05: Make risk assessments for practical work to identify and control risks
- 9TWSc.05: Carry out practical work safely, supported by risk assessments where appropriate
- 9TWSa.03: Make conclusions by interpreting results, explain the limitations of the conclusions and describe how the conclusions can be further investigated

Links to other topics

- Topics 1.1 and 1.2 in Stage 7 provide essential prior understanding for this topic. In those topics students will have learnt about excretion as a life process and how organs work together as organ systems.
- Activity 2.1 requires students to remember work from Topic 2.1 in Stage 8 (food test for proteins) in order to plan their own investigations.
- There are many opportunities in Stage 9 for students to write risk assessments for practical work. Students will need these skills in this chapter.

Using models

- Students should appreciate that the diagram of the excretory system in the Student's Book is a model.
- Questions 7 and 8 in the Student's Book provide students with the opportunity to consider the strengths and limitations of analogies used to represent kidney filtration.

  This could be extended by asking students to develop their own analogy for the kidney filtering system.

Science in context

Teaching ideas

If there is a dialysis centre near you, it may be worth contacting them for some information (e.g. the number of people that they see each week, why those people come, how long their dialysis takes). This can help to add some local relevance to the work.

Common misunderstandings

- Ensure that students are able to explain excretion in terms of removal of wastes that are made in the cells of the body. So, faeces is mainly undigested food and its elimination from the digestive system is not an example of excretion. See Background science above.

2.1a Starting point

Teaching and learning ideas

Write MRS GREN on the board and ask students what this means. Establish that the letters are a mnemonic to remember the seven life processes – the seven processes that happen in all living things. Ask students to say what each letter stands for, and then to explain what each of movement, reproduction, sensitivity, growth, respiration, excretion and nutrition mean.

Ask students to work in groups to write down an example of each life process in a plant and of each life process in a large animal. Ask a spokesperson from each group to read out one life process and an example from a plant and an example from a large animal. Ask other groups if they agree or not, and correct any misconceptions before writing a correct answer on the board.

Remind students of their work on viruses from Topic 3.1 in Stage 7. Ask them to state why viruses are not considered to be living things.

Explain to students that excretion refers to the removal of wastes produced inside an organism's cells, and that in this chapter they are going to look out how excretion of a substance called urea happens in humans.

Read through the Excretion section in the Student's Book, and ask students to answer questions 1 and 2. Challenge some students to answer question 3.

2.1b The excretory system

Students will learn:

- To describe the structure of the human excretory (renal) system
- To describe how the kidneys filter blood to remove urea, which is excreted in urine.
- To describe the uses, strengths and limitations of models and analogies

Glossary words

bladder, blood vessels, excretion, excretory system, kidney, organ system, renal system, urea,

ureter, urethra, urinate, urine

Resource preparation

Practical notes 2.1b

Worksheet 2.1b

Teaching and learning ideas

- Write the following words on the board: cells, tissues, organs, organ systems. Draw a single line between each word and the next one in the list. Then ask students to suggest a phrase that could be used to link the two words. For example: *cells* of the same type and grouped together form a *tissue.* Use this activity to remind students of their work in Stage 7 Chapter 1 about organs systems and introduce the excretory systems as containing the kidneys, ureters, bladder and urethra.
- Explain to students that the kidneys filter out waste substances from the blood (mainly a soluble substance called urea). Use a simple filtering demonstration (using standard filtering apparatus and fine sand) to demonstrate the idea. See Practical notes 2.1b. Point out to students that this is a model for what happens in the kidneys. Explain that the fine sand represents blood cells and that the yellow liquid represents the urea dissolved in blood plasma.

- Read through the section of the Student's Book entitled The excretory system, and ask students to answer questions 4–6.
- Ask students to look at figure 2.1 in the Student's Book, and explain that it is a model. Ask students to suggest a way in which the model is good (e.g. it shows all the organs and how they connect to one another) and one way in which the model is poor (e.g. it does not show exactly what the organs look like, it does not show the size of the organs, it does not show how they function).
- Ask students to think back to the filtering model about how the kidneys function (above) and to suggest ways in which it is a good model (e.g. it shows how insoluble things such as blood cells can be separated from soluble wastes) and ways in which it is a poor model (e.g. the kidneys do not use all the blood plasma/water that goes through them to make urine, the real organs are flexible).
- Remind students that physical objects and paper diagrams are both types of models, and that all models have points that make them useful but also aspects that are poor. Remind students that analogies are also models, where we compare something to an everyday object or idea. Then ask students to answer question 7 in the Student's Book. Challenge students to go on to answer question 8.
- Challenge students to look at figure 2.1 in the Student's Book and to say which of their kidneys is higher up the body than the other. The left one is higher, and students should appreciate that human organ drawings are done as though you are looking at the front of a person.
- Challenge students to draw up a summary table to show the models used in this topic, identifying one good point and one poor point for each.
- Finish by asking students to complete Worksheet 2.1b. Before printing/copying, decide how many of the words at the bottom of the sheet to remove, in order to match the level of difficulty to your students.

Differentiation

For less confident students, leave all of the words and phrases at the bottom of Worksheet 2.1b.

When using the filtering model for the kidneys for more confident students, leave out any explanation of what you are doing. Ask students to explain why you are showing them filtration apparatus and to identify what each of the components represents.

Extension ideas

Ask students to find out where urea is made, and why it is made. They should present their work as two or three bullet points (encouraging them to use their own words, rather than just copy out something from a book or website).

Homework suggestions

Worksheet 2.1b could also be used for homework.

Alternatively, ask students to draw a flow chart to explain how urea is removed from the blood and leaves the body.

2.1c Kidney problems

Students will learn:

- To describe the structure of the human excretory (renal) system
- To describe how the kidneys filter blood to remove urea, which is excreted in urine
- To describe examples where scientific understanding has been improved by the unexpected outcomes of scientific enquiries
- To plan methods used to test hypotheses
- To identify hazards and identify ways to minimise risks
- To carry out practical work in a safe manner to minimise risks
- To interpret results and form conclusions using scientific knowledge and understanding

Resource preparation

For the practical activity: Biuret solution, test tubes or spotting tiles, pipettes, different 'urine' samples for testing, labels, eye protection

Practical notes 2.1c

Worksheet 2.1c

Teaching and learning ideas

Important note: You may have students in your class who have, or may know someone who has, kidney disease. You should approach this topic carefully, briefly and from a scientific point of view. Take care not to go into medical details about treatment or prognosis of kidney disease. Be aware that if you do have students who have kidney disease they may want to share their experience with the class, but this may not be appropriate.

This is extension work so could be omitted altogether if likely to cause upset for your students.

- Explain to students that some people have kidneys that do not work. In some people, kidney disease damages the filters inside the kidneys, which means that proteins are allowed into the urine. Normally, the proteins are kept in the blood plasma.
- Read through the paragraph and table in the Student's Book, entitled Kidney problems. Encourage students to answer questions 9 and 10.
- Challenge students to answer questions 11 and 12.
- Look through the activity box in the Student's Book, entitled Detecting kidney disease. Encourage students to work together in groups to plan a method for this experiment and write a risk assessment. Encourage students to decide on which team member should perform which job, in the design of the experiment, in carrying it out and in recording the results. As part of their conclusion, students should agree on one further investigation that could be done. See Practical notes 2.1c for more details. If you do not wish students to plan their own practical work, instructions for students can be found on Worksheet 2.1c.
- Read through the remainder of the Student's Book page for this chapter, highlighting the fact that the idea for dialysis came about as the result of experiments on filtration.

Differentiation

For less confident students, consider using the instructions for the practical work on Worksheet 2.1c or guiding them through the process of planning, by looking back at the Student's Book material in Stage 8 Chapter 2 (which covers both the Biuret test for proteins and how to develop a risk assessment).

Ask more confident students to write their own plans for the experiment, together with a risk assessment, and then work together in groups to compare their plans and amend them as necessary.

Homework suggestions

Students answer the questions in Workbook Topic 2.1: *Producing urine*.

Chapter 2 Checking students' progress

End of chapter reflection

- To review the learning points of the chapter
- To test understanding through answering questions

Resources

- Student's Book [End of chapter review questions]
- Workbook [Self-assessment grid, End-of-chapter questions]

Approach

Use the 'Key facts' and 'Check your skills progress' boxes in the Student's Book as a way of recapping the main learning points.

Ask students to complete the Self-assessment checklist at the end of Chapter 2 in the Workbook. Encourage students to work together in small groups to discuss which aspects of the material are the most difficult to understand, and then have a class vote. You may wish to revisit some Learning Objectives again if students find any areas particularly challenging. Use the Student's Book text and questions to revisit these aspects. Go through the answers orally in class. Take note of whether giving the answer appears to resolve students' uncertainty or not, and work through further examples.

The Teacher's comments in the Workbook can be completed to give students personalised feedback on next steps. Students could be asked to answer specific End of chapter review questions in their books. Alternatively go through some or all of the End of chapter review questions in the Student's Book orally in class.

Finish by asking students to complete the End-of-chapter questions at the end of Chapter 2 in the Workbook. Provide feedback on students' answers, and work with students individually to identify areas that still need more work.

Chapter 3 Variation and inheritance

Chapter overview

In this chapter students will investigate variation within animals and plants. They will learn about species and how variation in characteristics is inherited (using ideas about chromosomes, genes, DNA and sexual reproduction). The chapter finishes with a look at how Darwin's theory of natural selection operates on the variations within a population, which can then lead to evolution.

Topic title	Number of 40-minute periods	Learning Objectives	Thinking and Working Scientifically Learning Objectives
3.1 Variation within a species	2–3	9Bp.03 Describe variation within a species and relate this to genetic differences between individuals	9TWSc.07 Collect, record and summarise sufficient observations and measurements in an appropriate form 9TWSa.03 Make conclusions by interpreting results, explain the limitations of the conclusions and describe how the conclusions can be further investigated 9TWSa.05 Present and interpret results, and predict results between the data points collected
3.2 Chromosomes, genes and DNA	1	9Bs.03 Know that chromosomes contain genes, made of DNA, and that genes contribute to the determination of an organism's characteristics 9Bp.03 Describe variation within a species and relate this to genetic differences between individuals	9TWSm.01 Understand that models and analogies reflect current scientific evidence and understanding and can change
3.3 Fertilisation and inheritance	2	9Bp.01 Describe the fusion of gametes to produce a fertilised egg with a new combination of DNA 9Bp.02 Describe the inheritance of sex in humans in terms of XX and XY chromosomes	9TWSm.02 Describe some important models, including analogies, and discuss their strengths and limitations 9TWSc.06 Make an informed decision whether to use evidence from first-hand experience or secondary sources
3.4 Fetal growth and development	2	9Bp.08 Discuss how fetal development is affected by the health of the mother, including the effect of diet, smoking and drugs	
3.5 Natural selection	1	9Bp.04 Describe the scientific theory of natural selection and how it relates to genetic changes over time	9TWSm.01 Understand that models and analogies reflect current scientific evidence and understanding and can change 9TWSa.05 Present and interpret results, and predict results between the data points collected 9TWSc.07 Collect, record and summarise sufficient observations and measurements, in an appropriate form
End of chapter questions	1	Check students' progress	

Background science

Variation within a species is caused by genes, the environment, or a combination of both.

The definition of species at this level should be the idea that it is a group of organisms that can breed to produce offspring that are also able to reproduce, and this is the definition that was met in Stage 7. It is not entirely accurate and species are now more commonly defined in terms of their DNA but this idea is not necessary at this level and can get very complicated.

Take time to ensure that students are clear about the definitions of deoxyribonucleic acid (DNA), genes and chromosomes, and ensure that they understand how they link together and how they link to previous work in Stage 7 (on cells and cell nuclei). This may be the first time that some of these key terms have been met and it is the time where many misconceptions can begin. DNA is the substance that makes up chromosomes (along with proteins). Each chromosome contains a single molecule of DNA, which is bound up with proteins that help preserve its structure. Each species has a certain number of chromosomes in the cell nuclei. For example, humans have 23 pairs of chromosomes (46 in total).

Some sections of DNA contain coded instructions for assembling a certain protein. A section such as this is called a gene. Every individual of the same species has the same genes, but usually has different types of these genes. A different version of a gene is called an allele. For example, all humans contain genes for eye colour but some of us have alleles that cause blue eyes and some have alleles that cause brown eyes. The vast array of different alleles means that there is a lot of variation within a species.

In any species, each individual's set of chromosomes carries the same genes but will show differences in the alleles they have (except for identical twins, who have identical alleles – they are genetically identical). When human gametes are formed, the 23 pairs of chromosomes are separated, meaning that one of each pair goes into every gamete formed. This means that when two human gametes fuse during fertilisation, a fertilised egg cell is formed that has 23 chromosomes from the egg cell and 23 chromosomes from the sperm cell, making a new cell with 23 *pairs* of chromosomes (46 in total). In this way, the new individual has inherited half of its genetic material from the mother and half from the father.

Darwin's theory of evolution by natural selection is based upon the idea that within a population of organisms of the same species, those that are better adapted to their environment are going to be more likely to survive when conditions are harsh. So, they will then be more likely to reproduce and therefore pass their genetic information on to the next generation.

For example, imagine a population of a certain species of bird on an island. Within this population there is variation among the birds. If, by chance, there were a sudden change in the environment, meaning that the only food source remaining for the birds was large nuts that they have to crack open with their beaks, the individuals that have the slightly larger, stronger beaks would be more likely to be able to crack the nuts and gain more food. This means that these individuals are more likely to survive and reproduce, and pass on their genetic information to the next generation. In the next generation slightly more of the population will therefore have the larger beaks. If this environment remains changed, with large nuts as the only available food source, birds with the larger, more powerful beaks will be selected and therefore over many generations there will gradually be more and more birds with larger beaks. Over many, many generations this one population of birds may change (evolve) so much that it becomes a different species.

3.1 Variation within a species

Topic overview

		Student's Book	Workbook	Downloadable material
3.1a	**Starting point** This short activity provides the opportunity for students to demonstrate their understanding of the word variation and to make observations on variation	Starting point table Questions 1–2		
3.1b	**Discontinuous variation** This activity introduces the term discontinuous variation and focuses on the enquiry skills of making observations (including measurements) and deciding on the best way to present data	Questions 3–4 Activity 3.1		
3.1c	**Continuous variation** This activity introduces the term continuous variation and focuses on the enquiry skills of making observations (including measurements) and deciding on the best way to present data	Questions 5–7 Activity 3.2		Practical notes 3.1c(1), 3.1c(2) Worksheets 3.1c(1), 3.1c(2)
3.1d	**Causes of variation** This activity examines the differences between variation caused by inheritance and variation caused by the environment.	Questions 8–13	Questions 1–7	

Learning Objectives

- 9Bp.03: Describe variation within a species and relate this to genetic differences between individuals
- 9TWSa.03: Make conclusions by interpreting results, explain the limitations of the conclusions and describe how the conclusions can be further investigated
- 9TWSa.05: Present and interpret results, and predict results between the data points collected
- 9TWSc.07: Collect, record and summarise observations and measurements in an appropriate form
- 9SIC.02: Describe how science is applied across societies and industries, and in research
- 9SIC.05: Discuss how the uses of science can have a global environmental impact

Links to other topics

- This topic builds on work from Stage 7 Chapter 3, in which variation between species was considered. Students now turn their attention to variation between members of the same species.

Science in context

Activity 3.2 in Topic 3.1 in the Student's Book provides the opportunity to cover the following learning objective:

- 9SIC.02 Describe how science is applied across societies and industries, and in research.

Teaching ideas

When students are doing their own work on variation in humans (possibly by measuring their heights), encourage them to think about what sort of industries use data of this type. Working out average heights of people, and how the changes in the number of people who have heights that are further and further from this average, is very important in clothing manufacture and retailing, and is also important in aspects of car design, building design, furniture design and planning, office and factory layouts. Students should be able to think of some other examples to add to this list.

Question 14 below the Science in context: Coral reef bleaching box in Topic 3.1 of the Student's Book covers the following learning objective:

- 9SIC.05 Discuss how the uses of science can have a global environmental impact.

Teaching ideas

Students should consider ways in which changes in patterns of environmental variations can be used as indicators of environmental change (e.g. bleaching of corals due to sea temperature rise).

Common misunderstandings

- The main misunderstandings come with the causes of variation and whether or not they are genetic or environmental. Explain to students that organisms inherit many of their characteristics such as eye colour but the environment will also have an effect, such as diet affecting an animal's size.

3.1a Starting point

Ask students to look at the pictures of the proboscis monkey (*Nasalis larvatus*) and the orangutan (*Pongo species*) in figures 3.1 and 3.2 in the Student's Book. Ask students if they can remember what a species is (from their work in Stage 7) and establish that it is a group of organisms that can reproduce to produce fertile offspring. Discuss the differences between the two species and remind students of the term variation. Ask students to suggest some variation that might occur between members of the same species (e.g. proboscis monkeys). Encourage students to answer questions 1–2.

3.1b Discontinuous variation

Students will learn:

- To describe how genetic differences between organisms can lead to variation within the species
- To use charts to display variation
- To collect and record observations and measurements appropriately
- To present results appropriately
- To interpret results and form conclusions using scientific knowledge and understanding

Glossary words

discontinuous variation, variation, whole number

Resource preparation

Squared or graph paper for drawing bar charts

Teaching and learning ideas

- Ask the students to see if they have bent little fingers or straight little fingers. Record the results in a table on the board. This should be done by a quick count of hands. Explain that this is variation with a distinct set of options called discontinuous variation. You can be in one group or another and not in between. Show the students how to present the results as a bar chart with spaces between the bars.
- Complete a quick survey of the class to find out everyone's shoe size. Show the students how to record this data in a tally chart (which is introduced more formally to students in 3.1c below). Then ask the students to complete a bar chart individually.
- Challenge students to collect data about eye colour in the class and present their data in a bar chart.
- Together with students, look at the section titled Discontinuous variation in the Student's Book. Encourage students to answer questions 3–4.

Differentiation

Less confident students may need to be provided with the scale on the *y*-axis of their bar charts.

More confident students will be able to draw bar charts without any help.

Homework suggestions

Ask students to complete Activity 3.1 *Investigating discontinuous variation* in the Student's Book.

3.1c Continuous variation

Students will learn:

- To describe how genetic differences between organisms can lead to variation within the species
- To use charts to display variation
- To collect and record observations and measurements appropriately
- To present results appropriately
- To interpret results and form conclusions using scientific knowledge and understanding

Glossary words

continuous variation, data, range, tally chart

Resource preparation

For activities: two sets of leaves from the same species – one from plants grown in shade, the other from plants grown in full light – at least five leaves in each set; 30 cm ruler; 1 m ruler; graph paper

Practical notes 3.1c(1), 3.1c(2)

Worksheets 3.1c(1), 3.1c(2)

Teaching and learning ideas

- Choose at least 10 students to come to the front of the class and arrange them in height order. Note: be careful to avoid selecting very tall or very short students if they may be uncomfortable having their height compared with others. Explain that this shows an example of continuous variation. This is where variation has any value within a range.
- Together with students, look at the section titled Continuous variation in the Student's Book. Encourage students to answer questions 5–13.
- There are two practical activities suggested here; one uses plant material and one uses the students themselves. It is suggested that you carry out one or the other.
- Activity – Individually or in pairs ask students to measure the lengths of different leaves from two groups of leaves from the same species; leaves that have grown in the shade and those that have grown in full sunlight. They should record their results in a table and calculate the average length of the leaves from each group – see Worksheet 3.1c(1) and Practical notes 3.1c(1). Some students may find parts of this activity challenging and will need more guidance than others.
- Activity – Individually or in pairs ask students to measure variation within the class. They should record their results in tables. There are different features that students can measure or record. The worksheet and Activity 3.2 in the Student's Book suggest ones that should not cause any controversy. Note: When completing Activity 3.2 you may have students who are reluctant/embarrassed to take off their shoes/socks. If this is the case you could ask the whole class to measure their height with their shoes on and then have a class discussion about the validity of the results and whether valid conclusions can be made. Weight, eye colour and hair and skin colour should be avoided – see Worksheet 3.1c(2) and Practical notes 3.1c(2). Students analyse their collected data on variation within the class by first producing appropriate charts and graphs. They should then make conclusions about the data collected – see Worksheet 3.1c(2).
- Challenge students to plan how they could test the question: do taller people have bigger feet? They should decide what they should measure, how many people they should sample, what control variable they might need (such as age group or gender). Ask students what sort of graph or chart they should draw to see if there is a correlation (relationship) between these two variables, and remind them if necessary that we use scatter graphs for this purpose.

Differentiation

Less confident students may need help presenting the data in the form of graphs and charts. They could be given the axes and just asked to plot the data. They may also need help with the mean (average) calculation. More confident students should be able to analyse the data without assistance.

Homework suggestions

Students answer questions 1–7 in Topic 3.1: Variation within a species in the Workbook.

Provide the students with a set of continuous data to copy down and draw tally charts and bar charts (without gaps between the bars).

3.1d Causes of variation

Students will learn:

- To describe how genetic differences between organisms can lead to variation within the species
- To present results appropriately

Glossary words

environmental variation, inherited variation

Resource preparation

Photos from the internet showing inherited variation in humans (e.g. pairs of people with different eye colours, natural hair colours, wavy/straight hair, lobed/lobeless ears, skin colour)

Teaching and learning ideas

- Look at figure 3.8 in the Student's Book and ask students why the tree grew in this shape. Then ask them to draw or explain what the tree may have looked like if the wind was not blowing. Establish the idea that the shape of the tree has been formed by the wind, and that other trees of the same species will not be this shape. This is an example of environmental variation – variation caused by the conditions around an organism.
- Tell students that humans also have characteristics that vary because of their surroundings (e.g. having a scar from a cut, hair length, speaking different languages). Point out to students that all humans also have characteristics in common (such as having two eyes, having red blood cells). However, there is some variation in these characteristics – some people have different coloured eyes, different people have different types of blood (called 'blood groups'). These variations are inherited from our parents – they are examples of inherited variation.
- Show students some photos downloaded from the internet of inherited variation in humans. See suggestions above.
- Together with students, read through the section of the Student's Book entitled Causes of variation. Encourage students to answer questions 8–13.
- Show students the photos again, and ask how each of the variations shown may be further changed by the environment (e.g. coloured contact lenses, hair dye, hair styling, earrings, exposure to the Sun).
- Ask students to suggest different ways in which there is variation in the characteristics of humans. Write these up on the board and ask students to classify each as 'inherited variation', 'environmental variation' or 'both inherited and environmental variation'. Encourage students to present this work as a table.
- Challenge students to suggest why there are small amounts of visible variation between two identical twins. Students should understand that identical twins have the same inherited variations, and that any differences between them are due to environmental variation.

Differentiation

Less confident students may need help in sorting variation into different groups. Point out to students that a lot of variation has both inherited and environmental causes. More confident students could come up with further examples of traits showing examples of environmental and inherited variation.

Homework suggestions

Students answer the questions in Topic 3.1: *Variation within a species* in the Workbook.

3.2 Chromosomes, genes and DNA

Topic overview

		Student's Book	Workbook	Downloadable material
3.2a	**Starting point** This is focused on the key terms for this topic	Starting point table		Worksheet 3.2a
3.2b	**Inheritance and variation** Students consider the underlying causes of inherited variation	Questions 1–5 Activity 3.3	Questions 1–5	

Learning Objectives

- 9Bs.03: Know that chromosomes contain genes, made of DNA, and that genes contribute to the determination of an organism's characteristics
- 9Bp.03: Describe variation within a species and relate this to genetic differences between individuals
- 9TWSm.01: Understand that models and analogies reflect current scientific evidence and understanding, and can change
- 9SIC.01: Discuss how scientific knowledge is developed through collective understanding and scrutiny over time
- 9SIC.04: Describe how people develop and use scientific understanding as individuals and through collaboration

Links to other topics

- This topic builds on work from Stage 7 Chapter 1, in which students first met cells and cell nuclei.

Using models

- This topic explores some different types of models, including physical models, and how they are used in the development of scientific understanding. Watson and Crick made a physical model when working on the structure of DNA. In Lesson 3.2b students are shown a physical model of a chromosome made by colouring in short sections on a piece of white cotton thread and then wrapping this in modelling clay/sticky tack (putty-like adhesive material) to form a chromosome shape.

Science in context

Questions 4 a–e below the Science in context: Discovering DNA box in Topic 3.2 of the Student's Book cover the following learning objectives:

- 9SIC.01 Discuss how scientific knowledge is developed through collective understanding and scrutiny over time.
- 9SIC.04 Describe how people develop and use scientific understanding as individuals and through collaboration.

Teaching ideas

Encourage students to think about ways in which an increasing knowledge of DNA and chromosomes is changing the way in which we live our lives. Students may have heard of police investigators collecting DNA from a place and using it to identify the people that had been in that place. Or they may have heard about using DNA analysis to identify the causes of someone's illness or using DNA to make medicines specifically for one person.

Common misunderstandings

- Some students may not realise that genetic material is found in all living things. In animals and plants, it is in the form of DNA.
- Some students think that chromosomes contain lots of short pieces of DNA (called genes). Ensure they understand that one very long molecule of DNA is coiled up and folded in a specific way by proteins to form a chromosome. This one long DNA molecule contains sections that are genes (a bit like a long book in which there are sections of random letters and then sections that make sense and can be read).
- Many students may forget that an individual has an exact copy of their chromosomes in the nucleus of almost every single cell in their body (exceptions include gametes, which have half the number of chromosomes).
- Students may hold the misconception that individuals are only different because of the differences in their DNA; it is important that students consider the impact of the environment as well. Beyond this, however, students should be clear that changes made to an organism during its lifetime will not affect the DNA of its gametes and therefore will not be passed on to the next generation. (Note that this is a bit of a simplification, and the field of epigenetics is throwing new light on our understanding of inheritance, but this is well beyond what students should learn at this level.)

3.2a Starting point

Resource preparation

Students will need access to secondary sources, including textbooks, dictionaries and the internet, to research the key terms.

Worksheet 3.2a

One of the most challenging aspects of this topic for students are the key terms – understanding what they mean and being able to understand the relationship between them.

Give students the key terms and ask students to work in groups to come up with a definition for each: chromosome, DNA, environmental variation, gene, genetic material, inherited variation, sexual reproduction, variation.

Ask students to write down what they think each of these terms means before doing any research. Once students have their ideas, give them 10 minutes to use the Student's Book and other secondary sources and develop or alter their original ideas to produce clear written definitions for all of these key terms. Students should show how they are changing and developing their original definitions, rather than just copying new definitions out of a secondary source.

As an alternative to this, students can complete Worksheet 3.2a, which gives students all of the key terms and definitions, and they have to match them. This could be done by drawing lines between the key terms and definitions or by cutting out all the boxes and sticking the key terms next to the appropriate definition. As a challenge for some students, remove one or two of the key terms and/or definitions before copying/printing the sheets and ask students to fill those in.

3.2b Inheritance and variation

Students will learn:

- To describe how a chromosome contains a molecule of DNA, some sections of which are genes
- To describe the role of genes in inherited variation
- To understand that models and analogies can change according to scientific evidence

Glossary words

chromosome, DNA, gene, genetic material

Resource preparation

Secondary sources including textbooks, internet access (optional: video material from the internet showing chromosomes)

Model of chromosome (made by colouring in short sections on a piece of white cotton thread and then wrapping this in modelling clay/sticky tack to form a chromosome shape)

Activity 3.3 *Researching work on genes* (extension activity)

Teaching and learning ideas

- To start, ask students to review the work on the key terms that they did in 3.2a and use this to write a few sentences, trying to use as many of the key terms as correctly as they can. You could call this the 'social media challenge' and limit students to just 140 characters if you wish, or simply limit it to three sentences. To review, ask students to share their 'posts' or sentences.
- Together with students, read through the DNA, Chromosomes, Genes and Discovering DNA sections in the Student's Book. Ask students to answer questions 1–4.
- It may help students to put all of the key terms from this topic together as a diagram. This is covered in question 5 and it is recommended that this is done as class, by adding to a diagram on the board. Draw the outline of an animal cell, and ask students where chromosomes would be found. Establish that this is in the nucleus and then draw a call-out from the nucleus to a magnified view of chromosomes. Then ask where DNA is found, and then where genes are found and add these to the diagram. Point out to students that this is a model that helps us to understand how everything fits together. There are some useful videos on various internet video storage sites that can be found using search terms such as 'What is a chromosome?'. Be aware, however, that many of these videos go beyond what students need to know at this level and it may be best to show the students this material without any sound, so that you can explain what is being shown.
- Show students the model of the chromosome (made of thread and modelling clay). Ask students what a chromosome is made of and then pull out the thread. Ensure that students realise that the thread represents the DNA molecule. Then ask students what the coloured sections of the thread represent (genes). Point out to students that this is an example of a physical model (one that you can touch) and ask them to think of an example of scientists using a physical model to develop their ideas (e.g. Watson and Crick).
- For an extension activity, use Activity 3.3 *Researching work on genes*. Students consider the Human Genome Project (HGP) and answer a series of short questions based on their findings. You could encourage students to work in groups, and to decide for themselves which team members answer which questions. The students then bring back their findings to the group and work together to create a report or poster to answer the questions. There are parallels to be drawn here in how teams of scientists work together, which is particularly true of big international projects such as the HGP.
- To finish, ask students to draw a model to show the relationship between cells, chromosomes, genes and DNA in a plant.

Differentiation

It may help less confident students if you provide them with the definitions of all the key terms so that they can use these to create their social media post or sentences (at the start of this learning episode).

Challenge more confident students to design a physical model to show the relationship between chromosomes, DNA and genes before showing the model.

Extension ideas

Activity 3.3 (see above). Students could research some ways in which a knowledge of DNA has changed our lives.

Homework suggestions

Workbook topic 3.2, questions 1–5

Activity 3.3 *Researching work on genes* could be completed as homework if not completed as part of the learning episode.

3.3 Fertilisation and inheritance

Topic overview

		Student's Book	Workbook	Downloadable materials
3.3a	**Starting point** This short activity revises the basic parts of a cell and reminds students of specialised cells	Starting point table		
3.3b	**Sexual reproduction** Students consider the structure and function of egg and sperm cells, labelling the different parts of each cell	Questions 1–4	Questions 1–4	Worksheet 3.3b
3.3c	**Fertilisation** Students first define fertilisation and consider the processes that are required to happen before fertilisation can occur. They then consider the structure and function of different parts of the male and female reproductive systems in humans	Questions 5–11	Questions 5–10	
3.3d	**Sex determination** Students consider how sex is inherited	Questions 12–18 Activity 3.4		

Learning Objectives

- 9Bp.01: Describe the fusion of gametes to produce a fertilised egg with a new combination of DNA
- 9Bp.02: Describe the inheritance of sex in humans in terms of XX and XY chromosomes
- 9TWSm.02: Describe some important models, including analogies, and discuss their strengths and limitations
- 9TWSc.06: Make an informed decision whether to use evidence from first-hand experience or secondary sources

Links to other topics

- This topic builds on work on specialised cells and organ systems from Stage 7 Chapter 1.

Using models

- Diagrams are used to clarify the types and number of chromosomes in human cells, and the roles played by the different parts of sperm and egg cells.
- Physical models can be used to illustrate the difference in size between sperm cells and egg cells. Note that the size difference between the models must be correct and appropriate. The sperm cell should be roughly 40 times smaller than the egg cell. These models could also be used to address Learning Objective 9TWSm.02, where students are given the opportunity to consider the strengths and limitations of the model. There are further opportunities to build physical models and to use animation as a way to model processes.

Common misunderstandings

- Some students may believe that copulation/sex/sexual intercourse and reproduction are the same things. It is important to clarify this with them so that they understand that sexual reproduction is the process by which offspring are produced using half the information from the mother and half from the father (via the use of gametes). The fusion of sperm and egg nuclei is fertilisation.
- Some students may believe that a tiny baby is inside the egg or the sperm and this then just grows inside the mother. It is important to make it clear to students that information from both the mother and the father (from the egg cell and the sperm cell) is needed to join together to make a fertilised egg cell. This then grows and divides, forming an embryo and then a fetus.

3.3a Starting point

Draw an animal cell on the board, and ask students what the parts are (revising cell membrane, mitochondria, nucleus and cytoplasm). Then challenge students to draw their own diagrams of plant cells.

Ask students why not all the cells in an organism are the same. Establish the idea that different cells have different functions and therefore have different features. Ask students to give examples of some specialised (reminding them of work in Stage 7 Chapter 1). Tell them that specialised cells called gametes are used by many organisms for reproduction (including plants and humans).

3.3b Sexual reproduction

Students will learn:

- To describe how fertilisation in humans occurs
- To explain why organisms inherit features from both their parents
- To describe the role of the XX and XY chromosomes in human sex determination
- To describe the uses, strengths and limitations of models and analogies

Glossary words

egg cell, embryo, fertilisation, fertilised egg cell, gamete, sex cell, sexual reproduction, sperm cell

Resource preparation

Micrograph images of sperm and egg cells (e.g. downloaded from the internet)
Tennis ball, poppy seed (or other items of a similar size difference)
Access to secondary sources of information (e.g. the internet)
Worksheet 3.3b

Teaching and learning ideas

- Explain to students that many organisms produce male and female cells that join together to produce an embryo, which then grows and develops into a new organism. This is how those organisms reproduce and is known as sexual reproduction. Ask students if they can remember the names of the male and female cells that allow flowering plants to reproduce from their previous studies. Remind them that these are the male pollen grain and the female egg cell. Explain that when a pollen grain gets into a flower, its nucleus combines with the nucleus of an egg cell to form an embryo inside a seed. This is how flowering plants reproduce sexually.
- Explain that humans also reproduce sexually, like all other mammals. Together with students, look at the section in the Student's Book on Sexual reproduction. Ask students to identify the nucleus, cytoplasm and cell membrane in the sperm and egg cells – parts of the cell that they will have studied in Stage 7. Go through the adaptations of the cells and discuss the functions of these. Encourage students to answer the questions.
- Ask students to draw a table to show the structures of the sperm cell and egg cell along with the functions of these. Challenge more confident students to do this without reference to the Student's Book. They can then check their answers with the Student's Book and make additions/corrections as necessary. For example:

Structure of an egg cell

Structure	Function
Food store	To provide nutrients to the fetus as it develops

- Show students micrograph images of sperm cells and egg cells and discuss with students the difference in size of the two cells. A human egg cell is 130–200 μm in diameter (about the size of a full stop on this page). A sperm cell is about 60 μm in length and its head is about 5 μm wide. So, in terms of size difference, the egg cell is about 40 times bigger than the head of a sperm cell.
- Ask students what could be used to help visualise the parts of the cells more easily and the size difference between the sperm and egg cells. Elicit the idea that this is where models can be useful. Point out how the diagrams of the sperm and egg cells in the Student's Book are simplified compared to what is seen in a micrograph. This makes it easier to see the main parts and understand what they do. To help visualise the size difference we could use a physical model (such as a poppy seed and a tennis ball – although you may be able to find objects that give a more accurate representation of the size difference).
- Try to show different types of images, including diagrams, computer-generated images and images from high-powered microscopes.
- Finish by asking students to complete Worksheet 3.3b, labelling the parts of the sperm cell and egg cell and considering their adaptations.

Differentiation

You may need to give less confident students lists of the labels and adaptations to use to complete Worksheet 3.3b. Or provide a table showing the functions of the different parts of the sperm cell and egg cell with structure names missing. Students then add the correct structure names next to the relevant function as a preliminary exercise, and then use their completed tables to complete the worksheet. For example,

Structure	Function
	To store the genetic information for the cell.

Challenge more confident students to find out whether bigger species of animals produce larger gametes. (They do not.)

Extension ideas

Some students could be encouraged to find out about other specialised cells, explaining what features they have and why those features are necessary to help the cells to carry out their specific functions.

Homework suggestions

Give students a micrograph of a sperm and/or an egg cell downloaded from the internet and printed out. Ask students to draw diagrams to model the different parts and label them with their names and functions.

3.3c Fertilisation

Students will learn:

- To describe how fertilisation in humans occurs
- To decide whether evidence from first-hand experience or secondary sources should be used

Glossary words

cervix, oviduct, penis, prostate gland, semen, sperm duct, testis, uterus, vagina

Resource preparation

Video clips showing fertilisation (i.e. from the internet; these must be watched prior to the lesson)

Recording devices (e.g. smartphones or tablets)

Teaching and learning ideas

- Together with students, look at the section in the Student's Book on Fertilisation in humans. Encourage students to answer questions 5–7.
- There are many good videos available on the internet to show students what happens as two gametes fuse to form a fertilised egg cell. These could be shown to students without sound (or subtitles) and students challenged to write a voice-over for the video. Students' voice-overs could be recorded and shared with the class if time and equipment permit.
- Encourage students to think about how knowledge of sexual reproduction in humans can help doctors to help people to become parents (e.g. IVF).

Differentiation

For less confident students, provide access to the Student's Book to help them complete the worksheet(s).

Ask more confident students to design their own worksheets to test knowledge and understanding of this topic. They can use the artworks on the worksheets to help them. They should prepare answers and mark schemes. If time allows they could try out their worksheets.

Extension ideas

Students should consider identical and non-identical twins. You could provide them with the following information:

There are two types of twins: identical twins and non-identical twins.

- One type of twins forms when two eggs are released from the ovaries and both are fertilised by different sperm.
- The other type is formed from just one egg that is fertilised by one sperm. The fertilised egg then divides and splits in two.

Students should use this information, along with secondary sources if required, to explain which type of twins are produced from which method and why this leads them to being identical and non-identical. Students will be introduced briefly to identical twins in the next section in the Student's Book.

Homework suggestions

Ask students to prepare test questions worth a total of 5 marks on the subject of fertilisation. They should present their questions together with a mark scheme. If time allows, questions could be answered by students in the following lesson, and feedback given on the structuring of the questions and mark schemes.

3.3d Sex determination

Students will learn:

- To describe the role of the XX and XY chromosomes in human sex determination
- To decide whether evidence from first-hand experience or secondary sources should be used
- To describe the uses, strengths and limitations of models and analogies

Glossary words

genetic material, inheritance, sex chromosome, sex determination

Resource preparation

For Activity 3.4: Strips of card or pieces of modelling clay in two different colours to represent chromosomes.
Camera (e.g. smartphone or tablet) and stand for optional stop–motion animation, or computer animation software

Secondary source material for research (possibly including internet access)

Teaching and learning ideas

- Ask students to find out the number of chromosomes in the body cells of five different animals of varying sizes. Challenge students to draw bar charts of their data and to use their data to say whether it is true that bigger animals have more chromosomes in their cells. (It is not true.)
- Ask students why all the animals in their research had even numbers of chromosomes in their cells. Establish that this is because chromosomes come in pairs. Both chromosomes in the pair are the same size and contain molecules of DNA that contain the same genes.
- With students read through the Human chromosomes section in the Student's Book.
- Ask students to suggest what it is about their cells that makes them either male or female. Ensure that students realise that this is also a form of inheritance.
- With students read through the Sex determination and Brothers, sisters and twins sections in the Student's Book.
- Activity 3.4 *Modelling chromosomes*. Students should work in groups to discuss how they might make a model to explain how each person gets half their chromosomes from their mother and half from their father. Ideas could include a single diagram, a series of diagrams or a physical model using modelling clay or pieces of coloured card. Some students might suggest a stop–motion animation or a computer animation of the process. Ask students to answer the activity questions A1–A3 in the Student's Book. If there is time and equipment students could make their models, but check each group's finished idea before allowing them to start.
- Finish by asking students to answer questions 12–18 in the Student's Book.

Differentiation

Some groups of students may be helped by being given more guidance in the activity, and being provided with the equipment that they need. This will need the teacher to decide what sort of model they are going to produce.

Extension ideas

Ask students to find out about sex determination in other animals.

Homework suggestions

Workbook Topic 3.3 questions 1–10

3.4 Fetal growth and development

Topic overview

		Student's Book	Workbook	Downloadable materials
3.4a	**Starting point** This short activity provides an opportunity to establish the extent of your students' prior knowledge about the development of a fetus	Starting point table		
3.4b	**The effect of diet, smoking, drugs and disease on the developing fetus** Students describe the effects that diet, various drugs and diseases can have on the developing fetus. They are also required to find information from secondary sources and present it in an appropriate way	Questions 1–12 Activity 3.5 Activity 3.6	Questions 1–10	

Learning Objective

- 9Bp.08: Discuss how fetal development is affected by the health of the mother, including the effect of diet, smoking and drugs

Links to other topics

- This topic builds on work on organ systems from Stage 7 Chapter 1 and work on balanced diets from Stage 8 Chapter 2.

Science in context

Question 1 following the Science in context: Fetal scanning box in Topic 3.4 of the Student's Book covers the following learning objective:

- 9SIC.02 Describe how science is applied across societies and industries, and in research.

Encourage students to think about how technology is used to monitor pregnancy (e.g. scanning).

Common misunderstandings

- Some students may not think that humans are alive until they are born. The fertilised egg cell is, though, a living cell.
- Many think that blood from the fetus and the mother mix in the placenta. This does not happen – the two different bloods never mix. Blood vessels from the mother and the fetus run very close to each other in the placenta and therefore substances can be exchanged between the two (substances diffuse from the mother's blood to the baby's blood and vice versa).

3.4a Starting point

Teaching and learning ideas

- Pose this question to your students: How does a fertilised egg become a baby?
- Encourage students to write as many ideas as they can. You can provide pointers, such as What does the developing baby need? How does it get these things? What is happening to the developing baby?
- Once students have written down their ideas they could share these with a partner or in a small group.
- Students should keep a written record of their thoughts to review at the end of this topic.

3.4b The effect of drugs on the developing fetus

Students will learn:

- To discuss the ways in which a mother's health and lifestyle choices (diet, smoking and drugs) can affect the development of a fetus
- To decide whether evidence from first-hand experience or secondary sources should be used
- To interpret results and form conclusions using scientific knowledge and understanding

Glossary words

drug, stillborn

Resource preparation

For Activities 3.5 and 3.6: Materials for poster making, secondary sources of information on fetal development, including any appropriate leaflets from healthcare centres and the internet

Teaching and learning ideas

- To start, look at the section Diet during pregnancy in the Student's Book. Encourage students to answer the questions.
- Activity 3.5 *Healthy eating during pregnancy*. Together with material from the Stage 9 Student's Book, students should look back at their work in Stage 8 Chapter 2 to think about what they would recommend for a pregnant women to have in her diet. Students could do further research using secondary sources, and should be encouraged to present their findings as a poster.
- Next, ask students to list some things (apart from food) that it may be dangerous for a mother to have. Write the ideas up on the board and elicit the idea that drugs (including medicines) can be harmful to a developing fetus.
- Together with students, look at the Drugs and pregnancy and the Tobacco smoke and pregnancy sections in the Student's Book. Encourage students to answer the questions.
- Activity 3.6 *Researching the effects of drugs on a fetus' development* in the Student's Book. Students will need to use secondary sources to research information about drugs and how they can affect the growth and development of a fetus. They are then asked to present this in a suitable format to give information to someone who is trying to become or who is pregnant.
- Finally, review the section Diseases and pregnancy in the Student's Book. Encourage students to answer the remaining questions in this topic.

Differentiation

Help less confident students with the activity by suggesting certain drugs for them to research (e.g. isotretinoin, tetracycline, warfarin, and possibly recreational drugs such as heroin and alcohol). You may also find a couple of suitable websites for them to use for their research or specific healthcare leaflets.

Ask more confident students to focus on explaining why using certain drugs cause development problems for the fetus (such as the concern that carbon monoxide, inhaled when smoking, reduces the oxygen carrying capacity of blood, and therefore reduces the amount of respiration that can happen and subsequently slows down the growth and development of the fetus).

Extension ideas

Ask students to research 'fetal alcohol spectrum disorder'. They should consider how it is caused, the levels of severity, problems associated with it, difficulties that children suffer with it and distinctive features of children that have it.

Homework suggestions

Workbook Topic 3.4 questions 9–11

3.5 Natural selection

Topic overview

		Student's Book	Workbook	Downloadable material
3.5a	**Starting point** A recap on adaptations of animals and survival rates	Starting point table		Worksheet 3.5a
3.5b	**Adaptation** Students look at how organisms are adapted to the environmental conditions in which they live	Questions 1–4 Activity 3.9		
3.5c	**Natural selection** Students consider the work of Darwin and the theory of natural selection	Questions 5–14 Activity 3.10	Questions 1–6	Worksheet 3.5c Practical notes 3.5c

Learning Objectives

- 9Bp.04: Describe the scientific theory of natural selection and how it relates to genetic changes over time
- 9TWSc.07: Collect, record and summarise sufficient observations and measurements, in an appropriate form
- 9TWSa.05: Present and interpret results, and predict results between the data points collected
- 9SIC.01: Discuss how scientific knowledge is developed through collective understanding and scrutiny over time
- 9SIC.04: Describe how people develop and use scientific understanding as individuals and through collaboration
- 9TWSm.01: Understand that models and analogies reflect current scientific evidence and understanding, and can change

Links to other topics

- This topic builds on work on ecosystems from Stage 8 Chapter 3.

Using models

- Activity 3.10 *Investigating natural selection* provides the opportunity to use a model to help students understand natural selection. In the model, forceps and tongs are used to represent bird beaks and different sizes of seed are used to represent prey. Students should try using the different tools to pick up the different seeds and link this to natural selection of birds with different beak shapes. They can also be asked to discuss the strengths and limitations of the model to address learning objective 9TWSm.02.

Science in context

Questions 10 a–e below the Science in context: Charles Darwin box in Topic 3.5 of the Student's Book covers the following learning objectives:

- 9SIC.01 Discuss how scientific knowledge is developed through collective understanding and scrutiny over time
- 9SIC.04 Describe how people develop and use scientific understanding as individuals and through collaboration.

Teaching ideas

Darwin used ideas from others (e.g. Thomas Malthus) and shared his ideas with other scientists (e.g. Alfred Russel Wallace) as he developed his theory. When he published his book, he had collected an enormous amount of evidence to support his ideas, which helped them to gain support quickly by many scientists.

Encourage students to consider plants and animals that are local to your area or country, and how they are adapted to where they live. They could also consider how some of these adaptations (such as thick or soft fur) make them useful to humans.

Ask students to think about any organisms that are now extinct but whose remains/fossils have been found in your country/local area. Challenge them to think about why the organisms became extinct.

Common misunderstandings

- Students can often think that natural selection and evolution are the same thing. It is important to emphasise that these are two different things. Natural selection can lead to evolution. Students must be clear that evolution (as applied to organisms) is a gradual change in characteristics of a species over many generations. Darwin (and Wallace) proposed the theory of evolution by natural selection to explain how this happens.

3.5a Starting point

Resource preparation

Worksheet 3.5a

Ask students to complete Worksheet 3.5a. This gives them the scenario of introducing populations of two different frog species into two different environments. Students are required to consider the survival chances of both species of frog in the different environments and give an explanation of which species of frog they think will be more likely to survive in each habitat and why they think this.

Review with students, sharing their ideas about how they think the population sizes will alter and why.

3.5b Adaptation

Students will learn:

- To describe how organisms are adapted to where they live

Glossary words

adaptation, ecosystem

Resource preparation

Access to secondary sources of information (including the internet)

Teaching and learning ideas

- First read the section Adaptation in the Student's Book and ask the students to answer questions 1–3. Ensure that students understand that an organism's adaptations will help it to survive. Ask students to give the names of some plants and animals (maybe local to your area or country) and how they are adapted to their environmental conditions.
- Activity 3.9 *Imaginative adaptations*. Ask students to complete this activity individually, and then to work through the questions with a partner (as outlined in the Student's Book). Students might include wide feet (to stop the animal sinking into the sand), a sandy colour (for camouflage) and a way to store water (like a camel does in its stomach). Challenge some students to go on to design a plant that is adapted for the same conditions. Another version of this activity is found in Stage 9 Chapter 4.
- Read the section Bird adaptations in the Student's Book and ask the students to answer question 4. Challenge students to list a few common birds in your area/country and to find out how their beaks are adapted for what they eat.
- Finish by asking students to write a definition for the term 'adaptation'.

Differentiation

Less confident students could be helped in the activity by giving them a set of descriptions for a desert habitat (e.g. very dry, sandy, hot in the day, cold at night). More confident students could find their own examples of adaptations.

Extension ideas

Ask students to find out how one plant and one animal are adapted to the same environmental conditions/habitat. For example, pine trees (*Pinus* species) and Artic foxes (*Vulpes lagopus*) (in the Arctic), jackrabbits (*Lepus* species) and cacti (in a desert), seaweeds and limpets (in tidal zones).

Homework suggestions

Activity 3.9 could be given as homework, if not already done in class.

Alternatively, ask students to write about one or two organisms from your local area/country and how they are adapted to help them survive.

3.5c Natural selection

Students will learn:

- To describe how genetic differences between organisms can lead to variation within the species
- To describe how genetic changes over time result in natural selection
- To collect and record observations and measurements appropriately
- To evaluate how the prediction is supported by the evidence collected
- To interpret results and form conclusions using scientific knowledge and understanding
- To present results appropriately
- To understand that models and analogies can change according to scientific evidence

Glossary words

evolution, natural selection, population

Resource preparation

Activity 3.10 *Investigating natural selection.* Rice, dried peas, haricot beans, forceps, small blunt knife, tongs, large tray/piece of white paper, timer. See teacher and technician practical notes for further support.

Worksheet 3.5c

Practical notes 3.5c

Teaching and learning ideas

- First read through the sections on Natural selection and Evolution in the Student's Book. Encourage students to the answer questions 5–9.
- Read through the Charles Darwin section in the Student's Book with students. Ask students to answer question 10 about Darwin and his work, discussing their ideas either in pairs or small groups. Encourage students to consider how Darwin came up with his idea about natural selection and the fact that he needed lots of observations and evidence that he could continually return to, as well as sharing ideas with peers, in order to formulate his theory. Ensure that students understand that 'evolution' (changes within populations over time) had already been observed by scientists; Darwin was trying to explain how this happened (Darwin did not describe evolution for the first time).
- Read through the sections on Natural selection and genetics and The peppered moth (*Biston betularia*) – natural selection in action. Encourage students to answer questions 11–13, and question 14 for a challenge.
- Activity 3.10 *Investigating natural selection.* In this activity, students use rice and beans to represent prey items which are the same species but different sizes, and then use different tools to pick these up – investigating which tool is the best for picking up the different-sized prey items. Students should link their findings to the idea of natural selection. Students should use Worksheet 3.5c during this activity as it gives the method and questions relating to this activity. See teacher and technician practical notes for more support and an equipment list. Challenge students to do the extension activity on Worksheet 3.5c. You could also challenge students to identify one strength of the model (e.g. it illustrates how selection occurs) and one limitation (e.g. the selection in this model is partly due to human choice, the model only demonstrates selection caused by a predator).
- Finish by playing a key term game with students using all of the key terms from this chapter. 'Taboo' would be a good game to play here; students should work in pairs and take it in turns to describe/define a key term to their partner without saying the word or a derivative of the word.

Differentiation

It may help some less confident students if they do Activity 3.10 *Investigating natural selection* before going through the theory of natural selection as it may help them to have a practical scenario to relate to.

Ask more confident students to explain how their own investigation demonstrates natural selection. Some students may be able to extend and develop the activity themselves, including additional tools that could represent different types of beak, or they could also be encouraged to develop a completely different activity that could be used to demonstrate natural selection.

Extension ideas

Students should use secondary sources to research different examples of natural selection. Encourage students to find out an example of evolution by natural selection in plants and in animals.

There are many short videos on the internet that explain the work of Darwin and how he came up with his theory of natural selection. It would be good for students to watch some of these once they have worked through the content in the Student's Book.

Homework suggestions

Workbook Topic 3.5, questions 1–6

Chapter 3 Checking students' progress

End of chapter reflection

- To review the learning points of the topic
- To test understanding through answering questions

Resources

- Student's Book [End of chapter review questions]
- Workbook [Self-assessment grid, End-of-chapter questions]
- Worksheets 3.2a, 3.3b, 3.4b(1), 3.4b(2), 3.5a [all of which may have already been used]

Approach

Use the 'Key facts' and 'Check your skills progress' boxes in the Student's Book for each topic in this chapter as a way of recapping on the main learning points.

Ask students to complete the Self-assessment checklist at the end of Chapter 3 in the Workbook. Encourage students to work together in small groups to discuss which aspects of the material are the most difficult to understand, and then have a class vote. You may wish to revisit some Learning Objectives again if students find any areas particularly challenging. Use the Student's Book text and questions to revisit these aspects. Go through the answers orally in class. Take note of whether giving the answer appears to resolve students' uncertainty or not, and work through further examples.

The Teacher's comments in the Workbook can be completed to give students personalised feedback on next steps. Students could be asked to answer specific End of chapter review questions in their books. Alternatively go through some or all of the End of chapter review questions in the Student's Book orally in class.

There are a lot of key words in this chapter, so consider playing a word definition game. Place at least 15 key words on the board. Students draw a 3 × 3 grid and choose nine key words to write into the spaces on the grid. The teacher reads out definitions of the words in a random order. The students cross out a word when they hear the teacher read out a definition for which it is the correct answer. They shout 'Line!' when they have crossed out a line of three words on the card. The line can be across, down or diagonally. Or they have to cross out all nine words before they can shout 'Finished!'.

Use the Context worksheet (Worksheet 3.6) as a framework for students to conduct their own research enquiries about evolution. It is suggested that students look at the roles that different people played in the development of scientific ideas about evolution.

Finish by asking students to complete the End-of-chapter questions at the end of Chapter 3 in the Workbook. Provide feedback on students' answers, and work with students individually to identify areas that still need more work.

Chapter 4 Population change

Chapter overview

In this chapter students will develop their knowledge and understanding of interactions between organisms and their environments. Students will consider the resources that organisms need to survive and the adaptations of some organisms that allow them to survive in certain environments. Factors affecting the population sizes of different organisms will be considered as well as the effects of an ever-increasing human population and what we can do to reduce the impact of this.

Topic title	Number of 40-minute periods	Learning Objectives	Thinking and Working Scientifically Learning Objectives
4.1 Populations and extinction	2	9Bp.03 Describe variation within a species and relate this to genetic differences between individuals 9Be.01 Describe what could happen to the population of a species, including extinction, when there is an environmental change	9TWSa.04 Evaluate experiments and investigations, including those by others, and suggest improvements, explaining any proposed changes 9TWSa.05 Present and interpret results, and predict results between the data points collected
End of chapter questions	1	Check students' progress	

Background science

When looking at the adaptations of living organisms for their habitats, be careful to emphasise to students that organisms need to be well adapted so that they can survive (meaning that they must get enough food, water and shelter) and reproduce. By reproducing, organisms are passing copies of their genetic information on to the next generation. Better adapted organisms are more likely to pass on their genetic information, especially when a resource (such as food) is in short supply. Better adapted organisms are sometimes said to be 'fitter', leading to the expression 'survival of the fittest', which is often used to describe natural selection (which students may have met in Stage 9 Chapter 3).

It is important to note that plants, and all other types of organisms, have adaptations to help them survive in certain habitats too; this concept is not simply limited to animals.

Population sizes of organisms can be affected by many different factors and this is explored in this chapter. Living factors that have an impact on organisms, such as competitors and diseases, are called biotic factors. Non-living factors that affect organisms, such as rainfall and temperature, are called abiotic factors.

When considering how predator–prey population sizes fluctuate, be careful to emphasise the idea that these are within the same ecosystem and do not necessarily apply to worldwide population sizes. Examples of simple predator–prey relationships are quite rare as predators usually have a choice of food sources, rather than just relying on the one prey organism.

Extinction occurs when there are no longer any living organisms of a certain species. There are various categories of endangerment (defined by the International Union for Conservation of Nature) that are highlighted before an organism becomes extinct in an attempt to prevent this from happening. Extinction can be caused by numerous factors, including new diseases, natural disasters, predators and competitors.

The drastic rise in human population has had a massive impact on ecosystems around the world. This impact is due to humans consuming natural resources, creating large amounts of waste, occupying large areas of land and producing a lot of pollution. Students are encouraged to consider how we humans can reduce our negative impact on the Earth while still continuing with economic development – this is part of sustainable development.

4.1 Populations and extinction

Topic overview

		Student's Book	Workbook	Downloadable material
4.1a	**Starting point** This short activity provides an opportunity to establish the extent of your students' prior knowledge of adaptations of organisms	Starting point table		Worksheet 4.1a
4.1b	**Resources and adaptation** This learning episode considers what animals and plants need in order to survive and reproduce, as well as adaptations of organisms	Questions 1–5	Questions 1–4	
4.1c	**Factors affecting the size of populations** This learning episode considers the factors that can affect the size of a population and how the size of one population can affect the size of another	Question 6–8	Questions 5–8	Worksheet 4.1c(1) Worksheet 4.1c(2)
4.1d	**Extinction** In this learning episode students consider the factors that can cause extinction and study examples of organisms that have become extinct due to specific factors	Questions 9–11 Activity 4.2 Activity 4.3	Questions 9–11	Worksheet 4.1d

Learning Objectives

- 9Bp.03: Describe variation within a species and relate this to genetic differences between individuals
- 9Be.01: Describe what could happen to the population of a species, including extinction, when there is an environmental change
- 9TWSa.04: Evaluate experiments and investigations, including those by others, and suggest improvements, explaining any proposed changes
- 9TWSa.05: Present and interpret results, and predict results between the data points collected
- 9SIC.03 Evaluate issues which involve and/or require scientific understanding
- 9SIC.05 Discuss how the uses of science can have a global environmental impact

Links to other topics

- This topic builds on ideas about populations, food chains/webs and environmental conditions in different habitats, met in Stage 7 Chapter 2 and Stage 8 Chapter 3. The idea of adaptation may also have been met in Stage 9 Chapter 3, but study of that chapter is not necessary before study of this chapter.

Science in context

Question 12 below the Science in context: Global pollution box in Topic 4.1 of the Student's Book covers the following learning objectives:

- 9SIC.03 Evaluate issues which involve and/or require scientific understanding
- 9SIC.05 Discuss how the uses of science can have a global environmental impact.

Teaching ideas

Encourage students to think about pollution in your local area/country and how this might be affecting organisms. What is being done about the problem? What difference can the students themselves make?

Common misunderstandings

- Students often think that organisms can develop an adaptation during their lives and then pass this adaptation on to their offspring. Organisms can develop certain behavioural adaptations throughout their lives, which they may be able to 'teach' to their offspring, but they cannot change their physiology and pass this on to their offspring in their genetic material. (This is a slight but important simplification at this level. In fact, certain changes in physiology may be able to be passed on to offspring but our understanding of these epigenetic processes (inherited phenotype changes that do not occur due to a change in the DNA sequence) is still at an early stage.)
- Many students will think that predator and prey populations are the same size. During this topic, students look at fluctuations of populations and it is important that within this you consider with students the fact that the prey populations are generally much larger than the predator populations. Students also need to understand that a change in the population of one species within a habitat will often impact on the size of a different population – students often believe that every species is independent of others.
- Some students will not understand that natural ecosystems are 'dynamic', which means that they are always changing. Within this, students may believe that once damage has been done that there is nothing that we can do about it. It is important to emphasise to students that there are always things that can be done to support an ecosystem and remove things that are having damaging effects.
- A few students may think that extinction of plant and animal species is of no concern to them or does not affect them. Discuss this with students and highlight the importance of species diversity and the role that every organism has on Earth as part of an ecosystem. Many organisms are also of great use to humans and the loss of organisms may mean that an organism that is useful to humans now or in the future is no longer available.

4.1a Starting point

Resource preparation

Worksheet 4.1a

Give students Worksheet 4.1a and ask them to add as much detail to it as possible, suggesting a habitat in which each of the organisms is found and what adaptations they have that enables them to survive in these habitats.

Students should work through this individually first and then share their ideas with a partner to further develop their answers. The completed worksheets can be looked at again in 4.1b.

4.1b Resources and adaptation

Students will learn:

- To explain ways in which organisms are adapted to their ecosystems

Glossary word

adaptation, ecosystem, environment, gene, inherited variation

Resource preparation

Secondary sources (e.g. textbooks, video clips on adaptation, dictionaries, access to the internet)

Pictures of different organisms that live in varying environments

Worksheet 4.1a completed by students in 4.1a

Teaching and learning ideas

- To start, give students five minutes to write a definition for 'adaptation'. If students have studied Stage 9 Chapter 3, this should be relatively straightforward. If not, encourage students to use secondary sources to find out or check what it means before writing a definition in their own words.
- Together with students, look at the Resources and adaptation, Adaptations to life in a desert and Adaptations to life in polar regions sections in the Student's Book. Discuss the requirements for survival of animals and plants with students and consider why these are slightly different. Ask students to work through the questions within these sections of the Student's Book.
- Show students pictures of various organisms from different ecosystems such as the Arctic, desert, rainforest and ocean. Ask them to work in pairs to identify the adaptations that each of the organisms has. Students could use secondary sources to support this. This activity could be made more interactive if you can print pictures of the different organisms and lay them at different points around the classroom. Students then have one minute at each picture to make their own notes about any adaptations before moving on to the next picture. Review all of the organisms once the activity has been completed.

- To extend this activity, students could complete a group challenge task where they are assigned a specific habitat and have to design an organism that could survive in it. Students should research their environment to find out the conditions and then consider the adaptations that their organism is going to have to have in order to survive in it. A similar activity is found in Stage 9 Lesson 3.5b.
- Finish by asking students to complete a review activity such as '3, 2, 1': three things they have learned, two things they have enjoyed doing and one question they still have. Students should write their ideas down individually and then share with a partner or the rest of the class. This activity can be used to gauge what students have learned from this learning episode.
- Return Worksheet 4.1a to students and ask them to add any extra information/ideas that they now have and change anything that they wish. Students should do this in a different coloured pen so that it is clear how their ideas have developed.

Differentiation

Less confident students may prefer the more structured and simple version of the activity in the Student's Book in Chapter 3.

Give more confident students more of a challenging habitat and a special feature of that habitat that the students' organisms need to be adapted for (e.g. the animal lives in the ocean but has to hunt for its food very deep in the ocean, a tropical rainforest in which the animal must avoid being eaten by jaguars).

Extension ideas

Students should consider the surface area to volume ratio of different organisms and how this may be an adaptation for surviving in different environments. For example, walruses are very large animals but in comparison to their very large volume they have a small surface area. Elephants, on the other hand, can be equally large but have large ears and wrinkled skin and therefore have a large surface area. The larger the surface area, the more heat energy can be transferred from the body. Ask students to prepare a list of three animals with small surface area to volume ratios and three with large surface area to volume ratios, and state how this is an adaptation in each case.

Homework suggestions

Workbook questions 1–4

Students could complete the 'design an organism' activity at home if not used in class.

4.1c Factors affecting the size of populations

Students will learn:

- To describe the effects of an environmental change on the population of a species (including extinction)
- To decide whether evidence from first-hand experience or secondary sources should be used

Glossary words

population, predator, prey

Resource preparation

Graph paper

Worksheets 4.1c(1), 4.1c(2)

Teaching and learning ideas

- To start, ask students to think about different population sizes of organisms in your country and then compare these to populations of the same species globally. It is best to prepare some examples before the lesson by looking up population sizes of some local organisms and the global populations of those same organisms, or populations of those organisms in other countries. Clarify with students that a population is a group of organisms of the same species that lives in a certain habitat and is often isolated from other populations of the same species in other habitats, which are not linked geographically.
- Together with students, look at the Factors affecting the size of populations section in the Student's Book. Discuss with students the term 'population'. Students must be clear that this refers to the number of individuals of one species in a distinct, localised habitat.
- Together with students, look at the Predator–prey interactions section in the Student's Book. Work through the graph showing the predator and prey populations and then ask students to work in pairs to explain to one another what exactly the graph is showing. Ask students to answer question 8 in the Student's Book, which requires students to give their own example of a predator–prey relationship.

- Students complete Worksheet 4.1c(1). This provides them with some data about the population numbers of Canadian lynx (*Lynx canadensis*) and snowshoe hare (*Lepus americanus*). Students are required to plot the data as a line graph (with the plotted points joined dot-to-dot), and then describe and state the pattern that is shown. It may be worth discussing with students why we use a line graph in this case (because it is showing the relationship between two continuous variables – time and population).
- Ask students to write a two-mark question on predator–prey relationships, and a mark scheme to go with it. In pairs they should try one another's questions and mark schemes. Encourage students to feed back to each other some strengths and some ideas for improvement for their questions and mark schemes. Give time at the end for students to write out neat, final copies of their questions and mark schemes.

Differentiation

Less confident students could use Worksheet 4.1c(2) instead of Worksheet 4.1c(1). This gives them support with drawing the line graph.

Challenge more confident students to write a paragraph to explain why predator–prey relationships of the type seen on Worksheet 4.1c(2) and in the Student's Book are very rare in reality. Encourage them to think back to their work with food webs in Stage 7 Chapter 2.

Homework suggestions

Ask students to sketch a predator–prey graph and to label it with key points, explaining why the numbers of prey and predators are increasing or decreasing at certain times.

4.1d Extinction

Students will learn:

- To describe the effects of an environmental change on the population of a species (including extinction)
- To decide whether evidence from first-hand experience or secondary sources should be used
- To interpret results and form conclusions using scientific knowledge and understanding
- To present results appropriately
- To evaluate experimental methods, explaining any improvements suggested

Glossary words

endangered, evolution, extinction, invasive species, natural selection

Resource preparation

Secondary sources of information, for example, the internet

Activity 4.3: Poster-making materials

Worksheet 4.1d

Teaching and learning ideas

- Begin by challenging students to name as many animals as they can that are extinct. Share ideas with the rest of the class and establish what students understand by the term extinct. Then ask students why the animals that they have named have become extinct. Write any different causes of extinction that come up in discussion on the board.
- Together with students, look at the Student's Book section on Extinction. Discuss the idea that extinction can be caused by many different factors, and add any factors that did not come up in your original discussion to the list on the board. Ask students to answer questions 9–11 in the Student's Book.
- Activity 4.2 *Investigating extinction*. Look at the activity box and ensure that students understand that scientists often go on searches for certain organisms in order to monitor their populations. If these searches do not find any of a certain organism, this is evidence that the organism no longer exists in that area. Ask students to use secondary sources to look at the list of animals in the activity box, and then to use secondary sources to find out the most likely cause for its extinction or endangerment. Students could fill in their findings on Worksheet 4.1d.
- Challenge students to use secondary sources to find further examples of animals that have become extinct or endangered in different ways to add to their worksheets. Less confident students or students who work more slowly could be supported by being given a list of animals to look up (e.g. Japanese sea lion (*Zalophus japonicus*), golden toad (*Incilius periglenes*), Tasmanian devil (*Sarcophilus harrisii*), Haast's eagle (*Harpagornis moorei*), Guam flycatcher (*Myiagra freycineti*), mosaic-tailed rat (*Melomys fraterculus*), black rhino (*Diceros bicornis*), woolly mammoth (*Mammuthus primigenius*)).
- Summarise some of the research findings by writing the list of headings from the left-hand column of Worksheet 4.1d on the board, and asking students to give some names of animals that they have found out about, so building up a class list of examples. If there is time, ask students to give their best/favourite/most interesting 'extra detail' that they have found out about one animal.

- With students, read the Science in context box in the Student's Book concerning global pollution. Encourage students to give some further examples of pollutants that affect many areas of our world if they can, and examples of pollutants that are a particular problem in your area of the world.
- Activity 4.3 *Plastic pollution poster.* Divide students into groups and challenge them to work together to produce a poster or a presentation to encourage people to be more careful about what they do with their plastic waste. When assessing students on this activity, consider how well they are working as a team (as well as the material that they produce). If necessary, help students to divide the different roles for the activity amongst themselves. Encourage students to make written evaluations of their work (in answer to question A4). Point out the importance of evaluation in science, including how scientists often evaluate another's work before it is published (as part of peer review) and afterwards (when they read about the work of other scientists).
- Finish by asking students to list all the ways in which extinction and endangerment can occur. Write these on the board, and then ask students to suggest steps to prevent extinction/endangerment happening in each way. Some are impossible to stop (e.g. catastrophic events such as meteorite strikes).

Differentiation

It may help less confident students to consider only the examples of extinction given in the Student's Book, rather than having to search for information on the internet.

Challenge more confident students to explain the wider impact of the extinction of the organism that they are researching (i.e. how the extinction affects other organisms in the same food web). Also ask students to consider which of the factors causing extinction is having the biggest impact and why they believe this to be so.

Extension ideas

Some students could be asked to complete further research about Macquarie Island – how and why so many organisms died out on the island, and what has more recently been done to restore the ecology of the island.

Homework suggestions

Workbook questions 9–11

Chapter 4 Checking students' progress

End of chapter reflection

- To review the learning points of the topic
- To test understanding through answering questions

Resources

- Student's Book [End of chapter review questions]
- Workbook [Self-assessment grid, End-of-chapter questions]
- Worksheets 4.1a, 4.1c(1), 4.1c(2), 4.1d, 4.2 [all of which may have already been used]

Approach

Use the 'Key facts' and 'Check your skills progress' boxes in the Student's Book for each topic in this chapter as a way of recapping on the main learning points.

Ask students to complete the Self-assessment checklist at the end of Chapter 3 in the Workbook. Encourage students to work together in small groups to discuss which aspects of the material are the most difficult to understand, and then have a class vote. You may wish to revisit some Learning Objectives again if students find any areas particularly challenging. Use the Student's Book text and questions to revisit these aspects. Go through the answers orally in class. Take note of whether giving the answer appears to resolve students' uncertainty or not, and work through further examples.

The Teacher's comments in the Workbook can be completed to give students personalised feedback on next steps. Students could be asked to answer specific End of chapter review questions in their books. Alternatively go through some or all of the End of chapter review questions in the Student's Book orally in class.

Ask students to work together to draw a concept map to show the material in this chapter. They should start by writing the title 'Population change' in the centre of a sheet of paper. Lines then link this to key points from the chapter. Encourage students to try to do this work without access to the Student's Book at first. Once they have finished their concept maps, allow them access to the Student's Book so that they can make corrections and additions to their maps as necessary. Then ask students to say what they had forgotten to include in their concept maps and/or what they needed to correct. Revise these topics.

Use the Context worksheet (Worksheet 4.2) as a framework for students to conduct their own research enquiries about the role of humans in the extinction of species. Students should be able to suggest some ways in which we can prevent some species from becoming extinct.

Finish by asking students to complete the End-of-chapter questions at the end of Chapter 4 in the Workbook. Provide feedback on students' answers, and work with students individually to identify areas that still need more work.

Chapter 5 The Periodic Table

Chapter overview

In this chapter students will learn more about the structure of an atom, building on content covered in Stage 8 Chemistry, Chapter 4. They will draw atoms to show electron arrangement and learn what atomic number and mass number tell us about an atom. Students will also look at trends and patterns within Group 1 of the Periodic Table and consider how the Periodic Table was developed.

Topic title	Number of 40-minute periods	Learning Objectives	Thinking and Working Scientifically Learning Objectives
5.1 Atomic structure and the Periodic Table	2–3	9Cm.01 Understand that the structure of the Periodic Table is related to the atomic structure of the elements and the Periodic Table can be used to predict an element's structure and properties	9TWSm.01: Understand that models and analogies reflect current scientific evidence and understanding and can change 9TWSm.02: Describe some important models, including analogies, and discuss their strengths and limitations 9TWSm.03 Use symbols and formulae to represent scientific ideas
5.2 Trends in the Periodic Table	2–3	9Cp.01 Understand that groups within the Periodic Table have trends in physical and chemical properties, using group 1 as an example	9TWSa.05 Present and interpret results, and predict results between the data points collected 9TWSc.01: Sort, group and classify phenomena, objects, materials and organisms through testing, observation, using secondary information, and making and using keys 9TWSa.02: Describe trends and patterns in results, identifying any anomalous results and suggesting why results are anomalous 9TWSm.01 Understand that models and analogies reflect current scientific evidence and understanding and can change
End of chapter questions	1	Check students' progress	

Background science

The current model of the atom taught to students at this level consists of a small, central nucleus consisting of positively charged protons, and neutrons, which have no charge. Negatively charged electrons orbit around the nucleus in regions called shells (also called energy levels). At this level it is useful to think of electrons orbiting the nucleus like planets orbit the Sun. However this is a simplification and it is impossible to know exactly how they are actually moving. The first electron shell can hold a maximum of two electrons, the second one can hold eight, and the third one (for the simple cases studied at this level) can hold eight. Electrons fill low-energy shells before they fill higher-energy ones. For the first 20 elements, electrons will fill the shells starting from the shell nearest the nucleus.

The atomic number of an element is the number of protons (and also electrons) its atoms have. Elements always have the same atomic number. Elements are ordered in the Periodic Table by their atomic number.

The mass number of an atom is the total number of protons plus neutrons. Students do not need to know, but the number of neutrons in the atoms of an element can vary; this gives rise to isotopes which have different mass numbers (but the same atomic number). The relative atomic mass as listed in the Periodic Table is a weighted average based on the relative abundance of all the isotopes of an element. This explains why it is often not a whole number. The existence of isotopes explains why arranging the elements in order of atomic weight does not always give a periodic pattern of properties.

In the Periodic Table, elements (other than the transition elements) are arranged in groups. The elements in each group all have the same number of electrons in their outer shell, for example, Group 1 elements have one electron in their outer shell. This results in elements of the same group having similar chemical properties. However, physical properties such as melting and boiling points change gradually with a predictable pattern within a group. The reason why the trends are in opposite directions, for example, for Groups 1 and 7, cannot be explained without some further knowledge about atomic structure and chemical bonding.

5.1 Atomic structure and the Periodic Table

Topic overview

		Student's Book	Workbook	Downloadable material
5.1a	**Starting point** Students draw a model of the atom	Starting point table		
5.1b	**Atomic structure** Students learn about what atomic number and mass reveal about the structure of the atom; they build models of atoms using different coloured sweets to represent protons, neutrons and electrons	Questions 1–2 Activity 5.1	Questions 1–3, 8	Worksheet 5.1b Practical notes 5.1b
5.1c	**The Periodic Table** Students learn how the structure of an element's atoms relates to its position on the Periodic Table.	Questions 3–6	Questions 4–7	Worksheet 5.1c

Learning Objectives

- 9Cm.01: Understand that the structure of the Periodic Table is related to the atomic structure of the elements and the Periodic Table can be used to predict an element's structure and properties
- 9TWSm.01: Understand that models and analogies reflect current scientific evidence and understanding, and can change
- 9TWSm.02: Describe some important models, including analogies, and discuss their strengths and limitations
- 9TWSm.03: Use symbols and formulae to represent scientific ideas

Links to other topics

- Students were introduced to the Rutherford model of the atom in Stage 8 Chapter 4 and learnt the charges of protons, neutrons and electrons. They learnt that protons and neutrons are found in the nucleus with electrons outside the nucleus.
- In Stage 9 Topic 6.1, they will learn about bonding.

Using models

- Students will use a simple model of the atom to learn what atomic mass and number show as well as electronic structure. They will build their own 2D models of the atom using different coloured sweets to represent the subatomic particles. Three different colours of sweets are needed to represent the electrons, protons and neutrons. Students should also be given the opportunity to discuss the strengths and limitations of their models to address learning objective 9TWSm.02.

Common misunderstandings

- Students may think that the elements on the Periodic Table are arranged by increasing atomic mass. In fact, the elements are arranged by increasing atomic number.
- When scientists were first organising the elements into a table they used atomic mass because atomic number was not known (protons, neutrons and electrons had not yet been discovered).
- Atomic mass is based on the mass of the entire atom and can vary within atoms of the same element. Atomic number is the number of protons (and electrons) which is unique to each element.

5.1a Starting point

Ask the students to draw a picture to show our current understanding of the structure of an atom.

Invite some students to share what they have drawn. Some will have drawn the Rutherford model of a central nucleus surrounded by electrons, which they learned in Stage 8. Some students may have additional knowledge and drawn the electrons in orbits, others may not have moved on from their understanding from Stage 7 and drawn atoms as solid spheres.

5.1b Atomic structure

Students will learn:

- To understand that models and analogies can change according to scientific evidence

Glossary words

electrons, neutrons, nucleus, protons, shells

Resource preparation

Worksheet 5.1b

Activity 5.1 *Build an atom.* A selection of different coloured sweets/counters/beans (at least three different colours); large piece of paper (A4 or A3); a Periodic Table that lists the elements with atomic number and relative atomic mass (to the nearest whole number)

Teaching and learning ideas

- Remind the class that the model of the atom changed over time based on evidence collected by scientists. Ask them to read through the section Models of the atom in the Student's Book. Discuss how the work of Bohr and Chadwick (who discovered the nucleus) highlighted weaknesses in the model, so it changed. The model of the atom we use in school looks like figure 5.2.
- Students answer question 1 in the Student's Book.
- Show the students a diagram of the Periodic Table (such as figure 5.5 in the Student's Book). Remind them that all the elements we know about are listed here. Point out the number next to the symbol of each element. Introduce this as the atomic number. Explain that each element has a unique atomic number and show how elements are arranged in order of this number from left to right on the Periodic Table.
- Ask students to read through the section on Atomic number and mass number in the Student's Book. Ask them to write down what the atomic number of an element tells us *(how many protons and electrons it has)* and also what the mass number tells us *(the total number of protons and neutrons in the nucleus).* Use nucleus representations and the Periodic Table to test students' understanding, for example:
  - What element has the atomic number 8? *(oxygen)*
  - How many protons does an atom of the element helium have? *(2)*
  - What is the symbol of the element with 19 electrons? *(K)*
  - A magnesium atom has a mass number of 24 and an atomic number of 12, how many neutrons does it have in its nucleus? *(12).*

  Make sure that students understand that not all elements have a mass number that is double its atomic number, although many do. The mass number is always higher than the atomic number.
- Students answer question 2 in the Student's Book independently to check their understanding.
- Students complete Worksheet 5.1b, where they draw the electronic arrangement for the first 20 elements. They can use the information on Electron arrangement from the Student's Book to help them.
- Students complete Activity 5.1 *Build an atom.* Students build models of atoms using coloured sweets, counters or beans to represent the neutrons, protons and electrons. Refer to Practical notes 5.1b for more information.

Differentiation

With less confident students use multiple examples to demonstrate to the class how to use atomic number and mass to work out the numbers of protons, neutrons and electrons in an atom.

You might like to complete the first few examples on Worksheet 5.1b to model how to draw an atom to show the electrons in the shells. More confident students should be able to complete Worksheet 5.1b without assistance.

Homework suggestions

Ask students to create their own list of questions about atomic number, mass number and electron arrangement along with the answers. For example, the atoms of which element have six electrons? (answer: carbon). They can use them to test a partner as a starter activity in the following learning episode.

Extension ideas

Ask students to produce a timeline showing how the model of atomic structure has been developed over time. They should consider how the work of scientists has been developed and how advances in technology have helped.

5.1c The Periodic Table

Students will learn:

- To understand how the position of an element in the Periodic Table can be used to predict its atomic structure and chemical and physical properties

Glossary words

period, Periodic Table

Resource preparation

Worksheet 5.1c

Teaching and learning ideas

- To review learning from the last learning episode, give students a piece of paper and ask them to draw a labelled diagram of an atom based on the model we use today. Ask students to swap drawings and mark each other's drawing using figure 5.2 in the Student's Book (or the diagram they drew into their books during the last learning episode). More confident students should also check that the numbers of protons and electrons are the same.
- Give each student a copy of Worksheet 5.1c. Ask them to complete tasks 1–4 using the Periodic Table (figure 5.5) in the Student's Book.
- Quiz the students by asking them questions about different elements – they have to write down the symbol of the element on a piece of scrap paper or mini-whiteboard and hold it up.
- Suitable questions can be:
  - Which element has the atomic number...?
  - Which element is in Group... Period...?
  - Which element is to the right/left/top/bottom of...?
- Ask the class to draw an atom of a given element to show the electrons in their shells. To support less confident students, model how to do this first before asking them to complete another by themselves. To extend the more confident, give them three different atoms to complete. Ask students to locate the element(s) on the Periodic Table using figure 5.5 in the Student's Book. Ask them to spot patterns connecting the structure of different atoms to their place on the Periodic Table. They should notice that the number of electrons in the outer shell shows the group the element is in and the number of electron shells shows the period.
- Ask students to read through the section 'The Periodic Table' in the Student's Book and answer questions 3–6.

Extension ideas

In preparation for the next learning episode, ask students to find out some properties of the Group 1 elements.

Homework suggestions

Workbook Topic 5.1, questions 1–8

5.2 Trends in the Periodic Table

Topic overview

		Student's Book	Workbook	Downloadable material
5.2a	**Starting point** Students complete a concept map to check their prior knowledge of the Periodic Table	Starting point table		Worksheet 5.2a
5.2b	**Trends in the Periodic Table** Students learn about Group 1 and how elements in this group are similar to each other and study trends in their properties	Questions 1–2 Activity 5.2	Questions 1–6	Practical notes 5.2b
5.2c	**Developing the Periodic Table** In this learning episode, students group elements using information about their physical and chemical properties in order to create their own version of the Periodic Table		Questions 7–8	Worksheet 5.2c

Learning Objectives

- 9Cp.01: Understand that the groups within the Periodic Table have trends in physical and chemical properties, using group 1 as an example
- 9TWSm.01: Understand that models and analogies reflect current scientific evidence and understanding, and can change
- 9TWSc.01: Sort, group and classify phenomena, objects, materials and organisms through testing, observation, using secondary information, and making and using keys
- 9TWSa.02: Describe trends and patterns in results, identifying any anomalous results and suggesting why results are anomalous
- 9TWSa.05: Present and interpret results, and predict results between the data points collected
- 9SIC.01 Discuss how scientific knowledge is developed through collective understanding and scrutiny over time
- 9SIC.04 Describe how people develop and use scientific understanding as individuals and through collaboration

Links to other topics

- Students were introduced to the Periodic Table in Stage 7 Chapter 4. They learnt about elements and how they are represented by symbols.

Science in context

Question at the end of the Science in context: Developing the Periodic Table box in Topic 5.2 of the Student's Book, covers the following learning objectives:

- 9SIC.01 Discuss how scientific knowledge is developed through collective understanding and scrutiny over time
- 9SIC.04 Describe how people develop and use scientific understanding as individuals and through collaboration.

Teaching ideas

You may want to develop ideas about the development of the Periodic Table further and ask students to research the work of other scientists from the past e.g. John Newlands and William Ramsay. You could also extend this to more recent discoveries such as the work of Yuri Oganessian.

Common misunderstandings

- Students may think that the Periodic Table in its present form is the way the elements have always been categorised. In this chapter they will learn that the Periodic Table has evolved, through the work of many scientists and consensus was not easily achieved.

5.2a Starting point

Use Worksheet 5.2a to set a task based on concept mapping – a useful method for exploring students' ideas about the Periodic Table and giving an opportunity to correct any misunderstandings before they are introduced to new information in this chapter. The Worksheet contains a concept map for the Periodic Table with links to key terms labelled with numbers. The students should write sentences to match the links. This activity can be used as an individual task to elicit understanding or as a group task to encourage discussion. Invite students to share their linking sentences with the rest of the class and use these answers as a basis for discussion.

5.2b Trends in the Periodic Table

Students will learn:

- To understand that trends in chemical and physical properties are seen within the different groups of the Periodic Table and describe these trends using the Group 1 elements as examples
- To describe trends and patterns shown in a set of results, identifying, and explaining, any anomalous results present

Resource preparation

Activity 5.2: Graph or squared paper, ruler, colouring pencils

Demonstration – Trends in properties of the alkali metals. Goggles/face shield for demonstrator, eye protection for students, forceps, filter paper, ceramic tile, scalpel or sharp knife to cut the metal, one or more large glass troughs (5 dm^3 capacity), at least two safety screens, universal indicator solution, small bottles of oil containing small pieces of the following metals: lithium (5 mm cubes), sodium (4 mm cubes), potassium (3 mm cubes)

Teaching and learning ideas

- Ask the class to write down a list of chemical and physical properties of elements. This reviews learning from Stage 7 Chemistry, Chapter 4.
- They read through the section Chemical and physical properties in the Student's Book to check their answers, and add any missing properties.
- Discuss with the class that chemical properties depend on the number of electrons in an atom's outer shell. Ask them why elements in the same group of the Periodic Table have similar chemical properties.
- Use the information from the section Groups in the Periodic Table in the Student's Book to introduce the class to Group 1.
- Demonstrate the reactions of the alkali metals with water. Use this to show the reaction of the alkali metals with water and the trend in reactivity as you go down the group. Refer to Practical notes 5.2b for more information.
- Students complete questions 1–2 in the Student's Book.

  Students complete Activity 5.2 *Physical property trends in Group 1* in the Student's Book. They will need graph or squared paper to draw a bar chart to show the melting points of the Group 1 elements and then answer questions about the trend.

Differentiation

During Activity 5.2 *Physical property trends in Group 1*, support less confident students by supplying them with ready-drawn axes or show them how to construct them. More confident students should be challenged to produce their own axes.

Homework suggestions

Students can add information and links to the concept map from the Starting Point activity.

Extension ideas

Ask students to write a prediction for the chemical and physical properties of rubidium and caesium. They should explain their answers using the observations made from the demonstration of the alkali metals with water.

5.2c Developing the Periodic Table

Students will learn:

- To explain how the Periodic Table was developed
- To understand that models and analogies can change according to scientific evidence
- To use observations, measurements, secondary sources of information and keys, to organise and classify organisms, objects, materials, or phenomena

Resource preparation

Worksheet 5.2c

Scissors

Teaching and learning ideas

- Introduce the idea of grouping, or classification, to the class using a familiar example such as the way products are arranged in a supermarket, for example, oranges are fruit so are placed in the fruit and vegetable aisle. Ask the class to explain how grouping products in this way is useful.
- Ask them to discuss in small groups how the way the elements are arranged in the Periodic Table is similar. They should recall from the last learning episode that elements with similar properties are placed in groups. Also, ask them to suggest why scientists developed the Periodic Table, that is, why is it useful to arrange elements in this way?
- Ask students to read through the Science in context box Developing the Periodic Table in the Student's Book.
- Explain that they are going to develop their own system of organising elements. This models the approach used by historical scientists including Mendeleev. Give small groups Worksheet 5.2c and ask them to cut out the cards. Each card contains information about an element (but the name is not given). Their task is to order the elements by atomic mass and then arrange them to group the elements according to their properties. Note that atomic mass/weight is the mass of one atom of the element (not to be confused with mass number, which is the number of neutrons and protons in the nucleus). Explain to the students that they are working from data just like scientists did when they first started to think of ways to organise the elements, that is, that they are working without knowledge of the structure of atoms. They can place elements that share similar properties under each other to form groups, but the order of atomic mass should remain the same from left to right.
- Reveal the names of each element: *X: potassium; R: copper; B: fluorine; P: sulfur; G: hydrogen; L: magnesium; Z: barium; F: nitrogen; W: oxygen; O: chlorine; S: lithium; Q: sodium; D: iron; E: calcium; M: gold; H: bromine.*

  Ask groups to compare their arrangements with another group's and then with a modern Periodic Table (figure 5.5 in the Student's Book). Ask them to explain what is similar and what is different and explain why. For example, there are no noble gases and there are other missing elements. This models how scientists in the middle of the 19th century worked – just as the students only have the data on the cards, scientists in the past only had the data for the elements that had already been discovered. Discuss how they used the concept of periodicity when they moved elements to form groups – the properties of the elements repeat, for example, lithium, sodium and potassium; fluorine, chlorine and bromine.

Homework suggestions

See extension idea below.

Extension ideas

Set students a challenge question. This can be completed for homework, perhaps using secondary sources.

'Describe how the modern Periodic Table is different to Mendeleev's table. Explain why it changed.'

Chapter 5 Checking students' progress

End of chapter reflection

- To review the learning points of the chapter
- To test understanding through answering questions

Resources

- Student's Book [End of chapter review questions]
- Workbook [Self-assessment grid, End-of-chapter questions]
- Worksheet 5.3

Approach

Use the Key facts and Check your skills progress boxes in the Student's Book for each topic in this chapter as a way of recapping on the main learning points.

Ask students to complete the Self-assessment checklist at the end of Chapter 5 in the Workbook. Encourage students to work together in small groups to discuss which aspects of the material are the most difficult to understand, and then have a class vote. You may wish to revisit some Learning Objectives again if students find any areas particularly challenging. Use the Student's Book text and questions to revisit these aspects. Go through the answers orally in class. Take note of whether giving the answer appears to resolve students' uncertainty or not, and work through further examples.

The Teacher's comments in the Workbook can be completed to give students personalised feedback on next steps. Students could be asked to answer specific End of chapter review questions in their books. Alternatively go through some or all of the End of chapter review questions in the Student's Book orally in class.

Use the Context worksheet (Worksheet 5.3) as a framework for students to conduct their own research enquiries about atomic structure. For example, students could focus on ideas about how nuclear power stations release energy from atoms or what happens inside a particle accelerator.

Finish by asking students to complete the End-of-chapter questions at the end of Chapter 5 in the Workbook. Provide feedback on students' answers, and work with students individually to identify areas that still need more work.

Chapter 6 Structure, bonding and the properties of matter

Chapter overview

In this chapter students will learn about covalent and ionic bonding, how atoms are arranged in matter as giant or simple structures and how to determine density. This chapter builds on ideas about atomic structure from Stage 9 Chapter 5 and particle theory, elements, compounds and physical properties from previous stages.

Topic title	Number of 40-minute periods	Learning Objectives	Thinking and Working Scientifically Learning Objectives
6.1 Chemical Bonds	2–3	9Cm.02 Understand that a molecule is formed when two or more atoms join together chemically, through a covalent bond 9Cm.03 Describe a covalent bond as a bond made when a pair of electrons is shared by two atoms (limited to single bonds) 9Cm.04 Describe an ion as an atom which has gained at least one electron to be negatively charged or lost at least one electron to be positively charged 9Cm.05 Describe an ionic bond as an attraction between a positively charged ion and a negatively charged ion	9TWSm.01 Understand that models and analogies reflect current scientific evidence and understanding and can change 9TWSm.02 Describe some important models, including analogies, and discuss their strengths and limitations 9TWSm.03 Use symbols and formulae to represent scientific ideas
6.2 Simple and giant structures	2–3	9Cp.04 Know that elements and compounds exist in structures (simple or giant), and this influences their physical properties	9TWSm.02 Describe some important models, including analogies, and discuss their strengths and limitations 9TWSp.03 Make predictions of likely outcomes for a scientific enquiry based on scientific knowledge and understanding 9TWSc.01 Sort, group and classify phenomena, objects, materials and organisms through testing, observation, using secondary information, and making and using keys
6.3 Density	3	9Cp.02 Describe how the density of a substance relates to its mass in a defined volume 9Cp.03 Calculate and compare densities of solids, liquids and gases	9TWSc.02 Decide what equipment is required to carry out an investigation or experiment and use it appropriately TWSc.07 Collect, record and summarise sufficient observations and measurements, in an appropriate form
End of chapter questions	1	Check students' progress	

Background science

The bonding and structure of substances determine their properties.

Bonding is how atoms are held together. In this topic students will learn about ionic and covalent bonds.

Ionic bonds normally (but not always) exist in compounds that contain metal and non-metal e.g. sodium chloride. An ionic bond is the electrostatic attraction between ions (charged atoms) with opposite charges and is caused by the transfer of electrons from the metal atom (causing a positive ion) to the non-metal atom (causing a negative ion).

Covalent bonds exist between non-metal atoms. They happen when atoms share electrons. The bond exists because of the electrostatic force between the shared electrons and the nucleus of each atom. Groups of non-metal atoms formed in this way are molecules.

The structure is the way the atoms are arranged relative to each other.

There are two major types of structure: giant and molecular.

Giant structures go on indefinitely, forming giant structures such as sodium chloride crystals. Molecular structures are made up of groups of atoms (molecules).

There are three groups of giant structures: ionic, covalent and metallic.

Molecular structures can be divided into simple (small molecules) and macromolecular (large molecules such as polymers). In this topic students will learn about simple molecular structure, with macromolecular being introduced at IGCSE.

There are strong covalent bonds between the atoms in a molecule (intramolecular force) and weaker forces of attraction (intermolecular) between molecules. This structure gives simple molecules low melting and boiling points.

In giant structures there are strong bonds between atoms/ions throughout the structure, which gives these substances high melting and boiling points.

Density also determines properties. Density is a ratio of a substance's mass to volume. So, materials with a high density have a high mass for their volume. The equation density = mass / volume is used to calculate density.

6.1 Chemical bonds

Topic overview

		Student's Book	Workbook	Downloadable materials
6.1a	**Starting point** In this short activity students classify substances in different ways. This allows you to see what they remember about elements and compounds and physical properties from previous stages.	Starting point table		Worksheet 6.1a
6.1b	**Covalent bonding** Students use dot-and-cross diagrams to model how covalent bonds are formed in molecules	Questions 1–3 Activity 6.1	Questions 1–6	
6.1c	**Ionic bonding** Students use models to show the structure of ions and how ionic bonds are formed. They use the charges of ions to determine the formulae of compounds.	Questions 4–5	Questions 7–8	

Learning Objectives

- 9Cm.02: Understand that a molecule is formed when two or more atoms join together chemically, through a covalent bond
- 9Cm.03: Describe a covalent bond as a bond made when a pair of electrons is shared by two atoms (limited to single bonds)
- 9Cm.04: Describe an ion as an atom which has gained at least one electron to be negatively charged or lost at least one electron to be positively charged
- 9Cm.05: Describe an ionic bond as an attraction between a positively charged ion and a negatively charged ion
- 9TWSm.01 Understand that models and analogies reflect current scientific evidence and understanding and can change
- 9TWSm.02: Describe some important models, including analogies, and discuss their strengths and limitations
- 9TWSm.03: Use symbols and formulae to represent scientific ideas

Links to other topics

- In Chapter 5 students learnt about atomic structure including electron configuration. They will use this knowledge to explain how an ionic bond forms in terms of ion formation.

Using models

- Students will use models to explain ionic and covalent bonding. They will evaluate the strengths and weaknesses of using a dot-and-cross model to represent a molecule.

Common misunderstandings

- A common misunderstanding is that molecules always contain atoms of different elements. In fact, elements can be made up of molecules e.g. oxygen and hydrogen.
- Many students only define bonding as a way of getting full shells (by electron sharing or electron transfer). It should be explained that bonding is due to electrostatic attraction and so there are other types of bonding, such as metallic bonding. This will aid progression to IGCSE, where students will need to consider why some bonds are stronger in terms of bond energies

6.1a Starting point

Resource preparation

Worksheet 6.1a

Scissors

Teaching and learning ideas

Give pairs or small groups of students a copy of Worksheet 6.1a and ask them to cut out the cards. They should put the substances into groups of their choosing e.g. they might group them as elements and compounds, metals or non-metals, state at room temperature. They should then choose a different way of grouping them.

After each pair has completed this, discuss as a class the different groupings each pair used. This task enables you to see what students remember about elements and compounds and physical properties from previous stages.

6.1b Covalent bonding

Students will learn:

- To understand that molecules are formed by covalent bonds
- To describe how a covalent bond is formed
- To describe the strengths and limitations of the dot-and-cross diagram as a model of a molecule

Glossary words

covalent bond, dot-and-cross diagram, molecule

Resource preparation

Molecular model kit (optional)

Teaching and learning ideas

- Ask students to use the formula on the cards from the Starting Point activity to draw the structures of oxygen, hydrogen and water.
- Students check their answers by referring to figure 6.1 in the Student's Book.
- Introduce the term 'molecule' as a group of 2 or more atoms joined together. Explain that the structures they have drawn are molecules.
- Students answer questions 1 and 2.
- **Challenge**: Students can also answer question 3.
- Introduce a covalent bond as what holds the atoms in a molecule together. Model how to draw a dot-and-cross diagram to show the covalent bond in a hydrogen molecule. Explain how the sharing of a pair of electrons results in each atom having a full outer shell and the electrostatic attraction between the positively charged nucleus of each atom and the negatively charged shared electrons is the covalent bond.
- Students complete Activity 6.1 *Drawing dot-and-cross diagrams* in the Student's Book. They follow the instructions to draw dot-and-cross molecules of chlorine and hydrogen chloride.

- If you have a molecular model kit you can ask students to also use this to build the molecules. Point out how the holes in the atoms represent how many covalent bonds they can form.
- Discuss the answer to A5 as a class. This is about the strengths and weaknesses of a dot-and-cross diagram as a model to represent a molecule.

Differentiation

To support less confident students in Activity 6.1 *Drawing dot-and-cross diagrams* draw chlorine atoms on the board for them to copy (step A1).

To extend more confident students ask them to answer challenge question 3. They can also complete more dot-and-cross diagrams to show methane (CH_4) and ammonia (NH_3).

Extension ideas

Show the class some more molecule models e.g. ball and stick, space fill. Discuss the strengths and weaknesses of each model.

Homework suggestions

Workbook questions 1–6.

6.1c Ionic bonding

Students will learn:

- To describe how negatively charged and positively charged ions are formed from atoms
- To describe an ionic bond
- Use models to show the structure of ions and ionic bonds
- To represent scientific ideas using recognised symbols or formulae

Glossary words

ion, ionic bond

Teaching and learning ideas

- Ask students to draw a lithium atom with atomic number 3 and mass number 7, showing its protons, neutrons and electrons (a review of learning from Chapter 5.1). Ask them why it has an overall neutral charge. They can check their answers by referring to the section Ions in the Student's Book.
- Remind the class that in a molecule the non-metal atoms achieve a full outer shell by sharing electrons. Ask them 'how could a sodium atom achieve this?', 'what will happen to its charge if it does this?'. Discuss that it could lose the electron on its outer shell, giving it an overall charge of +1. This means that it loses its outer shell, revealing the full shell underneath.
- Ask them to repeat this but with a fluorine atom with atomic number 9 and mass number 19. This time the fluorine atom will need to gain an electron, giving it a charge of -1.
- Students read through the section Ionic bonding in the Student's Book.
- They answer question **4**.
- Ask them to draw a similar diagram to figure 6.4 in the Student's Book to show what happens when a lithium atom reacts with a fluorine atom.

Differentiation

To extend more confident students ask them to answer challenge question 5 and the challenge task.

Extension ideas

Discuss that elements in the middle of the Periodic Table do not belong to a group – most of these elements have atoms that can form ions with different charges, and this is shown by numbers in brackets e.g. iron(II) forms Fe^{2+} ions and Fe(III) forms Fe^{3+} ions. They are known as transition metals. Provide students with the names of compounds containing transition metals and ask students to write the formula of compounds e.g. iron(II) chloride, copper(II) oxide, iron(III) bromide.

Homework suggestions

Workbook questions 7–8.

6.2 Simple and giant structures

Topic overview

		Student's Book	Workbook	Downloadable materials
6.2a	**Starting point** Students group substances into molecules, ionic compounds and other	Starting point table		
6.2b	**Giant structures** Students build structures of sodium chloride and diamond and explore the properties of giant structures	Questions 1–2	Questions 1 and 3	
6.2c	**Simple structures** Students consider why simple structures have low melting and boiling points and carry out an activity where they use data to determine the structures of mystery substances	Activity 6.2 Question 3	Question 4	

Learning Objectives

- 9Cp.04: Know that elements and compounds exist in structures (simple or giant), and this influences their physical properties
- 9TWSc.01: Sort, group and classify phenomena, objects, materials and organisms through testing, observation, using secondary information, and making and using keys
- 9TWSm.02: Describe some important models, including analogies, and discuss their strengths and limitations
- 9TWSp.03 Make predictions of likely outcomes for a scientific enquiry based on scientific knowledge and understanding

Links to other topics

- The ability of molten or dissolved ionic substances and metals as electrical conductors is discussed in the chapter. This has links to Stage 7 Chapter 9, where students were introduced to charge.

Using models

- Students will build models to show the structure of giant and simple structures. They use these to explain the properties of these substances (melting/boiling points, strength/hardness and electrical conductivity).

Science in context

- There are many opportunities to discuss how the structure of substances relates to its properties and application. For example, the use of diamond as a cutting tool because of its hardness or metals as electrical conductors.

Common misunderstandings

- Students may think that when substances with a simple structure melt it is the strong covalent bonds between the atoms in the molecules that are overcome. In fact, it is the weaker forces of attraction between the molecules that are overcome; the covalent bonds remain intact.

6.2a Starting point

Resource preparation

Cards cut from Worksheet 6.1a (from previous topic's Starting Point activity)

Teaching and learning ideas

- Provide each pair or small group with the cards and ask them to group the substances into three groups: ones that consist of molecules containing covalent bonds, ones which contain ions joined together by ionic bonds (ionic compounds) and ones which are neither. If they need help, remind them that molecules contain non-metals and ionic compounds consist of metal ions bonded to non-metal ions.
- Check that they have correctly classified them:

  Molecules: oxygen, water, ammonia, carbon dioxide, methane, chlorine, glucose, hydrogen, nitrogen dioxide

  Ionic compounds: magnesium oxide, sodium chloride, calcium carbonate, copper hydroxide

  Neither: helium, sodium, gold, copper, iron

6.2b Giant structures

Students will learn:

- To understand how the physical properties of an element or compound may be determined by it's structure (simple or giant)
- To use models to show the structure of giant structures

Glossary words

giant structure

Resource preparation

Salt crystals (sodium chloride) and magnifying glasses

Items to represent atoms e.g. grapes, modelling clay, marshmallows, plus cocktail or kebab sticks

Teaching and learning ideas

- Show the class an image of a magnified salt (sodium chloride) crystals or allow them to use magnifying glasses to look at salt so they can see that the crystals are cube shaped.
- Ask them to draw diagrams to suggest why they are this shape by thinking about how the ions in a salt crystal are arranged.
- Students read through the section Giant structures in the Student's Book and check their structure by using figure 6.5. Discuss why sodium chloride is a giant ionic structure and diamond is a giant covalent structure.
- Ask them why each sodium is next to a chloride ion and vice versa, and why the formula of sodium chloride is NaCl.
- Provide groups with items for them to build their own models of sodium chloride and diamond based on figure 6.5. (Make sure they know that in diamond each carbon atom forms four bonds).
- Show the class 3D models of giant structures on the internet.
- Students read through the section Properties of giant structures in the Student's Book and answer question 1.
- **Challenge**: Students can also answer question 2.

Extension ideas

Students can do their own research and find more examples of substances with giant structures.

Homework suggestions

Workbook questions 1 and 3.

6.2c Simple structures

Students will learn:

- To understand how the physical properties of an element or compound may be determined by it's structure (simple or giant)
- To use models to show the structure of simple structures
- To classify substances based on their structure using secondary information

Glossary words

simple structure

Resource preparation

Molecule building kit

Teaching and learning ideas

- Ask students to draw particle diagrams to show the molecules from the Starting Point activity. Students can refer to figure 6.1 in the Student's Book to see how water is drawn and then repeat for the others.
- View 3D molecules by visiting www.molview.org. You can search for molecules and rotate the models.
- Ask students to write down what happens to the particles when a solid substance melts and then boils (this is a review of learning from Stage 7).
- Draw the structure of ice on the board (see figure 6.7 in the Student's Book) and ask students to predict what happens to water molecules when ice melts and then boils. Check that students do not have the misunderstanding that the covalent bonds between the atoms are broken.
- Students can read through the sections Simple structures and Properties of simple structures in the Student's Book to check their answers and answer question 3.
- Students complete Activity 6.2 *Testing properties*. This checks their understanding of the properties of giant and simple structures. For your information, the mystery substances are: A – magnesium oxide; B – gold; C – silicon dioxide; D – bromine.

Differentiation

In Activity 6.2 *Testing properties*, less confident students can just answer questions A1 and A2.

To extend more confident students ask them to answer A1, A2, A3 and A4.

Extension ideas

Show students data about the length of alkane molecules and their boiling point. Ask them to suggest a reason for the pattern. They will see that as the length of the molecule increases so does the boiling point. This is because there are more attractive forces between the molecules that need to be broken, so more energy is required.

Homework suggestions

Workbook question 4.

6.3 Density

Topic overview

		Student's Book	Workbook	Downloadable material
6.3a	**Starting point** Students will learn about density and how it is different from weight. They realise that large items can have a small density	Starting point table		
6.3b	**Determining the density of a regular solid** Students revise how to calculate volume, they learn the formula to calculate density and they use it to calculate the density of a regular solid	Questions 1–2	Questions 1–3	Practical notes 6.3b Worksheets 6.3b(1), 6.3b(2), 6.3b(3)
6.3c	**Determining the density of an irregular solid** Students plan a practical to find the density of an irregular solid using a measuring cylinder and water	Activity 6.3	Question 6	Practical notes 6.3c Worksheet 6.3c
6.3d	**Determining the density of a liquid and a gas** Students plan a practical to determine the density of a liquid	Questions 3–4	Questions 4 and 7	Practical notes 6.3d Worksheet 6.3d

Learning Objectives

- 9Cp.02: Describe how the density of a substance relates to its mass in a defined volume
- 9Cp.03: Calculate and compare densities of solids, liquids and gases
- 9TWSc.02: Decide what equipment is required to carry out an investigation or experiment and use it appropriately
- TWSc.07 Collect, record and summarise sufficient observations and measurements, in an appropriate form

Links to other topics

- Students will revisit density when they learn about floating and sinking in Chapter 9.

Using models

- The particle model is used to explain the difference in density when a substance is a solid, liquid or gas.

Science in context

Teaching ideas

You can discuss the uses of materials with low or high densities. For example, scientists develop materials with very low densities such as aerographite. Students should be able to suggest suitable applications depending on the density of the substance, such as using much less dense aluminium instead of iron when building aircraft.

Common misunderstandings

- Students may think that larger objects are heavier than small objects – to remedy this, you could show them a piece of polystyrene and a smaller piece of wood or iron to compare their masses.
- For most substances they are more dense when in the solid state than liquid or gas. However, water is one of the few substances that is less dense as a solid than it is a liquid. This is because when water freezes, the molecules arrange themselves into a crystal pattern which spaces the molecules out more. However, since it is the most common substance that students will see as both a solid and a liquid, they may think that all substances are like that. You could use wax to show that solid wax sinks in liquid wax.

6.3a Starting point

Resource preparation

A sponge and a piece of wood of roughly equivalent size and shape

Large piece of polystyrene

Teaching and learning ideas

- You could show the students a sponge and a piece of wood of roughly equivalent size and shape, then ask them if they weigh the same or if the wood weighs more, or if the sponge weighs more. They will know that the wood weighs more but they might find it difficult to explain why. They could come up with the idea that the sponge has holes in it, so there is less matter in the same amount of space.
- Show the students a large piece of polystyrene and ask them if they think it will be heavy or light. When they say that it is light, ask them why it is light if it is so big. This is light for the same reason that a sponge is light, but there are less obvious holes in polystyrene, so it will be more difficult to come to that conclusion.

6.3b Determining the density of a regular solid

Students will learn:

- To describe the mass of a substance in a defined volume as density
- To use the mass and volume of solids, liquids and gases to calculate and compare their densities
- To collect and record observations and measurements appropriately
- To choose the appropriate equipment for an investigation and use it correctly

Glossary words

density, mass, volume

Resource preparation

For the practical: a balance, 30 cm ruler, blocks of different substances in cuboid shapes

Practical notes 6.3b

Worksheets 6.3b(1), 6.3b(2), 6.3b(3)

Teaching and learning ideas

- Practical 6.3b: determining the density of a regular solid. Students calculate the volume of a block by measuring its length, width and height. They then weigh the substances and calculate their densities.
- Worksheets 6.3b(1) and 6.3b(2) – students calculate the densities of regular objects.
- Students answer questions 1 and 2 in the Student's Book.
- Challenge: ask students why ships float if they are made mainly of steel (the answer is that they are not made of solid steel – the rooms contain only air, which is less dense than water so the density of the ship averages out).
- Use the PhET density simulation found at https://phet.colorado.edu/en/simulation/density to see if students can predict correctly whether a substance can float on water.

Differentiation

For less confident students use Worksheet 6.3b(3) with particle diagrams of substances and students can work out why one is more dense than the others. The sheet also has images of blocks made up of 1 cm^3 pieces so they can easily work out the volume and how to distribute the mass. More confident students just use Worksheets 6.3b(1) and (2).

Extension ideas

Students can rearrange the equation to calculate volume or mass:

Volume = mass/density

Mass = volume × density.

Suitable questions are:

Mercury has a density of 13.5 g/cm^3. Some mercury has a mass of 270 g. What volume of mercury do you have?

Steel has a density of 7.4 g/cm^3. A lump of steel has a volume of 5 cm^3. What is its mass?

Homework suggestions

Students complete questions 1–3 on this topic from the Workbook.

6.3c Determining the density of an irregular solid

Students will learn:

- To describe the mass of a substance in a defined volume as density
- To use the mass and volume of solids, liquids and gases to calculate and compare their densities
- To plan methods used to test hypotheses
- To choose the appropriate equipment for an investigation and use it correctly
- To decide how to obtain sufficiently reliable data from observations and measurements

Glossary words

density, mass, volume

Resource preparation

For the practical: a balance; irregularly shaped objects, such as stones; measuring cylinders, container of water

Practical notes 6.3c

Worksheet 6.3c

Teaching and learning ideas

- Show students an object that is not a regular shape. Ask students how they can calculate the volume of this object.
- Practical: investigating the density of an irregular solid. Worksheet 6.3c and Practical notes 6.3c provide full details. Students measure the volume of an object by measuring how much water it displaces. They then weigh the object to get its density.
- Students discuss as a class whether their method worked and any improvements they could make.
- Students complete Activity 6.3 *Measuring the density of everyday objects*. You could supply students with objects to use or let them choose them from the classroom.
- Challenge: present the students with a problem to see if they can work out whether an object is made from a pure metal.

Differentiation

Less confident students could use a regular solid to begin with. They could calculate the volume using length × width × height and then work out the volume using water. This way, they will work out that the second method also works.

You can show the students the PhET simulation for density (found at https://phet.colorado.edu/en/simulation/density). You could use this to show that when you completely submerge a block in the water, the volume of the water increases by the same amount as the volume of the block. More confident students may want to start with an irregular object as a challenge.

Extension ideas

Challenge: when doing the PhET simulation with density, you could put a block that floats in the water and ask why the volume does not go up by the volume of the block when the block is floating (it is because the block is only partially submerged).

Homework suggestions

Students answer question 6 in the Workbook.

6.3d Determining the density of a liquid and a gas

Students will learn:

- To describe the mass of a substance in a defined volume as density
- To use the mass and volume of solids, liquids and gases to calculate and compare their densities
- To plan methods used to test hypotheses
- To collect and record observations and measurements appropriately
- To interpret results and form conclusions using scientific knowledge and understanding

Glossary words

density, mass, volume

Resource preparation

For activity: measuring cylinder, liquids, balance

For demonstration: gas syringe or measuring cylinder, calcium carbonate, bung, delivery tube, tripod, Bunsen burner/spirit burner, heat-proof mat, conical flask, vat of water (if using a measuring cylinder instead of a gas syringe), eye protection

Practical notes 6.3d

Worksheet 6.3d

Teaching and learning ideas

- Ask the students to draw squares of the same size on a piece of paper. Tell them to fill the first square with particles to represent a solid. Then the second to show a liquid, and the third to show a gas. Ask them to write down how they think the density of a substance changes as it is melted, and then boiled.
- They can check their ideas by using the section Density and particle theory in the Student's Book.
- Practical: investigating the density of a liquid. Students plan an investigation into the density of a liquid by taking the mass of a container, taking the mass of the liquid + container, subtracting the container mass and then measuring the volume of the liquid. See Practical notes 6.3d.
- Discuss the problems with measuring the density of a gas.
- Students think of suitable ways for measuring the density of a gas.
- Students complete answers to Worksheet 6.3d on the densities of gases.
- Students could work out that liquids have different densities by filling two identical measuring cylinders with different liquids. Use the same volume and then weigh them. The liquids should have different masses.

Differentiation

Less confident students may need help to recap the particle model and the concept of density.

More confident students could be challenged to explain why some liquids have different densities or why some gases have different densities.

Extension ideas

Students could explain why liquids and gases with a higher density have a greater pressure using particle theory.

Students could explain how hot air balloons work using the idea of density.

Students can research why solid water (ice) floats on top of liquid water.

Homework suggestions

Students use the idea of density of liquids to explain why oil floats on water.

Students complete questions 3–4 in the Student's Book and questions 4 and 7 on this topic from the Workbook.

Chapter 6 Checking students' progress

End of chapter reflection

- To review the learning points of the chapter
- To test understanding through answering questions

Resources

- Student's Book [End of chapter review questions]
- Workbook [Self-assessment grid, End-of-chapter questions]
- Worksheet 6.4

Approach

Use the Key facts and Check your skills progress boxes in the Student's Book for each topic in this chapter as a way of recapping on the main learning points.

Ask students to complete the Self-assessment checklist at the end of Chapter 6 in the Workbook. Encourage students to work together in small groups to discuss which aspects of the material are the most difficult to understand, and then have a class vote. You may wish to revisit some Learning Objectives again if students find any areas particularly challenging. Use the Student's Book text and questions to revisit these aspects. Go through the answers orally in class. Take note of whether giving the answer appears to resolve students' uncertainty or not, and work through further examples.

The Teacher's comments in the Workbook can be completed to give students personalised feedback on next steps. Students could be asked to answer specific End of chapter review questions in their books. Alternatively go through some or all of the End of chapter review questions in the Student's Book orally in class.

Use the Context worksheet (Worksheet 6.4) as a framework for students to conduct their own research enquiries about how structure relates to properties. For example, students could focus on applications of high strength materials, low density materials or materials with very high melting points.

Finish by asking students to complete the End-of-chapter questions at the end of Chapter 6 in the Workbook. Provide feedback on students' answers, and work with students individually to identify areas that still need more work.

Chapter 7 Chemical changes

Chapter overview

In this chapter students will learn about how mass and energy are conserved during chemical reactions and describe them using word and symbol equations. They will also learn about the equipment, reactants and techniques used to prepare dry crystals of a soluble salt from the reaction of acids with metals and metal carbonates, about how to use reactivity to predict displacement reactions and the factors which affect the rate of a chemical reaction. This chapter builds on content about chemical changes and reactions from Stage 7 and 8 Chemistry as well as introducing content that will be revisited in Cambridge IGCSE or equivalent level.

Topic title	Number of 40-minute periods	Learning Objectives	Thinking and Working Scientifically Learning Objectives
7.1 Changes in chemical reactions	2–3	9Cc.05 Understand that in chemical reactions mass and energy are conserved	9TWSp.03 Make predictions of likely outcomes for a scientific enquiry based on scientific knowledge and understanding 9TWSc.04 Take appropriately accurate and precise measurements, explaining why accuracy and precision are important 9TWSc.05 Carry out practical work safely, supported by risk assessments where appropriate
7.2 Word and symbol equations	1–2	9Cc.01 Use word equations and symbol equations to describe reactions (balancing symbol equations is not required)	9TWSm.03 Use symbols and formulae to represent scientific ideas
7.3 Methods for making salts	4–5	9Cc.01: Use word equations and symbol equations to describe reactions (balancing symbol equations is not required) 9Cc.03 Describe how to prepare some common salts by the reactions of metals with acids, and metal carbonates with acids, and purify them, using filtration, evaporation and crystallisation	9TWSc.05 Carry out practical work safely, supported by risk assessments where appropriate 9TWSc.07 Collect, record and summarise sufficient observations and measurements, in an appropriate form 9TWSm.03: Use symbols and formulae to represent scientific ideas 9TWSp.05: Make risk assessments for practical work to identify and control risks 9TWSc.02 Decide what equipment is required to carry out an investigation or experiment and use it appropriately
7.4 Displacement reactions	2–3	9Cc.02 Identify examples of displacement reactions, and predict products (limited to reactions involving calcium, magnesium, zinc, iron, copper, gold and silver salts)	9TWSm.02 Describe some important models, including analogies, and discuss their strengths and limitations 9TWSp.03 Make predictions of likely outcomes for a scientific enquiry based on scientific knowledge and understanding 9TWSc.07 Collect, record and summarise sufficient observations and measurements, in an appropriate form 9TWSm.03 Use symbols and formulae to represent scientific ideas 9TWSa.02: Describe trends and patterns in results, identifying any anomalous results and suggesting why results are anomalous 9TWSc.05 Carry out practical work safely, supported by risk assessments where appropriate

7.5 Rates of reaction	3–4	9Cc.04 Describe the effects of concentration, surface area and temperature on the rate of reaction, and explain them using the particle model	9TWSp.04 Plan a range of investigations of different types to obtain appropriate evidence when testing hypotheses 9TWSc.07 Collect, record and summarise sufficient observations and measurements, in an appropriate form 9TWSa.02: Describe trends and patterns in results, identifying any anomalous results and suggesting why results are anomalous 9TWSp.05 Make risk assessments for practical work to identify and control risks 9TWSm.02 Describe some important models, including analogies, and discuss their strengths and limitations 9TWSp.03 Make predictions of likely outcomes for a scientific enquiry based on scientific knowledge and understanding 9TWSc.05 Carry out practical work safely, supported by risk assessments where appropriate 9TWSc.03 Decide when to increase the range of observations and measurements, and increase the extent of repetition, to give sufficiently reliable data
End of chapter questions	1	Check students' progress	

Background science

During a chemical reaction, atoms in the reactants are rearranged to form atoms in the products. During this process the chemical bonds between the atoms in the reactants are broken and new bonds are formed between the atoms in the products. Breaking bonds between the atoms in the reactants requires energy and the making of new bonds between atoms to form the products releases energy.

During a chemical reaction there is no overall change in the amount of energy because energy cannot be created or destroyed, only converted from one form to another. This is known as the conservation of energy. During the reaction there is no destruction or forming of atoms. The total number of atoms of each element in the reactants is equal to the total number of atoms of each element in the products. The atoms in the reactants are just rearranged to form different products. This means that there is no change in mass, and the mass of the reactants will be equal to the mass of the products formed. This is known as the conservation of mass. However, in an open system in which the product is a gas, the gas can move into the air and so it *appears* that mass has been lost. This observation was the basis of the phlogiston theory of burning, which proposed that when substances burnt a substance was lost.

Chemical reactions can be shown as balanced symbol equations, which show the symbols or formulae of the reactants and products. They are balanced because they show that the numbers of atoms of each element on both sides of the equation are equal. Note: Balancing symbol equations is not expected at Stage 9.

Ionic compounds are made up of positive and negative ions (which may be made from a single atom e.g. Fe^{2+} or a group of atoms e.g. CO_3^{2-}). The formulae of ionic compounds show the ratio of ions in the compound. The positive and negative charges on the ions cancel each other out to form a compound with no overall charge.

Acids react with metals and metal carbonates to produce a salt. Reactions with acids follow simple patterns and these can be shown by general word equations.

1. Metals higher than copper in the reactivity series will react with acids to form a salt and hydrogen.

   metal + acid → salt + hydrogen

   For example:

   zinc + hydrochloric acid → zinc chloride + hydrogen

2. Metal carbonates will react with acids to form a salt, water and carbon dioxide.

   metal carbonate + acid → salt + water + carbon dioxide

   For example:

   copper carbonate + sulfuric acid → copper sulfate + water + carbon dioxide

The salt produced is a compound containing a metal ion and a non-metal ion from the acid. If the salt is soluble it will dissolve to form a solution. There are different methods of preparing salts depending on whether the salt is soluble or insoluble. This chapter only looks at preparing soluble salts from an insoluble metal or metal carbonate. The preparation of an insoluble salt by precipitation is covered in courses beyond Stage 9.

To make pure crystals of a soluble salt, with no contamination from the acid, all of the acid must be used up in the reaction. If a metal or insoluble carbonate is used this can be achieved by adding an excess of the solid. Any unreacted metal/carbonate can be removed by filtration before the salt solution undergoes evaporation in order to crystallise the salt. The slower the evaporation process, the larger the crystals. Therefore it is preferable to first heat the solution using a Bunsen burner (spirit burner) to remove some of the water but then leave it in a warm place to evaporate the rest.

Metals can be written as a list from most reactive to least reactive to form a reactivity series. The reactivity series that appears in the Student's Book contains just a few metals. However, the list can be expanded to include many more metals.

The reactivity series can be used to make predictions about whether a chemical (displacement) reaction will occur. In courses beyond Stage 9, students will learn how the position of a metal in the reactivity series determines the method of extraction from the metal ore.

In a displacement reaction, the more reactive metal displaces the less reactive metal from a solution or in a solid compound (such as in the thermite reaction).

Temperature, concentration and surface area are three factors which can affect the rate of a chemical reaction.
In courses beyond Stage 9, students will learn about collision theory. In this theory, the rate of reaction depends on the rate at which reacting particles collide with enough energy to react (the activation energy). These collisions are known as 'successful collisions'.

The rate of a chemical reaction can be determined in different ways. For example, by measuring the volume of gas produced over time, by measuring the decrease in mass of the reactants over time or by measuring colour changes or opacity. Students do not calculate rates of reactions in this chapter, only compare rates in terms of more/less gas given off in a certain time period.

Increasing the temperature of the reaction increases the rate because the particles have more energy, so they move around faster colliding more frequently and with more energy. This increases the rate at which successful collisions occur.

Increasing the concentration of a reactant in solution increases the rate of a reaction because increased concentration means more particles in a given volume, which means more frequent collisions.

Increasing the surface area of a solid reactant increases the rate of reaction because there is a greater surface area for the particles to collide with, so more frequent collisions occur.

Students will also learn about catalysts in courses beyond Stage 9. A catalyst is a substance that increases the rate of a reaction without being changed chemically or in mass by the end of the reaction.

7.1 Changes in chemical reactions

Topic overview

		Student's Book	Workbook	Downloadable material
7.1a	**Starting point** Students use their prior knowledge to consider what happens to the mass of iron when it rusts	Starting point table		
7.1b	**Conservation of mass and energy** In this activity students consider why evidence from the burning of fuels led scientists in the past to come up with the phlogiston theory, and why this theory is no longer accepted because of what we now know about conservation of mass and energy	Questions 1–5		Practical notes 7.1b
7.1c	**Mass change when metals react with oxygen** Students collect their own evidence to test the hypothesis that the mass of metals increase when they are burned	Activity 7.1	Questions 1–7	Practical notes 7.1c Worksheet 7.1c

Learning Objectives

- 9Cc.05: Understand that in chemical reactions mass and energy are conserved
- 9TWSp.03: Make predictions of likely outcomes for a scientific enquiry based on scientific knowledge and understanding
- 9TWSc.04: Take appropriately accurate and precise measurements, explaining why accuracy and precision are important
- 9TWSc.05: Carry out practical work safely, supported by risk assessments where appropriate
- 9SIC.05: Discuss how the uses of science can have a global environmental impact

Links to other topics

- This topic builds on Stage 7 Chapter 5,1, where students were introduced to how compounds are formed during a chemical reaction. Students should be able to represent chemical reactions as particle diagrams.
- In Stage 8 Chapter 6.4 students learnt about exothermic and endothermic reactions, which is built on here when learning about conservation of energy.

Using models

- Students will use particle diagrams from the Student's Book to represent chemical reactions. In these diagrams different colours can be used to represent the different atoms. Students should note that the number of each differently coloured atom is the same at the start and end of the chemical reaction. This observation can then be used to introduce the idea that the number of atoms does not change and that they are just rearranged to form different products. This model is useful in explaining why mass is conserved.

Science in context

The question in the carbon capture box in Topic 7.1 of the Student's Book covers the following learning objectives:

- 9SIC.05 Discuss how the uses of science can have a global environmental impact.

Common misunderstandings

- Students may think that the mass of the products of a precipitation reaction will be higher than the mass of the reactants because they observe a solid being made. This is explored in 7.1a Starting point.
- Students may think that as a gas is produced in a reaction then the total mass will decrease. This misunderstanding may arise because students believe that the gas was part of the reactants released on heating, or atoms were destroyed on heating.

7.1a Starting point

Students have learnt about precipitation reactions in Stage 7 Chapter 5. To review this, explore their initial ideas about conservation of mass by demonstrating the reaction between solutions of silver nitrate and potassium iodide to form a yellow precipitate of silver iodide. Repeat the reaction, but this time use an electronic balance to measure the mass of the two reactants and ask the students to predict what they think the mass will be after you mix them together. Listen to their ideas why. Mix the reactants and show them the mass. It will be the sum of the mass of the two reactants. Ask the class if their prediction was correct.

7.1b Conservation of mass and energy

Students will learn:

- To understand that mass and energy are conserved in a chemical reaction
- To use scientific knowledge to make predictions

Glossary words

combustion, theory

Resource preparation

For the demonstration: eye protection, Bunsen burner/spirit burner, heat-resistant mat, gas jar filled with oxygen with about a 2 cm depth of limewater in the bottom, deflagrating (combustion) spoon, charcoal

Practical notes 7.1b

Teaching and learning ideas

- Ask the class what they know about carbon dioxide and its link to climate change. Consolidate their ideas through discussion that it is produced when fuels, like petrol, are burnt in air.
- Explain that the burning of fuels is an example of a combustion reaction. Demonstrate the burning of carbon in air and in oxygen and use limewater to test the gas formed. Practical notes 7.1b give more information.
- Ask students to use their observations from the reaction to write the word equation. Reveal the answer (carbon + oxygen $\rightarrow$ carbon dioxide).
- Tell the class that the mass of the carbon was 3 g. Ask them to predict if the mass of carbon dioxide produced (if all the carbon reacts) is higher, lower or the same and why.
- Ask the students to read through the sections Testing theories and Conservation of mass in the Student's Book. Tell them that they can change their predictions now if they wish.
- Reveal the correct answer – that the mass of the carbon dioxide will be higher than 3 g, because the carbon has joined with oxygen particles. No particles have been created or destroyed.
- Use the particle diagram (figure 7.4) in the Student's Book to discuss what happens during the reaction. Carbon is a solid made up of carbon atoms, oxygen is molecules containing two oxygen atoms. During the reaction the carbon atoms break apart and join with the oxygen molecules to form a carbon dioxide molecule.
- Use this model to discuss conservation of energy. Explain that carbon is a solid, so the carbon atoms in it are held together by strong forces of attraction and so energy is needed to break apart the carbon atoms. This energy is stored in the new bonds formed between the carbon and oxygen atoms in carbon dioxide. So no energy is created or destroyed.
- Students answer questions 1–5 in the Student's Book.

Differentiation

Less confident students may need additional support with the concept of mass conservation in reactions. This could be done by modelling the reaction using beads to represent the compounds and showing that the atoms are just rearranged during the reaction. Extend more confident students' understanding of how models change by discussing with them how the development of particle theory has changed our model of what happens when substances burn. They can also answer challenge question 2 in the Student's Book.

Extension ideas

Give students the hypothesis 'Metals gain mass when they are heated'. Ask them to write down if they think this is true and why.

Homework suggestions

Ask students to write a method for an experiment where they could prove that mass is conserved during a chemical reaction.

7.1c Mass change when metals react with oxygen

Students will learn:

- To understand that mass and energy are conserved in a chemical reaction
- To collect accurate and precise measurements
- To carry out practical work in a safe manner to minimise risks
- To identify hazards and identify ways to minimise risks

Resource preparation

For the demonstration: eye protection, heat-resistant mat, iron/steel wool, Bunsen burner/spirit burner

For the activity (each pair): eye protection, magnesium ribbon, crucible with lid, tongs, pipe clay triangle, Bunsen burner/spirit burner, tripod, heat-resistant mat, access to balance

Practical notes 7.1c

Worksheet 7.1c

Teaching and learning ideas

- Ask the class 'does iron burn?' They will probably say no, as they will not consider that metals can burn.
- Demonstrate to the class the burning of iron wool. Wear eye protection. Place a small amount of iron wool on a heat-resistant mat and tease it apart to allow air to circulate. Ask students to describe the appearance of the iron (shiny and grey). Hold a Bunsen burner/spirit burner by the base and use it to heat one part of the iron wool; it will glow red and spread throughout the rest of the wool until it has all reacted. Ask them to observe the iron wool now – it will be a dull, dark grey colour and crumble easily.
- Ask students to predict what happens to the mass of metals when they are burnt and write a short plan that they can use to test this idea. They should use their scientific knowledge and understanding.
- Students calculate the change in mass when magnesium is burnt by following the instructions on Worksheet 7.1c. Practical notes 7.1c gives more information.
- Discuss with the students how they made sure their measurements were as accurate as possible, and why this is important. They can also mention any sources of error e.g. not all the magnesium oxide was weighed, some was lost when they lifted the lid, not all the magnesium reacted.
- Students complete Activity 7.1 *Heating magnesium* in the Student's Book (see the Student's Book answers).

Differentiation

The reaction could be modelled using beads with less confident students to show why the mass increased.

More confident students can be challenged to answer the challenge question on Worksheet 7.1c.

Extension ideas

Provide students with a series of questions that ask them to use conservation of mass to calculate the mass of reactants.

Example: iron and sulfur react to form iron sulfide: iron + sulfur → iron sulfide

32 g of sulfur reacted with iron to form 88 g of iron sulfide. What was the mass of iron?

Homework suggestions

Students answer questions 1–7 in the Workbook.

7.2 Word and symbol equations

Topic overview

		Student's Book	Workbook	Downloadable material
7.2a	**Starting point** Students describe chemical reactions as word equations	Starting point table		
7.2b	**Word and symbol equations** Students follow steps as they learn to construct balanced symbol equations, including writing the formulae of some ionic compounds. This is extension work as balancing symbol equations is not required at Stage 9	Questions 1–6	Questions 1–10	Worksheet 7.2b

Learning Objectives

- 9Cc.01: Use word equations and symbol equations to describe reactions (balancing symbol equations is not required)
- 9TWSm.03: Use symbols and formulae to represent scientific ideas
- 9SIC.02: Describe how science is applied across societies and industries, and in research
- 9SIC.04: Describe how people develop and use scientific understanding as individuals and through collaboration

Links to other topics

- Students were introduced to writing word equations in Stage 8 Chapter 6.1. In this chapter they will revisit this, as well as extend their learning by writing symbol equations. In Stage 9 Chapter 6 students were introduced to the concept of ions and ionic compounds. In this chapter they will be using the charges on ions to write formulae of ionic compounds.

Using models

- The particle models used in Topic 7.1 can be recapped and used to build an understanding of why symbol equations need to be balanced.

Science in context

The question at the bottom of the Science in context: Symbol equations box in Topic 7.2 of the Student's Book covers the following learning objectives:

- 9SIC.02 Describe how science is applied across societies and industries, and in research
- 9SIC.04 Describe how people develop and use scientific understanding as individuals and through collaboration.

Common misunderstandings

- When writing formulae of compounds, students are often confused about what the subscript numbers represent. A subscript number *after* an element symbol shows how many atoms of that element are in the compound if it is greater than one. For example, H_2O tells us that the molecule contains 2 atoms of hydrogen – not that the number of atoms of oxygen is multiplied by 2. The use of brackets in formulae can also be confusing. Students may think the subscript number applies only to the element immediately to the left of the bracket rather than to all the element symbols inside the brackets. Students will get practice at writing and interpreting formulae during the activities.
- Superscript numbers are also used to represent the number of charges on ions, for example, Fe^{2+}. Care should be taken that students understand the difference between the use of superscript and subscript numbers in representing ions and formulae.
- When balancing equations (extension work), students should understand that they cannot change the subscript numbers in the formulae in order to balance equations (as this would change the actual formula of the substance). Numbers are added before formulae only. For example, to balance: $Fe + O_2 \rightarrow FeO$ you cannot change the product to FeO_2 but instead should write $2Fe + O_2 \rightarrow 2FeO$. This is extension work as balancing symbol equations is not required at Stage 9.

7.2a Starting point

To check their understanding on how to write the word equations, ask students to read through the section Writing word equations in the Student's Book and answer question 1.

Read out some descriptions of other chemical reactions, or show videos of examples. Ask the students to write down word equations of each reaction.

7.2b Word and symbol equations

Students will learn:

- To describe chemical reactions using word or symbol equations (not including balancing)
- To represent scientific ideas using recognised symbols or formulae

Glossary words

formula, ion, product, reactant, symbol equation, word equation

Resource preparation

Worksheet 7.2b

Scissors, modelling clay in two different colours

Teaching and learning ideas

- Show the class a video clip of a rocket or space shuttle taking off. Inform them that the fuel the rocket uses is hydrogen. The fuel tanks contain hydrogen and oxygen, which are mixed together and ignited to form an explosive reaction, producing water. Ask them to write down the word equation for this reaction.
- Tell the class that to get the most explosive reaction it is important that the correct volume of oxygen is used so all the hydrogen reacts. Tell the class that they are going to work out what this is.
- Provide the class with modelling clay in two different colours. Tell them to make balls of modelling clay and build molecules of hydrogen (H_2) and oxygen (O_2). Remind them that when reactants react they split apart to form individual atoms, and these combine to form the products. The students model this using their models to make water (H_2O). They will see that they have one oxygen atom left over. Ask them what they need to do to use it up (make another hydrogen molecule so two molecules of water are formed).
- Show the class how to turn this into a balanced symbol equation: $2H_2 + O_2 \rightarrow 2H_2O$
- Ask them what the ratio of hydrogen to oxygen needs to be for an explosive reaction 2 hydrogen : 1 oxygen (or double the amount of hydrogen than oxygen).
- Ask the students to read the section Symbol equations in the Student's Book and answer question 2.
- Return to the word equation they wrote in question 1: sodium + chlorine → sodium chloride. Ask them to write a balanced symbol equation. This is extension work as balancing symbol equations is not required at Stage 9. Discuss that they need to work out the formulae of sodium chloride. Give each student Worksheet 7.2b, which will help them to understand how to work out the formulae of compounds from their constituent ions. They cut out and use cards to construct compounds and write down their formulae. You may like to discuss with the class that compounds on the sheet are all ionic compounds and that the formulae of molecules like H_2O and CO_2 cannot be worked out this way as these compounds are not formed from ions. Also, the charges on the cards do not match the normal convention for how they are written, that is, charges on the left for non-metals and each charge shown separately.

- Students answer questions 3–5 in the Student's Book. They can use the preceding information for guidance if they wish. Check answers to make sure students do not have any misconceptions surrounding the use of subscript numbers in the formulae. If students are finding this difficult, show the class how to answer some of the questions.
- Challenge: Students can also answer question 6.

Differentiation

Less confident students may find using modelling clay to model the reactions helpful.

Extension ideas

Extend more confident students by asking them what is the link between the charge on an ion of an element and the position of the element in the Periodic Table. They will notice that metal elements form ions with a positive charge and that Group 1 elements have a single charge and Group 2 elements have a double charge. Metals in the middle section (the transition metals) often form different ions with different charges, for example, iron can form Fe^{2+} or Fe^{3+} ions. Non-metal elements form negative ions. Those in Group 6 have a double negative charge and those in Group 7 have a single negative charge.

Homework suggestions

Workbook topic 7.2, questions 1–10.

7.3 Methods for making salts

Topic overview

		Student's Book	Workbook	Downloadable material
7.3a	**Starting point** Students review their knowledge of acids from Stage 7 by answering questions	Starting point table		
7.3b	**Reactions between metals and acids** In this activity students react metals with acid and perform the hydrogen 'pop' test. They also learn how to name the salts made during a reaction	Question 1	Question 2	Practical notes 7.3b Worksheets 7.3b(1), 7.3b(2)
7.3c	**Reactions between acids and carbonates** In this activity students react a range of carbonates with acid and perform the test for carbon dioxide. They use this test as well as flame tests to identify mystery substances They also plan an investigation to test an idea	Questions 2–4 Activity 7.2	Questions 1, 3 and 6	Practical notes 7.3c Worksheet 7.3c
7.3d	**Making a soluble salt** Students use filtration, evaporation and crystallisation to make a soluble salt – copper sulfate.	Question 5	Questions 4, 5, 7 and 8	Practical notes 7.3d Worksheet 7.3d

Learning Objectives

- 9Cc.01: Use word equations and symbol equations to describe reactions (balancing symbol equations is not required)
- 9Cc.03: Describe how to prepare some common salts by the reactions of metals with acids, and metal carbonates with acids, and purify them, using filtration, evaporation and crystallisation
- 9TWSm.03: Use symbols and formulae to represent scientific ideas
- 9TWSp.05: Make risk assessments for practical work to identify and control risks
- 9TWSc.02: Decide what equipment is required to carry out an investigation or experiment and use it appropriately
- 9TWSc.05: Carry out practical work safely, supported by risk assessments where appropriate
- 9TWSc.07: Collect, record and summarise sufficient observations and measurements, in an appropriate form
- 9SIC.02 Describe how science is applied across societies and industries, and in research
- 9SIC.05 Discuss how the uses of science can have a global environmental impact.

Links to other topics

- Students were introduced to acids in Stage 7 Chapter 4.2. In this chapter they will build on this by exploring the reactions of acids with metals and metal carbonates, and how to name the salts made.

Science in context

The additional questions in the Science in context: Feeding the world box in Topic 7.3 of the Student's Book cover the following learning objectives:

- 9SIC.02 Describe how science is applied across societies and industries, and in research
- 9SIC.05 Discuss how the uses of science can have a global environmental impact.

Teaching ideas

There are opportunities in this topic to explore the many industrial uses of salts, and how they are made in industry. One example is fertilisers.

Common misunderstandings

- When a salt solution is formed from the reaction of a metal and acid the products are a salt and hydrogen. Students may not understand how the salt formed may be a solution in water, if water is not a product. It is important that they understand that acids are solutions (this is the reason they are referred to as dilute or concentrated acids). The salt produced may therefore dissolve into the water present, even though all acid particles may have reacted. Remember however that not all salts are soluble.

7.3a Starting point

To review understanding of acids from Stage 7 Chapter 4, write a series of questions or comments on the board. Ask students to choose one, three or five to answer.

- What is an acid?
- How can you tell if a liquid is an acid?
- Name one acid you might find in a chemistry lab.
- Name one acid you might find at home.
- Describe the hazards and how you should reduce risks when using acids

Invite students to share their answers.

7.3b Reactions between metals and acids

Students will learn:

- To describe the methods used to prepare and purify salts from the reactions between acids and metals, acids and metal carbonates
- To choose the appropriate equipment for an investigation and use it correctly
- To carry out practical work in a safe manner to minimise risks

Resource preparation

For practical (each pair): eye protection, dilute hydrochloric acid (about 0.5 M), test tube and rack, rubber bung, magnesium ribbon (about 1 cm long), wooden splint

For challenge activity: 4 test tubes, magnesium ribbon, zinc pieces, iron nail, copper pieces, balances, filter paper, spatula

Practical notes 7.3b

Worksheets 7.3b(1), 7.3b(2)

Teaching and learning ideas

- Tell the class that most metals react with acid. Ask them to explain what they might observe when a metal is added to an acid. They may answer a colour change, bubbles, a solid being formed, etc.
- Students react magnesium and hydrochloric acid and test the hydrogen produced by following the instructions on Worksheet 7.3b(1). They write down their observations. Discuss the signs that a chemical reaction has taken place. Practical notes 7.3b give more information.
- Ask the students to read the sections Making salts and Reaction of acid with metals in the Student's Book. Ask them to name the salt that they made (magnesium chloride) and write a word equation for the reaction. Discuss that the salt is formed as a solution, because the acid they used was a solution.

- As further practice at writing word equations, students complete Worksheet 7.3b(2). They cut out the cards showing reactants and match them up to show different word equations.
- Students complete question 1 in the Student's Book.

Differentiation

Students can also write balanced symbol equations for the reactions (extension work).

Extension ideas

Students can complete the challenge task on Worksheet 7.3b(1) and plan a way to find out how different metals react with acid. You may wish for them to perform the practical. Practical notes 7.3b give more information.

Homework suggestions

Ask students to use secondary sources to research the uses of some common salts. Suitable salts they can research include sodium chloride (flavouring/preservative), copper sulfate (fungicide), ammonium nitrate (fertiliser), calcium sulfate (plaster), silver bromide (photography), barium sulfate (X-rays of gut).

Students can also write the formula of each compound.

7.3c Reactions between acids and carbonates

Students will learn:

- To describe the methods used to prepare and purify salts from the reactions between acids and metals, acids and metal carbonates
- To choose the appropriate equipment for an investigation and use it correctly
- To carry out practical work in a safe manner to minimise risks
- To present results appropriately

Resource preparation

For the demonstration: eye protection, piece of marble, limestone or chalk, tile, dilute hydrochloric acid

For the activity (each pair): eye protection, test tubes, test-tube rack, delivery tube with bung, magnesium carbonate, zinc carbonate, copper(II) carbonate, bicarbonate of soda (sodium hydrogen carbonate), dilute hydrochloric acid, limewater (calcium hydroxide solution), spatula, mystery substances A–D (hydrated potassium carbonate, magnesium sulfate, calcium carbonate and sodium chloride)

Practical notes 7.3c

Worksheet 7.3c

Teaching and learning ideas

- Demonstrate a reaction to students by adding drops of hydrochloric acid onto a piece of marble, chalk or limestone. Tell them that the rock contains the chemical calcium carbonate. Ask them to say if there is any evidence of a chemical reaction (bubbles).
- Ask them to use the Reactions between acids and carbonates section in the Student's Book to find and write down the word equation for the reaction just demonstrated.
- Students can answer questions 2–3 in the Student's Book. Students react a range of carbonates with hydrochloric acid and test the carbon dioxide produced by following the instructions on Worksheet 7.3c. Students record their observations in a suitable table. They should identify any patterns in their results. Practical notes 7.3c give more information.
- Students complete Activity 7.2 *What is the powder?* Students use information about the reactions with acids to identify powders as being metals, carbonates or neither.
- Go through the answers with the class (see Answers document).
- Students can also carry out a practical version of this activity. Present the class with four mystery powders labelled A–D. Ask them to test each to see if it is a carbonate. Practical notes 7.3c give more information.

Differentiation

Students can complete challenge question 4 in the Student's Book and the challenge activity on Worksheet 7.3c, which asks them to write the word equations for the reactions they performed.

Extension ideas

Discuss with the class that adding acid to samples and testing to see if carbon dioxide is produced is an analytical technique used to identify carbonates, but this test does not show what metal is present. Further tests such as flame tests can be used.

Homework suggestions

Ask students to research into the effects of acid rain on limestone and marble structures.

7.3d Making a soluble salt

Students will learn:

- To describe the methods used to prepare and purify salts from the reactions between acids and metals, acids and metal carbonates
- To choose the appropriate equipment for an investigation and use it correctly
- To carry out practical work in a safe manner to minimise risks
- To present results appropriately

Glossary words

crystallisation, evaporation, filtration

Resource preparation

Worksheet 7.3d for Practical: Making a soluble salt. Each pair will need: eye protection, copper(II) carbonate powder, dilute sulfuric acid, beaker, measuring cylinder, conical flask, filter funnel, filter paper, Bunsen burner/spirit burner, tripod, gauze, glass rod, spatula, evaporating basin, universal indicator paper

Practical notes 7.3d

Teaching and learning ideas

- Tell the class that they will be carrying out reactions involving acids. Discuss the hazards associated with the dilute acids they use in class (they are irritants and can cause irritation to skin and eyes). Ask them to write down a list of ways of controlling the risks when using acids. Examples include: wear eye protection, avoid spills, clean up small spills using a paper towel, tell the teacher if there is a larger spill, if acid is spilled on skin wash it off immediately, wash hands after using acids, put the lid back onto the bottle when you have finished using the acid. Go through the examples, making sure all students understand them so they can follow them when performing the experiments later in the learning episode.
- Show the class the equipment for the separating techniques: filtration and evaporation. This can be as actual equipment set up or as photographs/images (not scientific diagrams). Ask them to work in pairs, and divide the pairs in half – half working on filtration and half on evaporation. Pairs should draw the equipment as a labelled scientific diagram and describe why they might use this technique.
- Ask pairs from each half to show their drawings to the rest of the class and read out their description.
- Students perform a reaction to make copper(II) sulfate crystals using the reaction between copper(II) carbonate and sulfuric acid. Give each student Worksheet 7.3d. Refer to the practical notes for more information.
- Students complete question 5 in the Student's Book.

Differentiation

To support less confident students, model how to carry out each stage of the method and ask them to copy before you move onto the next.

Extension ideas

Ask students to choose a variable that could affect the size of salt crystals formed (for less confident students you can supply this from: temperature of the acid when the solid is added, amount of stirring when performing the reaction, the length of time the solution is heated for using the Bunsen burner/spirit burner, the temperature the solution is left at during crystallisation. For more confident students, ask them to choose a variable; they can use the internet to research one. They should then write a method that they could follow to see how this variable affects the size of the crystals. This should include how they could measure the size of the crystals.

Homework suggestions

Workbook, Chapter 7 Topic 3 questions 1–8.

7.4 Displacement reactions

Topic overview

		Student's Book	Workbook	Downloadable material
7.4a	**Starting point** Students revisit the reactivity series, and explain displacement reactions of metals using a simple sporting analogy	Starting point table		
7.4b	**Displacement reactions** Students learn about displacement reactions and apply their knowledge of the reactivity series; they make predictions on whether displacement reactions will or will not happen, based on the respective reactivity of two metals	Questions 1–4 Activity 7.3	Questions 1–6	Worksheets 7.4b(1), 7.4b(2) Practical notes 7.4b(1), 7.4b(2)

Learning Objectives

- 9Cc.02: Identify examples of displacement reactions and predict products (limited to reactions involving calcium, magnesium, zinc, iron, copper, gold and silver salts)
- 9TWSm.02: Describe some important models, including analogies, and discuss their strengths and limitations
- 9TWSp.03: Make predictions of likely outcomes for a scientific enquiry based on scientific knowledge and understanding
- 9TWSc.05: Carry out practical work safely, supported by risk assessments where appropriate
- 9TWSa.02: Describe trends and patterns in results, identifying any anomalous results and suggesting why results are anomalous
- 9TWSc.07 Collect, record and summarise sufficient observations and measurements, in an appropriate form
- 9TWSm.03 Use symbols and formulae to represent scientific ideas
- 9SIC.02 Describe how science is applied across societies and industries, and in research.

Links to other topics

- Students know from Stage 7 *Changes to materials* how to use the particle model to describe chemical reactions.
- Students learned how to write word equations in Stage 8 Topic 6.1 *Using word equations*.
- This topic gives students opportunities to develop their knowledge and understanding of using word equations and the particle model to represent and explain chemical reactions.

Using models

- The particle model used in topics 7.1 and 7.2 can be used again here and linked specifically to the example of the displacement reactions.

Science in context

The question in the Science in context: Global dimension box in Topic 7.4 of the Student's Book covers the following learning objective:

- 9SIC.02 Describe how science is applied across societies and industries, and in research.

Teaching ideas

Look out for opportunities to show students everyday examples of displacement. These need not be chemical examples, for example, new governments coming into power after a general election or year groups moving up at the start of a new school year.

Common misunderstandings

- Students often think that all metals react in the same way. It is important that students understand that there is an order of how reactive metals are, which doesn't change. The practical activities in this topic should help students to understand this idea.

7.4a Starting point

Remind students of the reactivity series and what it shows. Introduce displacement reactions and use a sporting analogy to explain them. For example, two basketball players with two different abilities trying to get possession of the ball. Explain that the players represent two different metals, and the ball represents a non-metal. Ask the students what a player with the ball represents (*a compound*). Ask the students what would happen if the better player tried to take the ball from the weaker player (*the better player would get the ball*). Ask the students what taking the ball from the other player represents (*a* chemical *reaction*). Ask the students what would happen if the weaker player tried to get the ball from the better player (*the better player would keep the ball*). Ask the students to identify the strengths and weaknesses of this analogy. (*Strengths: it is a simple and easy to understand analogy that describes how one metal can take the place of another. Weaknesses: the players are not 'joined' to the ball (as in a compound); sometimes a weaker player can successfully tackle a better player, but this cannot happen with metals.*)

7.4b Displacement reactions

Students will learn:

- To identify displacement reactions, and predict their products (involving zinc, magnesium, calcium, iron, copper, gold and silver salts only)
- To describe the uses, strengths and limitations of models and analogies
- To use scientific knowledge to make predictions
- To carry out practical work in a safe manner to minimise risks
- To describe trends and patterns shown in a set of results, identifying, and explaining, any anomalous results present

Glossary words

displacement reaction, inert, reactivity series, reactivity, salt

Resource preparation

Worksheet 7.4b(1) Making a reactivity series

Worksheet 7.4b(2) Using the reactivity series to make predictions

Practical notes 7.4b(1) Demonstration – Nail and coin demonstration

Practical notes 7.4b(2) Practical – Making a reactivity series

Activity 7.3: *Comparing the reactivity of metals*

For the demonstration (iron nail and coin): copper coin, iron nail, copper sulfate solution, iron(II) sulfate solution, forceps, eye protection

For the practical (making a reactivity series): eye protection, samples of zinc, iron, copper and magnesium, solutions of zinc sulfate, iron(II) sulfate, copper sulfate and magnesium sulfate (in dropping bottles or in small beakers with dropping pipettes), spotting tile, forceps/tweezers

Secondary resources:

- A video clip showing the extraction of copper from copper oxide using charcoal in a lab

Teaching and learning ideas

- To check students' understanding of displacement reactions following the basketball players and ball analogy, challenge students to predict what might happen when an iron nail is placed in copper sulfate solution and a copper coin is placed in an iron sulfate solution. This would work well as a 'think–pair–share' activity. Take ideas from the class and then demonstrate that iron will displace copper from copper sulfate solution but copper will not displace iron from iron sulfate solution. (Please note that many 'copper' coins are now only copper-plated steel and are magnetic, but these will work fine. Non-magnetic 'copper' coins are likely to be cupronickel but will also work fine as both copper and nickel are less reactive than iron.) Practical notes 7.4b(1) give more information about the demonstration.
- Students complete Activity 7.3 *Comparing the reactivity of metals* in the Student's Book, analysing some predictions made by students and explaining why the prediction was right or wrong. They also produce and justify a reactivity series for four metals, based on results provided for a series of displacement reactions.

- Worksheet 7.4b(1) gives the students a chance to carry out the displacement reactions described in Activity 7.3 themselves, and to construct a reactivity series from their results. Practical notes 7.4b(2) give more information.
- In Worksheet 7.2b(2) students use the reactivity series to predict the outcome of a series of displacement reactions between other metals and solutions to those in Activity 7.3: *Comparing the reactivity of metals*. Challenge students to complete the table in pairs using the reactivity series in the Student's Book to inform their decisions.
- Students complete questions 1–4 in the Student's Book.
- Show students the Science in Context box about how carbon is used to extract iron from iron oxide.

Differentiation

The students may observe some bubbling during the displacement reactions, especially those involving magnesium due to heating by the reaction. Make sure that students focus on the colour changes and appearance of a coating on the metals.

Extension ideas

Explain to students where carbon would appear in the reactivity series shown in the Student's Book. Ask them to discuss briefly in pairs why they think that a non-metal could be included in the reactivity series of metals and take feedback. If possible, show the students a video of the extraction of copper in the laboratory using charcoal. Explain that carbon will displace copper from copper oxide because carbon is more reactive than copper. Challenge students to work out the word equation for the reaction.

Homework suggestions

Workbook Topic 7.4, questions 1–6

Students research how scrap iron can be used in a displacement reaction to extract copper from a solution.

7.5 Rates of reaction

Topic overview

		Student's Book	Workbook	Downloadable material
7.5a	**Starting point** Challenge students to think about some ways they could make magnesium react with acid more quickly, in a think–pair–share activity	Starting point table		
7.5b	**What is the rate of a reaction?** Students consider that not all reactions happen at the same speed, and learn how they can compare rates of reaction	Questions 1–5 Activity 7.4	Questions 1–2	Worksheet 7.5b Practical notes 7.5b
7.5c	**Investigating the effect of temperature on rate of reaction** Students investigate how altering the temperature of the reaction mixture affects the rate of a reaction	Question 6	Questions 6 and 11	Worksheet 7.5c Practical notes 7.5c
7.5d	**Investigating the effect of concentration on rate of reaction** Students investigate how altering the concentration of a reactant in solution affects the rate of a reaction	Question 7	Questions 3, 7–9	Worksheet 7.5d Practical notes 7.5d
7.5e	**Investigating the effect of surface area on rate of reaction** Students investigate how altering the surface area of a solid reactant affects the rate of a reaction	Question 8 Activity 7.5	Questions 4, 5, 10	Worksheet 7.5e Practical notes 7.5e

Learning Objectives

- 9Cc.04: Describe the effects of concentration, surface area and temperature on the rate of reaction, and explain them using the particle model
- 9TWSm.02: Describe some important models, including analogies, and discuss their strengths and limitations
- 9TWSp.03: Make predictions of likely outcomes for a scientific enquiry based on scientific knowledge and understanding
- 9TWSc.05: Carry out practical work safely, supported by risk assessments where appropriate
- 9TWSa.02: Describe trends and patterns in results, identifying any anomalous results and suggesting why results are anomalous
- 9TWSc.03: Decide when to increase the range of observations and measurements, and increase the extent of repetition, to give sufficiently reliable data
- 9TWSc.07 Collect, record and summarise sufficient observations and measurements, in an appropriate form
- 9TWSp.04 Plan a range of investigations of different types to obtain appropriate evidence when testing hypotheses.
- 9TWSp.05 Make risk assessments for practical work to identify and control risks.

Links to other topics

- Students know from Stage 7 *Changes to materials* how to use the particle model to describe chemical reactions.
- Students learn to write word equations in Stage 8 Topic 6.1 *Using word equations.*

Using models

This topic provides the opportunity for students to consider the strengths and limitations of analogies used to explain the effect of temperature, concentration and surface area on the rate of reaction. One example of an analogy could be dodgem cars or children in a playground.

For the children in a playground analogy:

- The higher the concentration, the more children are in the playground and the more likely it is that they will collide. The more collisions there are, the higher the rate of reaction
- The higher the temperature the more energy the children have and the faster they run, meaning that they are more likely to collide. The more collisions there are, the higher the rate of reaction
- To model higher surface area the children could have their arms and legs spread out as far as possible when moving around. The higher the surface area the more potential places on the children there are that can collide and the more likely collisions are. The more collisions there are, the higher the rate of reaction.
- To address learning objective 9TWSm.02 students could be given the chance to discuss the strengths and limitations of the analogy used.

Science in context

Teaching ideas

Look out for opportunities to show students everyday examples of controlling rates of reaction by choosing suitable temperatures. For example, refrigerating food and vaccines to extend their useful life, and using heating to cook food.

Common misunderstandings

- Students often think that reactions proceed at a constant rate, whereas the rate of a reaction usually decreases as the reaction takes place. Plotting a graph of mass lost or volume of gas produced over time will help show (qualitatively) that the rate is fastest at the start but decreases.
- Some students may confuse reaction time and reaction rate. The two variables are inversely proportional to each other: as the reaction rate increases, the reaction time decreases. An analogy that may help understanding is that of a car journey. The faster the car goes, the sooner you reach your destination. This will be true however far the journey is or whatever car is being driven (analogous to different reactions).
- Students may think that the reaction rate increases as the number of collisions increases, rather than as the rate of collisions increases. This idea is incorrect because there will be many collisions in a very slow reaction if it is given enough time. In terms of the dodgem analogy, this is like driving round very slowly for a long time on the ride.
- Students may think that the energy of collisions increases as the concentration increases, rather than that the rate of collisions increases. This idea is incorrect because increasing the concentration of a solute only increases the number of particles in a given volume.
- Students may think that the number of collisions increases if the surface area of a solid reactant is increased. This is correct over a given time, so the rate of collisions increases. In terms of the dodgem analogy, this is like the owner of the ride placing traffic cones on the floor. The cars are more likely to collide with a cone if the cones are spread around, rather than being clumped together in one place.

7.5a Starting point

Remind students of the reactions of metals with acids they have seen. For example, magnesium reacting with dilute hydrochloric acid. If possible, show them the reaction again. Ask students to think about some ways they could make the magnesium react more quickly. This would work well as a 'think–pair–share' activity with students initially in pairs.

7.5b What is the rate of a reaction?

Students will learn:

- To describe and explain how the rate of reaction is affected by changes in concentration, surface area and temperature
- To plan investigations used to test hypotheses
- To describe the uses, strengths and limitations of models and analogies
- To carry out practical work in a safe manner to minimise risks
- To interpret results and form conclusions using scientific knowledge and understanding

Glossary words

rate of reaction

Resource preparation

Worksheet 7.5b PMI chart

Practical notes 7.5b Demonstration – What is the rate of reaction?

Activity 7.4 *Which reaction is the fastest?* Optional: graph paper

For the demonstration (What is the rate of reaction?): eye protection, 100 cm^3 conical flask, measuring cylinder (50 or 100 cm^3), 1 M hydrochloric acid, 10 g of fine marble chips, balance (±0.01 g), cotton wool, stopwatch

or eye protection,100 cm^3 conical flask, measuring cylinder (50 or 100 cm^3), 1 M hydrochloric acid, 0.4 g of fine marble chips, 100 cm^3 gas syringe, bung and rubber tube to act as a delivery tube, clamp stand, boss and clamp, stopwatch

or eye protection, 100 cm^3 conical flask, measuring cylinder (50 or 100 cm^3), 1 M hydrochloric acid, approx. 0.1 g of magnesium ribbon as a loose coil, 100 cm^3 measuring cylinder, shallow water trough, bung and tube to act as a delivery tube, clamp stand, boss and clamp, stopwatch

Teaching and learning ideas

- Give students examples of three chemical reactions that happen at very different rates. For example, a firework, magnesium ribbon reacting with acid and a nail rusting. Ask students in pairs to put the reactions in order of fastest to slowest, and to add any other reactions they can think of in between. Take feedback from student pairs and then ask them to write down ways that might alter the rate of the reactions.
- As an introduction to monitoring the rate of a reaction, demonstrate the reaction of marble chips with dilute hydrochloric acid on a balance *or* with a gas syringe to collect the gas produced, dependent on resources available. Alternatively, demonstrate the reaction between magnesium and hydrochloric acid, collecting the hydrogen over water. The procedures for all three experiments are described in Practical notes 7.5b.
- Place students in groups of about four to discuss the advantages and disadvantages of different methods of monitoring the rate of reaction supported with a PMI chart (Plus/Minus/Interesting) in Worksheet 7.5b. A PMI chart is a type of chart that contains three columns that students can use to organise the pluses, minuses and interesting things associated with a method, idea or decision. If necessary, prompt the discussion with questions such as: Which method(s) work when a gas is evolved?; Which methods work well when the gas evolved is soluble in water and why?; What other methods can they can think of?; Are there any examples where the methods looked at would not work?.
- In Activity 7.4 *Which reaction is the fastest?* in the Student's Book, students are provided with a brief description of a practical and a table of results. They have to give the variables involved and analyse the data to state when the reaction was fastest. Students could also plot a graph of this data. Go through the answers with the class and explain how the data shows that the rate of reaction changes.
- Challenge students to explain why the rate of reaction decreases as the reaction takes place.
- Introduce the idea that for a chemical reaction to happen, the reactant particles *must* collide with each other, and they must do this with enough energy. Since reaction rate is the number of reactions in a given time, the reaction rate will be higher if the reactant particles collide more often. Consider using an analogy such as dodgem cars at a fun fair. Dodgem (bumper) cars are a fairground ride where small cars are powered electrically from the floor or ceiling. The aim of the ride is to drive the car into as many other cars as possible. Challenge students to suggest what factors can affect the rate of a chemical reaction and why (the effects of temperature, concentration and surface area will be explained in 7.3c, 7.3d and 7.3e).
- Students complete questions 1–5 in the Student's Book.

Differentiation

Less confident students may need more emphasis on understanding the concept of rate being change over a specified amount of time. They may need to be provided with a value for rate.

More confident students could calculate the rate of reaction themselves using the volume of gas over time.

Extension ideas

Ask students to research how and why temperature affects the rate of reaction, in preparation for the next lesson.

Homework suggestions

Workbook Topic 7.5, questions 1–2.

7.5c Investigating the effect of temperature on rate of reaction

Students will learn:

- To describe and explain how the rate of reaction is affected by changes in concentration, surface area and temperature
- To plan investigations used to test hypotheses
- To describe the uses, strengths and limitations of models and analogies
- To carry out practical work in a safe manner to minimise risks
- To use scientific knowledge to make predictions
- To describe trends and patterns shown in a set of results, identifying, and explaining, any anomalous results present
- To interpret results and form conclusions using scientific knowledge and understanding

Resource preparation

Worksheet 7.5c How does temperature affect rate of reaction?

Practical notes 7.5c Practical – Investigating the effect of temperature on rate of reaction

For the practical (Investigating the effect of temperature on rate of reaction): eye protection, Bunsen burner/spirit burner, tripod, gauze and heat-resistant mat, thermometer, dilute hydrochloric acid, sodium thiosulfate solution, 100 cm^3 conical flask, cross drawn on a piece of paper, stopwatch, 10 cm^3 measuring cylinder, 50 or 100 cm^3 measuring cylinder, tongs, stop bath (details in Practical notes)

Teaching and learning ideas

- Through questioning, check students recognise that particles of a substance (in any state) move faster as the temperature increases. Now challenge the students to work in pairs to imagine what would happen in the dodgem car analogy if the cars moved quickly rather than slowly. The cars would collide more often and with more energy. Remind students that chemical reactions happen when particles collide, and extend the challenge to predict, in terms of rate of collision and energy of particles, what would happen if the temperature of a reaction mixture increased. Students then make predictions of what they think would happen to the reaction rate if the temperature were increased.
- Introduce the reaction between sodium thiosulfate solution and dilute hydrochloric acid by mixing 25 cm^3 of sodium thiosulfate with 25 cm^3 of hydrochloric acid in a 100 cm^3 conical flask. Place the conical flask on a piece of paper with a black cross and show that the cross disappears as the solution becomes opaque, if possible, using a visualiser or webcam. Explain to students that they will carry out an investigation into the effect of temperature on the reaction. Provide them with Worksheet 7.5c and ensure they are clear about the procedure. Emphasise they will measure the *time* for the cross to disappear: a short time for the cross to disappear means a fast rate of reaction. Practical notes 7.5c support this investigation.
- Ask students to read through the information about temperature in How do the factors affect the rate of reaction? in the Student's Book to check their predictions. Take answers from the class and correct any misunderstandings.
- Students answer question 6 in the Student's Book.

Differentiation

Less confident students could have the axis for graph of temperature against time provided.

More confident students should be able to plot the graph themselves and could also calculate the rate and plot a graph of temperature against rate.

Homework suggestions

Workbook Topic 7.5, questions 6 and 11.

Write up the practical, including a conclusion and particle diagrams to explain the effect of temperature on rate of reaction.

7.5d Investigating the effect of concentration on rate of reaction

Students will learn:

- To describe and explain how the rate of reaction is affected by changes in concentration, surface area and temperature
- To plan investigations used to test hypotheses
- To describe the uses, strengths and limitations of models and analogies

- To carry out practical work in a safe manner to minimise risks
- To use scientific knowledge to make predictions
- To describe trends and patterns shown in a set of results, identifying, and explaining, any anomalous results present
- To interpret results and form conclusions using scientific knowledge and understanding

Glossary word

concentration

Resource preparation

Worksheet 7.5d How does concentration affect rate of reaction?

Practical notes 7.5d Demonstration/Practical – Investigating the effect of concentration on rate of reaction

For the demonstration (Investigating the effect of concentration on rate of reaction): eye protection, magnesium ribbon (2 × 3 cm strips), hydrochloric acid (2 M, 1 M and 0.5 M), two boiling tubes, stopwatch

For the practical (Investigating the effect of concentration on rate of reaction): eye protection, 100 cm^3 conical flask, hydrochloric acid of different concentrations, sodium thiosulfate solution, cross drawn on a piece of paper, stopwatch, boiling tubes, 25 or 50 cm^3 measuring cylinders × 2

Teaching and learning ideas

- Through questioning, check students recognise that particles of a substance become more crowded as the concentration increases. Now challenge the students to work in pairs to imagine what would happen in the dodgem car analogy if more cars were moving during the ride. The cars would collide more often but the energy of the collisions would remain the same. Remind students that chemical reactions happen when particles collide, and extend the challenge to predict, in terms of rate of particle collisions, what would happen if the concentration of a reactant in a reaction mixture were increased. Students then make predictions of what they think would happen to the reaction rate if the concentration were increased.
- Demonstrate the reaction between 25 cm^3 of 1 M hydrochloric acid with a 3 cm strip of magnesium ribbon in a boiling tube and time the reaction. Now challenge students to predict what will happen if you repeat the reaction with the same amount of 0.5 M and 2.0 M hydrochloric acid. Allow them to discuss their answers briefly in pairs before taking feedback. Student may make quantitative predictions such as double the concentration will halve the time of reaction. (After demonstrating the reaction, it will be apparent that this is not the case. This is because the reaction is exothermic, and the greater the concentration of acid the more the reaction mixture increases in temperature. In addition, the concentration of acid decreases as the reaction proceeds.
- Revisit the method used in Worksheet 7.5c (the disappearing cross) and ask students to plan a method to investigate the effect of changing the concentration of hydrochloric acid on the rate of reaction between sodium thiosulfate and dilute hydrochloric acid. Students then carry out the practical described in Worksheet 7.5d to test their predictions. Practical notes 7.5d support this investigation.
- Ask students to read through the information about concentration in How do the factors affect the rate of reaction? in the Student's Book to check their predictions. Take answers from the class and correct any misunderstandings.
- Students answer question 7 in the Student's Book.

Differentiation

Less confident students could have the axis for graph of concentration against time/volume of gas provided.

More confident students should be able to plot the graph themselves and could also calculate the rate and plot a graph of concentration against rate.

Extension ideas

Ask students to write a test question and answer on the effects of temperature and concentration on the rate of reaction.

Homework suggestions

Workbook Topic 7.5, questions 3, 7–9

Write up the practical, including a conclusion and particle diagrams to explain the effect of concentration on rate of reaction.

7.5e Investigating the effect of surface area on rate of reaction

Students will learn:

- To describe and explain how the rate of reaction is affected by changes in concentration, surface area and temperature
- To plan investigations used to test hypotheses
- To describe the uses, strengths and limitations of models and analogies
- To carry out practical work in a safe manner to minimise risks
- To use scientific knowledge to make predictions
- To describe trends and patterns shown in a set of results, identifying, and explaining, any anomalous results present
- To interpret results and form conclusions using scientific knowledge and understanding

Glossary word

concentration

Resource preparation

Worksheet 7.5e How does surface area affect rate of reaction?

Practical notes 7.5e Demonstration/Practical – Investigating the effect of surface area on rate of reaction

Activity 7.5 *Investigating the effect of surface area*

For exploring surface area (optional) each pair of students will need: 8 cubes of side length 1 cm, for example, plastic linking cubes as used in Mathematics lessons

For the demonstration (dissolving sugar): sugar candy crystals, fine grain sugar (for example, caster sugar), 100 cm^3 beaker × 2, stirring rod × 2, stopwatch

For the demonstration (Investigating the effect of surface area on rate of reaction): eye protection, magnesium ribbon, magnesium turnings (small shavings of magnesium ribbon), 1 M hydrochloric acid, boiling tube × 2, stopwatch, balance (±0.01 g)

For the practical (Investigating the effect of surface area on rate of reaction): eye protection, 100 cm^3 conical flasks, small, medium and large marble chips, spatula, bung with rubber tubing, gas syringe, dilute hydrochloric acid, 25 or 50 cm^3 measuring cylinder, balance (±0.1 g), clamp stand, boss and clamp. Optional: graph paper

Teaching and learning ideas

- Provide pairs of students with eight cubes of side length 1 cm. Challenge them to work out the surface area of the cubes when arranged in one cube of side length 2 cm, and as eight cubes of side length 1 cm. Explain that in each case the volume of the material is the same but halving the cube size doubles the surface area. Explain that when a solid reacts, it is only particles on the surface which can react and so surface area is important.
- Check students' understanding of surface area and reaction rate by asking: will crystals of sugar candy or a spoon of powdered sugar dissolve faster in water? Dissolving is not a chemical reaction, but the idea of size and surface area of particles can be linked to chemical reactions and collisions. If possible, carry out the demonstration. If sugar candy crystals are not available, use sugar crystals for coffee.
- Demonstrate the effect of particle size on reaction time by using magnesium ribbon and turnings with hydrochloric acid. Measure 25 cm^3 of acid into two boiling tubes. Find the mass of a 3 cm strip of magnesium ribbon and measure out the same mass of magnesium turnings. Ask students to predict the outcome before you add the magnesium to each tube at the same time.
- Remind students of the work they did (or will do) in Stage 9 Chapter 7 about neutralising acids, in the context of indigestion remedies. Ask students to complete Activity 7.5 *Investigating the effect of surface area* in the Student's Book, based on planning an investigation of the reaction of indigestion tablets with dilute hydrochloric acid, making predictions about whether 'large' pieces, 'small' pieces or powdered tablets will have the fastest rate of reaction with the acid.
- Students then use ideas from Activity 7.5 to make predictions about an analogous practical supported by Worksheet 7.5e. Show students samples of the three different sized marble chips. Ask students in pairs to construct a short, written explanation to explain how increasing the surface area of a solid reactant affects the rate of reaction. Students then make predictions about what they think would happen to the reaction rate if they used 'large' pieces of marble chip compared with 'small' pieces of marble chip. Each student pair can then compare their ideas with another pair, and refine their ideas accordingly before carrying out the practical as described on Worksheet 7.5e. Practical notes 7.5e support this investigation. Students could also plot a line graph of their data to interpret the changing rate of reaction as the reaction proceeds, and to compare the rates of reaction.

- Ask students to read through the information about surface area in How do the factors affect the rate of reaction? in the Student's Book to check their predictions. Take answers from the class and correct any misunderstandings.
- Students answer question 8 in the Student's Book.

Differentiation

Less confident students may need a framework to be provided for the predictions about the analogous practical and graph axis for their results for the data from Worksheet 7.5e.

More confident students should be able to calculate their own rates of reaction and design their own graphs to use.

Extension ideas

Apply ideas about surface area and rate of reaction to explain why a wet towel dries quickly when it is hung up, but not if folded up on the floor.

Homework suggestions

Workbook Topic 7.5, questions 4, 5, 10

Write up the practical, including a conclusion and particle diagrams to explain the effect of particle size on rate of reaction.

Chapter 7 Checking students' progress

End of chapter reflection

- To review the learning points of the topic
- To test understanding through answering questions

Resources

- Student's Book [End of chapter review questions]
- Workbook [Self-assessment grid, End-of-chapter questions]
- Worksheet 7.6

Approach

Use the 'Key facts' and 'Check your skills progress' boxes in the Student's Book for each topic in this chapter as a way of recapping on the main learning points.

Ask students to complete the Self-assessment checklist at the end of Chapter 7 in the Workbook. Encourage students to work together in small groups to discuss which aspects of the material are the most difficult to understand, and then have a class vote. You may wish to revisit some Learning Objectives again if students find any areas particularly challenging. Use the Student's Book text and questions to revisit these aspects. Go through the answers orally in class. Take note of whether giving the answer appears to resolve students' uncertainty or not, and work through further examples.

The Teacher's comments in the Workbook can be completed to give students personalised feedback on next steps. Students could be asked to answer specific End of chapter review questions in their books. Alternatively go through some or all of the End of chapter review questions in the Student's Book orally in class.

Use the Context worksheet (Worksheet 7.6) as a framework for students to conduct their own research enquiries about the extraction of metals from their ores. In principle, all metals may be extracted from their compounds by passing electricity through molten mixtures or solutions of metal salts. However, a metal that is less reactive than carbon may be extracted more cheaply by heating one of its compounds with carbon. In these reactions, industrial chemists adjust factors such as temperature and surface area of solid reactants to produce reasonable amounts of metal in a reasonable time. Students may decide to focus on globally significant metals, such as iron and copper. They could extend their research to discover why some metals (such as tungsten) cannot be extracted by heating with carbon, even though they are less reactive than carbon.

Finish by asking students to complete the End-of-chapter questions at the end of Chapter 7 in the Workbook. Provide feedback on students' answers, and work with students individually to identify areas that still need more work.

Chapter 8 Energy

Chapter overview

In this chapter students will build on their previous knowledge from Stage 7 about energy transfers, that in every energy transfer some energy is dissipated and about energy as something that cannot be created or destroyed. They will also meet the term 'potential energy' and the joule as the unit of energy.

They will learn about the principle of conservation of energy and use this to solve quantitative problems. They will also learn about the three types of energy transfer – conduction, convection and radiation, and how evaporation by cooling works.

Topic title	Number of 40-minute periods	Learning Objectives	Thinking and Working Scientifically Learning Objectives
8.1 Energy conservation	1–2	9Pf.03 Know that energy is conserved, meaning it cannot be created or destroyed	9TWSm.02 Describe some important models, including analogies, and discuss their strengths and limitations 9TWSp.03 Make predictions of likely outcomes for a scientific enquiry based on scientific knowledge and understanding 9TWSc.06 Make an informed decision whether to use evidence from first-hand experience or secondary sources
8.2 Heating and cooling	2	9Pf.02 Describe the difference between heat and temperature 9Pf.04 Know that thermal energy will always transfer from hotter regions or objects to colder ones, and this is known as heat dissipation	9TWSm.01 Understand that models and analogies reflect current scientific evidence and understanding and can change 9TWSp.03 Make predictions of likely outcomes for a scientific enquiry based on scientific knowledge and understanding 9TWSc.07 Collect, record and summarise sufficient observations and measurements, in an appropriate form 9TWSa.03 Make conclusions by interpreting results, explain the limitations of the conclusions and describe how the conclusions can be further investigated 9TWSp.05 Make risk assessments for practical work to identify and control risks 9TWSc.05 Carry out practical work safely, supported by risk assessments where appropriate
8.3 Conduction, convection, radiation and evaporation	3–4	9Pf.05 Describe thermal transfer by the processes of conduction, convection and radiation 9Pf.06 Explain cooling by evaporation 9Pf.04 Know that thermal energy will always transfer from hotter regions or objects to colder ones, and this is known as heat dissipation	9TWSp.01 Suggest a testable hypothesis based on scientific understanding 9TWSp.04 Plan a range of investigations of different types to obtain appropriate evidence when testing hypotheses 9TWSp.05 Make risk assessments for practical work to identify and control risks 9TWSc.07 Collect, record and summarise sufficient observations and measurements, in an appropriate form 9TWSm.02 Describe some important models, including analogies, and discuss their strengths and limitations

			9TWSp.03 Make predictions of likely outcomes for a scientific enquiry based on scientific knowledge and understanding 9TWSc.04 Take appropriately accurate and precise measurements, explaining why accuracy and precision are important 9TWSc.05 Carry out practical work safely, supported by risk assessments where appropriate 9TWSa.03 Make conclusions by interpreting results, explain the limitations of the conclusions and describe how the conclusions can be further investigated
End of chapter review	1	Check students' progress	

Background science

Students may already recognise energy as something that cannot be created or destroyed and be aware of the concept of energy conservation from their previous studies.

Note: As energy has become more central to our modern lives, our scientific ways of talking about it have been changing, too. The Institute of Physics (IoP) has made some recommendations for science education about energy – a 'stores and transfers model' – which they consider has advantages over the traditional approach that describes various 'energy types'. The new Cambridge Primary and Lower Secondary Science curriculum framework has deliberately steered clear of adopting any specific model for teaching energy, preferring to allow teachers and text book authors the choice of model, and indeed the opportunity to highlight to students the fact that a model is being used, and to discuss with them how useful the model is.

So, in this course we have moved away from naming energy types, and instead focused on energy sources, energy transfers and the various ways in which energy can be stored. This, we hope, will enable you as teachers to exercise the freedom that the curriculum framework designers have intended you to have, choosing your own energy model and approach. In making that choice you will want to consider the language and ideas your students are likely to meet when they move on to prepare for Cambridge IGCSE or equivalent and other higher qualifications.

Let us consider what is different and possibly valuable in the 'stores and transfers' model for teaching at this Lower Secondary level. 'Potential energy' is a term that students often find a rather odd and confusing. 'Potential for what?' they may wonder, or, 'Is this actual energy or just a possibility?' So, in the IoP's recommended approach, this is first described to them as 'stored energy' or an 'energy store'. This stored energy could be *elastic* – stored in a stretched or squashed material, or *gravitational* – stored by moving a mass upwards against the force of gravity, and able to be recovered by letting the mass fall back down. (Because gravity is dealt with in a later chapter in this book, we have been cautious about using the word 'gravitational', and have just talked about energy stored in an object due to its height.)

Other 'energy stores' in the new 'stores and transfers' model include *kinetic* energy, *thermal* energy, *chemical* energy and *nuclear* energy. In the old language, they were all 'energy types'. IGCSE also introduces the term *internal* energy, which is a 'catch-all' for the total energy stored in a system, an important idea, but something which cannot be directly measured. Only *changes* in internal energy can be determined. So, getting your students used to talking about *changes* in energy is helpful preparation for more advanced learning later.

In both the old and the new language, *heating* and *doing work* are ways of transferring energy. *Heating* occurs when *thermal* energy is either the input or the output of the transfer. *Work is done* whenever a force moves though a distance. Students at this level need to recognise that we can store energy by applying a force and making something move; and that this stored energy can be recovered if an object then pushes back and begins to move itself – either due to its elasticity or just due to its weight making it fall. (They do not yet need to talk about force and displacement as being vector quantities.)

Heating does not always cause a temperature rise. Sometimes the thermal energy transferred into a material is used to do work: for example, the work of separating and rearranging molecules to turn them from solid to liquid state – ice melting at 0 °C without a temperature rise – or the work of expanding the volume of a gas by heating it. (Students will not explore such examples as these in any detail until later in their studies.)

What is quite different between the new 'stores and transfers' model and the old 'energy types' approach, is the way in which they describe energy in wave motion. Waves, like *sound* and *light*, carry energy from one object and place to another, when they are simple travelling waves. This is how students at this level usually experience sound and light. So, in this book, we have used the 'stores and transfers' approach of describing *sound* and *light*, along with a flow of *electric current*, as ways of transferring energy. This will hold good for students in their later studies, although they will also learn how standing waves can be stores of energy. By contrast the old language for teaching about energy calls electrical, light and sound energy 'energy types'. That is something we have avoided doing in this book; nor have we called these things 'energy sources'. Rather, we have pointed, beyond the means by which energy is being transferred, to a source from which it is coming – e.g. light from the Sun. This seems to fit better with the language and ideas about energy that students will meet at IGCSE or equivalent and beyond.

You can read more about the arguments for choosing a clear model and helpful language for teaching about energy at http://practicalphysics.org. However, because of the intention of the Cambridge Primary and Lower Secondary Science curriculum frameworks at Stages 1 to 9 not to specify a specific model, you do have a choice. However, note that the Learning Objectives 9Pf.04 and 9Pf.05 do refer to thermal energy and thermal transfer.

In this chapter we are looking at thermal energy. There are a number of key concepts involved.

Firstly, students need to be secure in their understanding that energy is conserved. There is always the same amount of energy in a system, it is just transferred from one place to another. During transfers thermal energy is always dissipated into the surroundings where it is effectively 'wasted' because it spreads out and cannot be captured and transferred usefully.

Secondly, we have the fact that thermal energy is not the same thing as temperature. The temperature of an object is a measure of the average amount of thermal energy each particle has. Therefore, the total amount of thermal energy in an object depends both on its temperature and the number of particles it contains.

Tied in with this, we have the important concept that thermal energy is always transferred from hotter objects (or areas) to cooler objects (or areas). This also helps to emphasise that objects (or areas) become cooler because they lose thermal energy as it is transferred elsewhere.

The final section of this chapter concentrates on thermal transfer mechanisms: conduction, convection and radiation.

Conduction happens mainly in solids because it involves energy transfer by particle vibration. In solids, the particles are in a fixed position and close together so transfer between particles is easier than in liquids and gases. Metals are better thermal conductors than non-metals. This is not because their particles are closer together but because of their atomic structure. Metals have free electrons. Free electrons are not tied to specific nuclei but form a sort of cloud surrounding the nuclei in the metals. These free electrons can move easily through the metal. The movement of these free electrons helps thermal conduction take place. (Incidentally, the existence of free electrons also explains why metals are good electrical conductors.)

Convection happens in fluids (liquids and gases). It cannot happen in solids. This is because it relies on the free movement of particles through the fluid. This movement is driven by density differences caused by particles that have gained thermal energy moving further apart (and therefore reducing density). Hot (less dense) fluids rise and cold (more dense) fluids fall. Note, it is the hot fluid (not 'heat') that rises. To say that heat rises is incorrect.

Radiation is different from convection and conduction in as much as it involves thermal transfer in the form of waves (infra-red radiation) and not particles. Surfaces which are hotter than their surroundings emit infra-red radiation into the surroundings and thus transfer thermal energy. The type of surface determines how well it emits infra-red radiation. Good emitters are dark with a matt finish. The lighter and shinier a surface is, the less infra-red radiation it emits. Similarly all surfaces that are cooler than their surroundings absorb infra-red radiation. Again, the nature of the surface determines how well it does this. Dark, matt surfaces are the best absorbers of infra-red radiation. Shiny, light coloured surfaces are very poor absorbers but are good at reflecting infra-red radiation back into the surroundings.

Evaporation can also be considered as a method of transferring thermal energy. When a liquid evaporates it removes thermal energy from its surroundings and uses this energy to change phase (or state) from liquid to gas. Because the surroundings have had thermal energy removed, their temperature decreases (the average energy of each particle in the liquid decreases). Also, the most energetic particles in the liquid evaporate first because they need the smallest amount of additional energy to escape the liquid. The rate of evaporation of a liquid depends on the surface area of the liquid, the temperature of the liquid, the temperature of the surroundings, the humidity of the surroundings and the movement of the surrounding air. Think of washing drying on a clothes line: evaporation is responsible for this. Clothes dry best on a warm, dry, windy day and when they are spread out, providing a large surface area for the water to evaporate from.

8.1 Energy conservation

Topic overview

		Student's Book	Workbook	Downloadable material
8.1a	**Starting point** This activity reviews students' prior knowledge and understanding in preparation for the new ideas they will meet in this chapter	Starting point table		Worksheet 8.1a(1) Worksheet 8.1a(2)
8.1b	**Pendulum** A teacher demonstration of energy transfers in a swinging pendulum introduces the principle of the conservation of energy. It also helps reinforce the meaning of potential energy Student practical using bouncing balls is used to illustrate energy being dissipated	Questions 1–2 Activity 8.1 Activity 8.2	Questions 1–6	Practical notes 8.1b
8.1c	**Measuring energy changes** Students are introduced to the joule as the unit of energy and explain what happens to energy during energy transfers, keeping in mind the conservation of energy throughout	Questions 3–8	Questions 7–9	

Learning Objectives

- 9Pf.03: Know that energy is conserved, meaning it cannot be created or destroyed
- 9TWSm.02: Describe some important models, including analogies, and discuss their strengths and limitations
- 9TWSp.03: Make predictions of likely outcomes for a scientific enquiry based on scientific knowledge and understanding
- 9TWSc.06: Make an informed decision whether to use evidence from first-hand experience or secondary sources
- 9SIC.02: Describe how science is applied across societies and industries, and in research
- 9SIC.03: Evaluate issues which involve and/or require scientific understanding
- 9SIC.05: Discuss how the uses of science can have a global environmental impact

Links to other topics

- Many topics in Lower Secondary Science use the idea of energy transfer:
  - how food can be used as a store of energy by animals and how energy is transferred in food chains (Stage 7 Biology Topic 2.2)
  - how the stored energy is released by respiration (Stage 8 Biology Topic 1.2)
  - how plants absorb energy from light, through photosynthesis, to release and store energy in the form of carbohydrates (Stage 9 Biology Topic 1.1)
  - how waves transfer energy from one object and place to another (Stage 7 Chapter 8 *Sound*, Stage 8 Chapter 9 *Light*)
  - how chemical reactions transfer energy between the reaction mixture and the surroundings (Stage 8 Topics 6.3 and 6.4)
  - how renewable and non-renewable energy resources are used to generate electricity (Stage 8 Topic 11.3)

Using models

- To address learning objective 9TWSm.01 and 9TWSm.02 students should be given the opportunities to discuss how any models use reflect scientific evidence and the strengths and limitations of these models.
- As explained above in 'Background science', there are different models for teaching energy, and teachers can choose which model to use. You could highlight to students when a particular model is being used, and challenge them to reflect on the use of such models and whether the one you are using helps them. Including diagrams or analogies could help this discussion.
- A visual analogy such as pouring a liquid from one watertight container to another could be used to explain the 'stores and transfers' model, in which energy is described as being in a store (e.g. a kinetic energy store, an elastic energy store, a thermal energy store). This energy may be transferred and when it does so it will end up in another store.
- Ask if there are any weaknesses or drawbacks in such a model. You could point out the danger that it might make them think of energy as a 'substance' in itself rather than an idea that helps us describe and predict the outcomes of interactions (i.e. forces, motions...) between objects.

Science in context

Question 8 below the Science in context: All-electric vehicles box in Topic 8.1 of the Student's Book covers the following learning objectives:

- 9SIC.02 Describe how science is applied across societies and industries, and in research
- 9SIC.03 Evaluate issues which involve and/or require scientific understanding
- 9SIC.05 Discuss how the uses of science can have a global environmental impact.

Teaching ideas

Look out for opportunities to discuss energy transformations and how much energy is dissipated during these. You could discuss with students how the uses of science can have a global environmental impact: for example, the environmental impacts of energy wastage when electricity is generated or when buildings are not well insulated.

Common misunderstandings

- Students may try to talk about the total amount of energy 'in' an object – for example, the amount of energy stored in a battery. But we can only measure energy *changes*, not totals. For example, when energy from a battery is used to move an electric vehicle, the quantity of energy being transferred into kinetic energy (giving it speed) and into increasing energy stored due to height too if the vehicle is climbing a hill, is called the 'work done'. Meanwhile, the energy being transferred because of electrical resistance and friction, both of which result in a temperature increase, is called 'heating'. Both the amounts of work done and heat transferred can be measured. From those energy changes (due to work being done and due to heating), we can deduce there must now be *less* energy stored in the battery, but we have no way of measuring and knowing the absolute total amount of energy in it.
- The concept of potential energy is difficult. The fact that an object has potential energy because of its height can be reinforced by discussing the fact that energy was used to place the object in its position and that this energy was transferred to the object. In the case of the pendulum, it is also helpful to stress that the kinetic energy the pendulum bob has when it is moving must have been transferred from somewhere and the only place it could have come from is the potential energy stored in the bob before it was released: emphasising the fact that if the bob hadn't been lifted up to a height, it would not have been able to move.

8.1a Starting point

This activity reviews students' prior knowledge and understanding in preparation for the new ideas in this chapter. Worksheets 8.1a(1) and 8.1a(2) can be used to assess prior knowledge. Worksheet 8.1a(2) is more challenging than 8.1a(1).

Students could work through the activity individually, in pairs or in small groups.

8.1b Pendulum

Students will learn:

- To understand that energy cannot be created or destroyed
- To use scientific knowledge to make predictions
- To describe the uses, strengths and limitations of models and analogies

Glossary words

conservation of energy, kinetic energy, potential energy

Resource preparation

For Activity 8.1: a pendulum with a long string and a massive bob for class demonstration and a large piece of white card; students should have diagrams of a pendulum

Practical notes 8.1b

For Activity 8.2, each group requires a metre rule and a tennis ball (or similar)

Teaching and learning ideas

- Refer to Activity 8.1 *Observing the motion of a pendulum* in the Student's Book. This activity works best using a pendulum with a long string and a massive bob. It is therefore probably best done by the teacher as a class demonstration. See Practical notes 8.1b.
- Demonstrate the swinging pendulum and question students about the types of energy the bob has at different points in the swing.
- Ask students to explain when each type of energy is at a maximum and why. Similarly, when is each energy type at a minimum?
- Demonstrate the equilibrium position and ask why a stationary bob at that point does not start moving.
- Displace the bob by a small amount, and release it from a point next to your nose.
- Now displace it by a much larger amount, putting your nose next to it before you release the mass. Ask the students to predict whether it will be safe for you to release it now, or will it hit your nose on its swing-back? Do not push the pendulum on release. (This can be repeated at different displacements.)
- Ask the students to mark on a diagram of the pendulum where the bob has its maximum and minimum values of potential energy. Where does it move fastest? Does it ever stop?
- Discuss air resistance, and let the pendulum swing for a longer time with a white card sheet behind; keep marking at intervals the furthest points of the swing.
- Challenge students to think about what has happened to the missing – 'wasted' – energy.
- The ideas met in this demonstration can be reinforced by asking students to do Activity 8.2 *Energy changes when bouncing a ball* where they look at kinetic and potential energy along with energy dissipation in the context of bouncing a ball. Predicting what the students expect to happen before they start this investigation is an important part of this practical.

Differentiation

- Less confident students should benefit from revisiting the ideas met during the pendulum demonstration after completing Activity 8.2.
- More confident students can discuss how and where the energy is dissipated during Activity 8.2 and/or plot a graph of bounce height against number of bounces.

Extension ideas

Students could investigate other examples of potential to kinetic energy transfers (or vice versa).

Homework suggestions

Questions 1–2 in the Student's Book. Students can provide answers to the oral challenge question: Where is the 'wasted' energy now that the pendulum has stopped swinging?

Workbook questions 1–6

8.1c Measuring energy changes

Students will learn:

- To understand that energy cannot be created or destroyed
- To decide whether evidence from first-hand experience or secondary sources should be used

Glossary word

joule

Teaching and learning ideas

- Ask students about the units in which energy can be measured, using food package labelling as an example – bring in examples from home, or search the internet for 'food package nutrition labels'. In many countries labels still include kilocalories or Calories with a big 'C', which is the same thing. Explain that joules and kilojoules and megajoules are the modern scientific unit for energy.
- Introduce the idea that we can only measure changes in energy; we don't ever know how much energy in total is in something.
- Always get all data converted into joules before students use it in calculations.

Differentiation

More confident students could research the amount (or percentage) of energy wasted by a range of machines or different types of ways of generating electricity.

For less confident students you could provide the relevant data for them.

Extension ideas

- This book deliberately avoids mention of power, since that is not included in the Cambridge Lower Secondary Science curriculum framework. However, you may wish your students to do research into the energy used in everyday electrical devices, by investigating around their homes or by using secondary information sources. If so, they will find most of the data is for power in watts (W). Energy use data, for example on electricity meters or bills, is normally in kilowatt hours (kWh). So, you would have to provide simple ways of converting both of these to joules:
  - 1 watt = 1 joule used every second. 1 kW = 1 kJ used every second. So, work either with the amount of energy used in one second, or alternatively calculate the amount used in one hour or in one day.
  - 1 kWh = 3 600 000 joules.

Homework suggestions

Questions 3–8 in the Student's Book. You could also ask students to research the electrical appliances in their homes, reading the labels to see how much energy power each uses – but see the note above (in 'Extension ideas') about explaining power and energy use units.

Workbook questions 7–9

8.2 Heating and cooling

Topic overview

		Student's Book	Workbook	Downloadable material
8.2a	**Starting point** This is a short opportunity to recap on prior knowledge and understanding that is needed for students to progress through this topic	Starting point table		
8.2b	**Heating and cooling** There are a lot of important ideas covered in this section. Students learn the difference between thermal energy (heat) and temperature. They then consider the direction of thermal transfer	Questions 1–5 Activity 8.3	Questions 1–5	Practical notes 8.2b(1) Practical notes 8.2b(2) Worksheet 8.2b

Learning Objectives

- 9Pf.02: Describe the difference between heat and temperature
- 9Pf.04: Know that thermal energy will always transfer from hotter regions or objects to colder ones, and this is known as heat dissipation
- 9TWSm.01: Understand that models and analogies reflect current scientific evidence and understanding, and can change
- 9TWSc.07: Collect, record and summarise sufficient observations and measurements, in an appropriate form
- 9TWSa.03: Make conclusions by interpreting results, explain the limitations of the conclusions and describe how the conclusions can be further investigated
- 9TWSp.05: Make risk assessments for practical work to identify and control risks
- 9TWSc.05: Carry out practical work safely, supported by risk assessments where appropriate
- 9TWSp.03: Make predictions of likely outcomes for a scientific enquiry based on scientific knowledge and understanding

Links to other topics

- In Stage 7 Chemistry students will have addressed objective 7Cm.06: Describe the three states of matter as solid, liquid and gas in terms of the arrangement, separation and motion of particles.

Using models

- The particle model of matter is used in the Student's Book to illustrate the difference between thermal energy and temperature.

Science in context

Teaching ideas

Look out for opportunities to discuss situations where we want thermal energy to transfer into or out of a building or object and also situations where we want thermal energy to be retained (for example, in an oven) or to stop it entering an object (for example, in a freezer). There is also the opportunity to investigate how theories about the nature of thermal energy (heat) have developed over time.

Common misunderstandings

- Students may use the terms 'heat' and 'temperature' interchangeably. Heat is the total amount of energy in a body, but temperature is a measure of the 'average' amount of heat energy in something. For this reason, an ocean has a colossal amount of heat in it, but since it is spread throughout a massive amount of water, its temperature is low and so it is cold.
- Students may confuse energy transfer words with ways of storing energy. For example, they might misuse the word 'heat' as a noun to refer to the energy stored in a warm object, which is correctly called 'thermal energy'.
- Students sometimes think that warm objects get colder because cold enters them. It is important to emphasise that objects get cooler because they lose thermal energy and not because they gain 'cold'.

8.2a Starting point

Take each statement in the Starting Point table and ask students to give an example that illustrates that statement. This could be a written exercise (individual or in small groups) or a whole class oral activity.

8.2b Heating and cooling

Students will learn:

- To describe how heat and temperature are different
- To understand how heat dissipation refers to the movement of thermal energy
- To understand that models and analogies can change according to scientific evidence
- To collect and record observations and measurements appropriately
- To identify hazards and identify ways to minimise risks
- To carry out practical work in a safe manner to minimise risks
- To use scientific knowledge to make predictions
- To interpret results and form conclusions using scientific knowledge and understanding

Glossary words

temperature, thermal energy

Resource preparation

Two beakers, tripod, gauze, heat-proof mat, Bunsen burner/spirit burner, timer, thermometer, water, eye protection

For Activity 8.3 you will need to supply containers of water at about 10 °C, room temperature (or about 20 °C) and about 45 °C. You may choose to have all students using the same large container at each temperature or use smaller containers, one per group. Practical notes 8.2b(2) relate to this activity

Practical notes 8.2b(1)

Teaching and learning ideas

- Discuss the difference between thermal energy and temperature (see Student's Book). Use the particle model illustrated in the Student's Book to explain the difference.
- Introduce the class practical shown on Worksheet 8.2b to emphasise the difference between thermal energy and temperature. Students should make a prediction of the outcome before undertaking this.
- Alternatively you could start this section with the practical work and then move on to the theory relating to the difference between thermal energy and temperature. This would be a good approach with more confident students.
- Discuss the idea of thermal energy flowing from hotter objects (or areas) to cooler objects (or areas) until their temperatures have equalised. You could illustrate this by asking students what would happen if you added some boiling water to ice. This is something you could demonstrate if needs be (see Differentiation).
- Move on to the practical in Activity 8.3 *Thermal transfer*. The aim of this practical is to be able to discuss why your hand feels warmer or cooler in each bowl of water, emphasising that this is because of thermal transfer either from your hand to the water or from the water to your hand.
- You can then move on to looking at dissipation of thermal energy. The idea of conservation of energy is important here so it is worth emphasising that no energy is actually 'destroyed'; it just becomes less useful as it spreads out into the surroundings.
- You can round off this section by looking at the Science in Context example to illustrate how scientific understanding develops and theories change over time. See the Extension ideas section below. If you pursue this it is worth noting that by the late 1700s, caloric was thought to be a massless substance.

Differentiation

- Less confident students could benefit from being given help with the risk assessment element of the practical shown on Worksheet 8.2b. They might also benefit from a demonstration that adding boiling water to ice results in water at that is hotter than 0 °C but cooler than 100 °C. You would need to take the temperature of the boiling water, the ice and the resultant warm/hot water. Before doing Activity 8.3 *Thermal transfer* it might be useful to remind some students that body temperature is about 37 °C.
- More confident students could be asked to read about the difference between thermal energy and temperature themselves and then report back on what they have learned. Also see the alternative approach to starting the work on temperature and thermal energy outlined in the Teaching and learning ideas section above.

Extension ideas

Students could be encouraged to think about why the caloric theory of heat was discarded. What evidence was there to suggest that it wasn't correct?

Homework suggestions

Questions 1–5 in the Student's Book

Ask student to give some real-life examples of thermal energy transferring from hot objects to cooler ones in order to reduce the temperature of the hot object and also examples where the transfer is intended to increase the temperature of the cooler object.

Workbook questions 1–5

8.3 Conduction, convection, radiation and evaporation

Topic overview

		Student's Book	Workbook	Downloadable material
8.3a	**Starting point** Students are introduced to the four types of heat transfer	Starting point table		Practical notes 8.3a
8.3b	**Conduction** Students find out about conduction They investigate which metals are the best conductors	Questions 1–4 Activity 8.4	Question 1–6	Practical notes 8.3b(1), 8.3b(2) Worksheet 8.3b
8.3c	**Convection** Students learn about convection They see it in action	Questions 5–8		Practical notes 8.3c(1), 8.3c(2) Worksheet 8.3c
8.3d	**Radiation** Students learn how radiation works They compare the three methods of energy transfer studied in the topic	Questions 9–11 Activity 8.5		Practical notes 8.3d(1), 8.3d(2) Worksheet 8.3d
8.3e	**Cooling by evaporation** Students find out how evaporation causes cooling They carry out an investigation into how different methods of insulation affect cooling	Questions 12–13	Questions 7–8	Practical notes 8.3e(1), 8.3e(2) Worksheet 8.3e

Learning Objectives

- 9Pf.04: Know that thermal energy will always transfer from hotter regions or objects to colder ones, and this is known as heat dissipation
- 9Pf.05: Describe thermal transfer by the processes of conduction, convection and radiation
- 9Pf.06: Explain cooling by evaporation
- 9TWSp.04: Plan a range of investigations of different types to obtain appropriate evidence when testing hypotheses
- 9TWSc.07: Collect, record and summarise sufficient observations and measurements, in an appropriate form
- 9TWSm.02: Describe some important models, including analogies, and discuss their strengths and limitations
- 9TWSp.01: Suggest a testable hypothesis based on scientific understanding
- 9TWSp.03: Make predictions of likely outcomes for a scientific enquiry based on scientific knowledge and understanding
- 9TWSp.05: Make risk assessments for practical work to identify and control risks
- 9TWSa.03: Make conclusions by interpreting results, explain the limitations of the conclusions and describe how the conclusions can be further investigated
- 9TWSc.05: Carry out practical work safely, supported by risk assessments where appropriate
- 9TWSc.04: Take appropriately accurate and precise measurements, explaining why accuracy and precision are important

Links to other topics

- In Stage 7 Chemistry Chapter 4 students carried out a simple investigation to find out which materials are the best conductors of heat.
- Density is studied in Stage 9 Chemistry Chapter 6, together with the differences in density between solids, liquids and gases and an explanation of density in terms of the particle model. Students can apply their knowledge of density and the particle model to understanding convection currents.
- Convection is also studied in Earth and space (in the context of explaining the movement of tectonic plates) in Chapter 12.
- In Stage 8 Chemistry Topic 6.4 students classified exothermic and endothermic processes based on the temperature change observed, and noted that evaporation is an endothermic process.

Using models

- Students should be familiar with the particle model of matter and should be able to apply this to explain convection, conduction and evaporation, and to explain why convection and conduction cannot take place in a vacuum. These models are shown in the Student's Book.

Science in context

Teaching ideas

Look out for opportunities to identify and discuss everyday examples relating to the use of conductors and insulators to either transfer thermal energy or to prevent thermal transfer. Also look for opportunities to discuss convection: for example, why water temperatures equalise faster if hot water is poured into a container before cold, rather than cold before hot. There will also be opportunities to discuss the effect of colour on how well surfaces emit, absorb and reflect radiation: for example, why silvered blankets are used to wrap people who are suffering from hypothermia and why the outside of freezers are normally coloured white. Examples of cooling by evaporation can also be discussed: for example, the role of evaporation in perspiration: – removing thermal energy from your body to lower your temperature.

Common misunderstandings

- Students might confuse heat radiation with radiation from nuclear substances. Explain that radiation refers to electromagnetic waves. Light is a type of electromagnetic radiation. X-rays and gamma rays are also types of electromagnetic radiation. These are the types that are associated with nuclear radiation.
- When considering thermal transfer by radiation, students find it difficult to understand that dark-coloured objects are both good emitters and good absorbers of thermal energy. Carrying out relevant practicals and demonstrations can help with this.
- Students may think that infra-red radiation is the same as heat radiation. In fact, other forms of radiation are capable of heating objects. Also, warm objects emit more types of radiation than just infra-red radiation.
- Students will confuse the terms boiling and evaporating. Evaporating is a process where the particles in a liquid that have the most energy escape from the surface of the liquid. It does not require the addition of energy to the liquid. Boiling occurs when you actively heat a liquid to give the particles more energy and turn them all into a gas.
- Students may have trouble understanding that insulation is good at keeping things cold as well as keeping things hot. For example, putting a cold liquid in a vacuum flask keeps the liquid cold. Show students that thermal energy transfer is a two-way process and that insulation prevents thermal energy from getting in or out.
- When applying the particle model to understand convection currents, students may get confused between the bulk (macroscopic) motion of air as an air current or wind and the random, sub-microscopic motions of individual air molecules.

8.3a Starting point

Resource preparation

For the demonstration: Bunsen burner/spirit burner, metal spoon or rod, flashlight, water

Practical notes 8.3a

This starting point is to demonstrate the four types of heat transfer and ask the students to explain what is happening. This will allow you to discover whether they know the words conduction, convection, radiation and evaporation. It will also allow you to see if they can link these thermal energy transfers with particle theory.

As a demonstration, light a Bunsen burner/spirit burner. Hold a metal rod or small spoon in the flame. Eventually, the end held away from the flame will feel warm. Give this to the students to feel. They have experienced conduction.

If you shine a light on the Bunsen burner or candle so that it casts a shadow on the wall, you should be able to see the convection currents rising up from the flame.

Ask students to hold their hands horizontally to the flame. They should feel thermal energy coming off it. This is radiation. This cannot be convection because the thermal energy is being transferred sideways and the thermal energy transferred by convection is only transferred upwards.

Ask students to drop a little water on their skin. Ask them how it feels after a few minutes. Their skin should feel cool. The water has evaporated and taken some energy with it.

8.3b Conduction

Students will learn:

- To understand how heat dissipation refers to the movement of thermal energy
- To describe the movement of thermal energy during conduction
- To carry out practical work in a safe manner to minimise risks
- To collect and record observations and measurements appropriately
- To interpret results and form conclusions using scientific knowledge and understanding
- To describe the uses, strengths and limitations of models and analogies

Glossary words

conduction, insulator, temperature, thermal energy

Resource preparation

For class conduction demonstration: a ball (any size)

For conduction practical: rods made of different metals, tripod, Bunsen burner/spirit burner and heat-proof mat, petroleum jelly, small metal objects such as drawing pins, timer, eye protection

For conduction in liquids demonstration: boiling tube, tongs, piece of ice small enough to fit in the boiling tube, water, Bunsen burner/spirit burner and heat-proof mat, piece of metal small enough to fit in the boiling tube, eye protection

Practical notes 8.3b(1), 8.3b(2)

Worksheet 8.3b

Teaching and learning ideas

- Discuss some examples of where conduction occurs in the home. For example, thermal energy is conducted from a cooker hob through the pan to the food inside the pan. Thermal energy is conducted from an iron to clothes, and the heat can remove wrinkles from the fabric. If you dry clothes on a heater, thermal energy from the heater is conducted to the clothes fabric and the water evaporates, so that the clothes dry.
- Some cooking pans have wooden handles which do not conduct thermal energy – they are insulators. This means that the thermal energy from the pan will not travel along the handle so that you can safely pick the pan up using the handle. A woollen hat or scarf acts as an insulator against the surrounding cold air.
- Explain that conduction happens when particles are close together and the thermal energy gets passed from particle to particle.
- Discuss how the particle model can be used to explain conduction using figure 8.6 in the Student's Book. Students may find it hard visualising the particle explanations for conduction. Students can role play this by lining up close together and passing a ball down the line. The individuals remain stationary, and represent the molecules or atoms in a solid, Then separate them and ask them if it is easier or harder to pass the ball down the line. The ball

represents thermal energy and the students are the particles. When they are close together, they are acting like the particles in a solid. When they get further apart, they are acting like the particles in liquids and gases.

- Students can carry out the practical on Worksheet 8.3b. See Practical notes 8.3b(1). They investigate how different metals conduct thermal energy. Students will probably find that the drawing pin fell off the copper rod quickest, showing that copper is the best conductor of heat. Their bar chart should have metal along horizontal axis and time along vertical axis.
- Demonstrate conduction in liquids using Practical notes 8.3b(2). Students can then answer questions 1–4. Question 4 is a challenge question. You might want to set this as a group discussion and then take feedback after a set time.
- Activity 8.4 *Conductors and insulators at home* is about conductors and insulators at home. You could set this as a homework task.
- You could ask students to come up with a list of different substances. They could ask a partner whether they would be good conductors or good insulators. You could take feedback after a set time to challenge any misconceptions.

Differentiation

- Less confident students may need help to identify independent and dependent variables for the bar graph in Worksheet 8.3b and with choosing a suitable scale for the vertical axis. More confident students should be able to identify the variables and choose suitable scales for themselves.

Extension ideas

- Students could extend the model of conduction described here to give a simple molecular account of conduction in solids, including the vibration of particles and transfer by electrons. This would help explain why metals are better conductors than non-metals.

Homework suggestions

- Activity 8.4 in the Student's Book.

8.3c Convection

Students will learn:

- To understand how heat dissipation refers to the movement of thermal energy
- To describe the movement of thermal energy during convection
- To carry out practical work in a safe manner to minimise risks
- To collect and record observations and measurements appropriately
- To interpret results and form conclusions using scientific knowledge and understanding
- To describe the uses, strengths and limitations of models and analogies

Glossary word

convection

Resource preparation

For class convection demonstration: a ball (any size)

For convection in a liquid practical: beaker, Bunsen burner or candle, tripod and gauze, potassium permanganate crystals, tweezers; large beaker, Bunsen burner or candle, wax, eye protection

For convection in air demonstration: a specially made box, piece of paper, candle

Practical notes 8.3c(1), 8.3c(2)

Worksheet 8.3c

Teaching and learning ideas

- Introduce convection as a type of energy transfer that occurs in gases and liquids. Talk about convection currents that occur when a heater is in a room using a simple diagram like this.

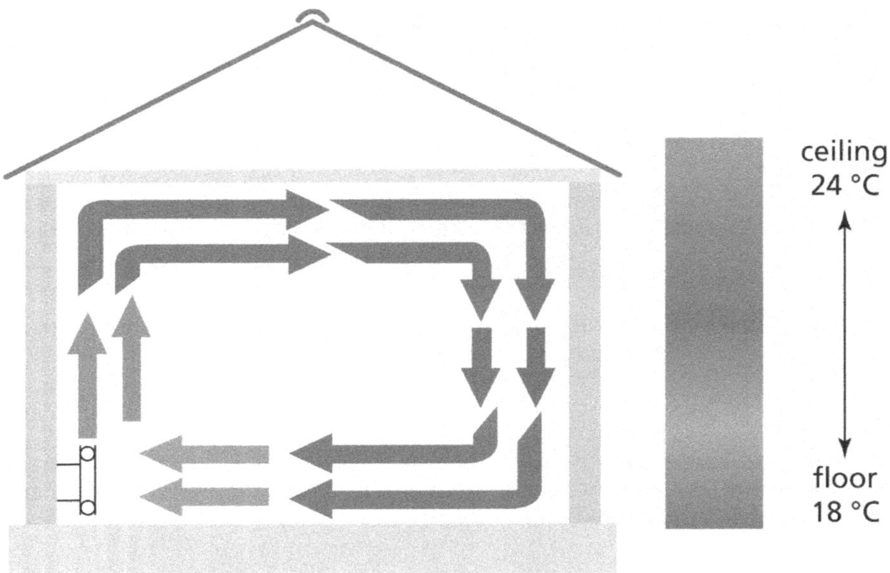

- Students could model the particles for convection. They do this by standing in a group on one side of a room. Say that one side of the room represents the surface of the liquid and the other side the bottom of the liquid. When you give students near the bottom of the liquid a ball (which represents thermal energy), they walk to the other side, then drop the ball and walk back. This shows that when they are hot they move up and when they are cold they move down.
- Students investigate convection in liquids. Worksheet 8.3c and Practical notes 8.3c(1) give full instructions.
- Discuss how particle theory can be used to explain convection. Also discuss how the theory can be used to explain why convection happens in liquids and gases but not solids. This diagram may be helpful. There is an explanation using this diagram below. You could give less confident students this text with some key words missing so that they simply have to fill in the blanks.

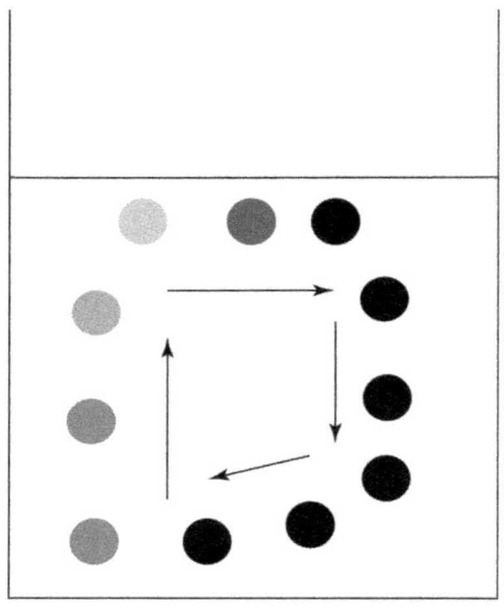

  - As the particles are heated, they move more, and become more spaced out.
  - This makes the substance less dense and the particles rise.
  - As they get to the top, they cool, get closer together and the substance becomes more dense.
  - Then the particles sink.

Ask students to discuss how well the particle model works when explaining convection and convection currents.

- You can demonstrate convection in air. See Practical notes 8.3c(2) for details.
- Students can make a Venn diagram of conduction and convection and compare the two.
- Discuss where convection currents are found using the Convection section in the Student's Book. Students can answer questions 5–8 in the Student's Book.

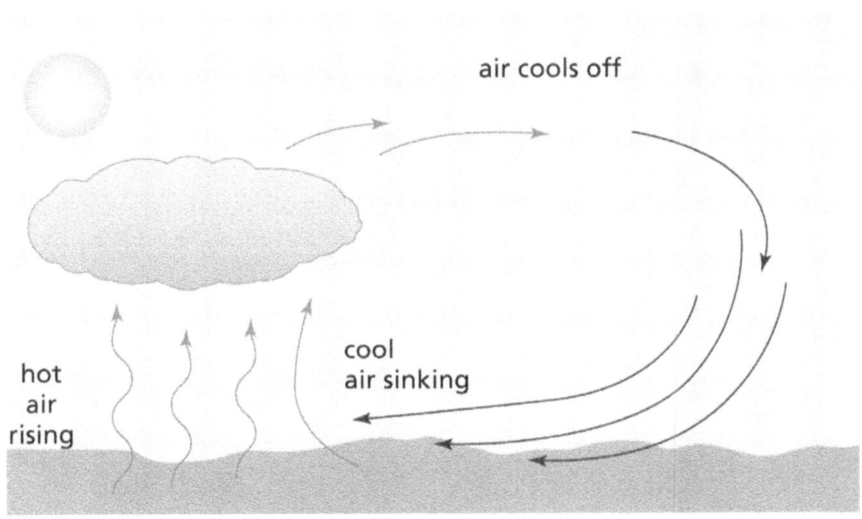

Differentiation

Less confident students should benefit from discussing more examples of convection currents: for example, in relation to where the heating element should be in an electric kettle and why air conditioning units are placed high on internal walls. More confident students can be given the challenge of using convection to explain how and why windy conditions occur. The diagram above may help with this.

Extension ideas

You could ask students to find out why the direction of sea breezes is usually from the sea towards the land during the day and reverses at night. (This involves knowing about the different specific heat capacities of land and water.)

Homework suggestions

Questions 5–8 in the Student's Book

Look at examples in the home where convection is considered when deciding on the position of objects.

8.3d Radiation

Students will learn:

- To understand how heat dissipation refers to the movement of thermal energy
- To describe the movement of thermal energy during radiation
- To carry out practical work in a safe manner to minimise risks
- To collect and record observations and measurements appropriately
- To interpret results and form conclusions using scientific knowledge and understanding
- To describe the uses, strengths and limitations of models and analogies

Glossary words

absorb, emit, radiation

Resource preparation

For radiation practical: shiny metal cans, metal cans painted black, kettle, water, two thermometers, a lid made of insulating material such as wood or bubble wrap, measuring cylinder or beaker, stopwatch. The shiny cans and the matt black cans should be as similar as possible to make this a fair test.

For radiation demonstration: a Leslie's cube (a metal cube with one side painted black and one side painted silver), heat-proof mat, kettle, thermometer (optional), infra-red sensor (optional)

Practical notes 8.3d(1), 8.3d(2)

Worksheet 8.3d

Teaching and learning ideas

- Discuss radiation. Start by talking about examples of where we need it. The Sun radiates energy to Earth. A camp fire radiates thermal energy to those who sit near it. You can prevent radiation reaching an object by wrapping it in shiny materials which reflect the radiation away from the object.
- Explain that thermal radiation occurs when thermal (infra-red) energy is transferred by waves.
- It does not need particles to move so it can travel through space. It can travel in a vacuum (an area where there is no matter). This diagram shows the differences between the methods of heat transfer.

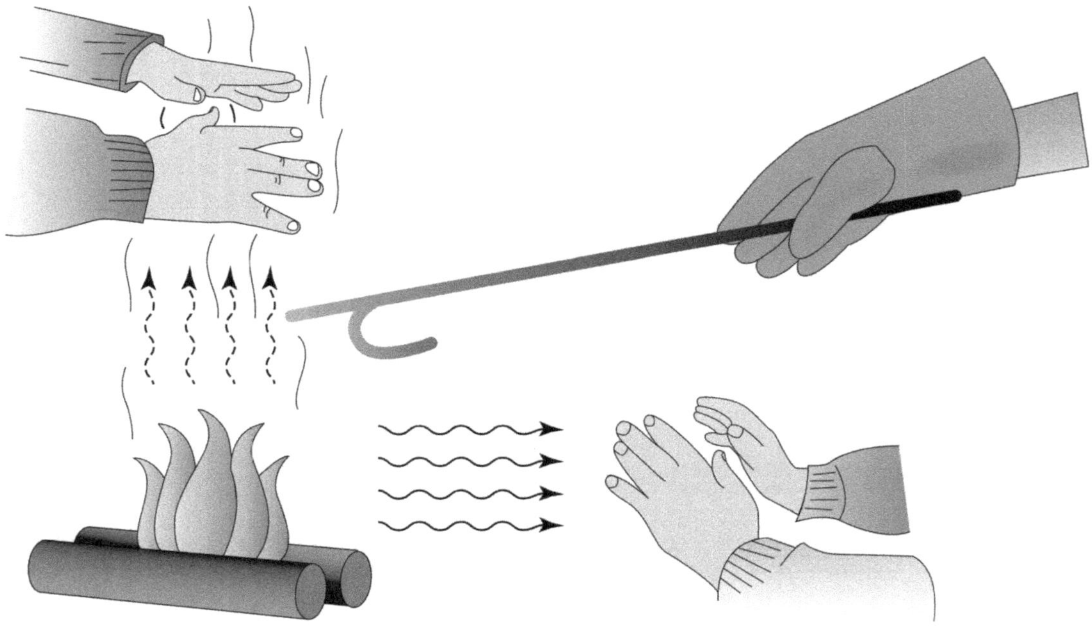

- Students should investigate radiation from different types of surface. Full instructions are given on Worksheet 8.3d and Practical notes 8.3d(1).
- You can also demonstrate radiation from different surfaces using a Leslie's cube. See Practical notes 8.3d(2). Tell students to put their hands near the surfaces and feel which one is the warmest. Explain that the warmest surface is the best radiator (it is emitting energy fastest).
- Ask students to design an experiment that would show that dark matt surfaces are better absorbers of thermal radiation than shiny white/silver surfaces. This is described in Activity 8.5 *How well do different coloured surfaces absorb thermal radiation?* in the Student Book. This could be the basis for a demonstration or, if there are enough infra-red heaters available, a class practical.
- Students can answer questions 9–11 in the Student's Book. Question 10 is a challenge question. You may want students to answer this in small groups and then take feedback after a time. You could use these statements for assessment. Students signal whether the statement applies to conduction, convection or radiation. The statement could apply to one, two, three or no methods of heat transfer.
  - Which method of heat transfer does this apply to? It only occurs in liquids and gases. (*Convection*)
  - Which method of heat transfer does this apply to? It mostly occurs in solids. (*Conduction*)
  - Which method of heat transfer does this apply to? It can travel through a vacuum. (*Radiation*)
  - Which method of heat transfer does this apply to? It requires particles to travel. (*Conduction and convection*)
  - Which method of heat transfer does this apply to? Works best with dull, black surfaces. (*Radiation*)
  - Which method of heat transfer does this apply to? Metals are the best at this type of heat transfer. (*Conduction*)
  - Which method of heat transfer does this apply to? It can be demonstrated with water and potassium manganate(VII) (potassium permanganate). (*Convection*)
- As a challenge, ask students why radiators are not painted black. The answer is that it is because they heat a room by convection rather than radiation. If they were painted black, heat would radiate out, but then it would only make the immediate area warm, rather than the whole room.
- Students can consider how thermal energy is lost in their house through conduction, convection and radiation. They can make suggestions about how these energy losses could be reduced. For those who live in very hot countries they could discuss this in the context of unwanted thermal energy entering buildings.

Differentiation

- Less confident students should benefit from detailed discussion of the vacuum flask before or whilst answering question 11. It would help to have a dismantled flask available so students could see the silvering on both sides on the glass.
- More confident students could design a basic solar panel designed to heat water.

Extension ideas

The following extension outcomes could be explored:

- Identify infra-red radiation as part of the electromagnetic spectrum.
- Investigate how the amount of radiation emitted also depends on the surface temperature and surface area of a body.

You could ask students why a hot piece of food cools faster when it has been cut up into smaller pieces. (It is because cutting it into smaller pieces increases the surface area and so there are more ways it can emit thermal energy.)

Homework suggestions

Questions 9–11 in the Student's Book

Activity 8.5 *How well do different coloured surfaces absorb thermal radiation?* in the Student's Book could be used as a homework task.

Workbook questions 1–6 on this topic cover conduction, convection and radiation.

8.3e Cooling by evaporation

Students will learn:

- To understand how heat dissipation refers to the movement of thermal energy
- To describe the movement of thermal energy during conduction, convection, and radiation
- To explain how evaporation causes cooling
- To use scientific knowledge and understanding to suggest a testable hypothesis
- To plan investigations used to test hypotheses
- To collect accurate and precise measurements
- To carry out practical work in a safe manner to minimise risks
- To collect and record observations and measurements appropriately
- To interpret results and form conclusions using scientific knowledge and understanding
- To describe the uses, strengths and limitations of models and analogies

Glossary word

evaporation

Resource preparation

For evaporation demonstration: thermometer, cotton wool, water

For evaporation practical: 100 cm^3 beaker, bubble wrap, thermometer, stopwatch, kettle

Practical notes 8.3e(1), 8.3e(2)

Worksheet 8.3e

Teaching and learning ideas

- Discuss the differences between evaporation and boiling. The section on Evaporation and boiling in the Student's Book will help with this.
- Demonstrate cooling by evaporation. Put a thermometer on a piece of damp cotton wool to demonstrate that evaporation reduces the temperature of it. See Practical notes 8.3e(1).
- Discuss the uses for cooling by evaporation. Point out that when we produce sweat, the sweat evaporates and cools the skin. Evaporation is also used in refrigeration systems and in air conditioning to decrease the temperature inside a building.
- Explain to students how evaporation works using particle theory and figures 8.16 and 8.17 in the Student's Book. When a liquid absorbs heat, the particles in a liquid start to move faster. The particles that move the fastest rise up into the air.
- Evaporation happens without liquids being heated, for example at room temperature.
- In preparation for their investigation, remind students how a vacuum flask prevents loss of thermal energy. Use the diagram (figure 8.14) in the Student's Book to help.
- Students should answer these questions in small groups.
  - Does part of the vacuum flask prevent thermal energy loss through conduction, convection or radiation (it might be more than one)?
  - How does it prevent thermal energy loss?

- Students should plan and carry out an investigation into how different methods of insulation affect cooling. Worksheet 8.3e gives full instructions. Practical notes 8.3e(2) give more guidance on how to run this investigation and expected results.
- As an assessment activity, students could write down two things that went well with the investigation and one improvement.
- As a challenge, students could explain why bubble wrap is a better insulator (due to trapped air) or why people recommend that you wear layers to keep warm (the trapped air between the layers means that there is reduced conduction and convection).
- Students can use their knowledge from the last few lessons to design a device that can cool something down, such as a phone or computer. Show them some pictures of cooling fins, a fan and a series of pipes carrying coolant that can help them. Talk about the features of the devices shown. They can finish this for homework.
- Students can use the idea of surface area for their cooling devices – the greater the surface area, the faster the thermal energy will be lost.
- To assess learning, ask students to write down one or two points on which they are least clear. This could be on any method of transfer of thermal energy – conduction, convection, radiation or evaporation.

Differentiation

You may need to assist less confident students with identifying independent and dependent variables. Some students may benefit from working in groups where they can discuss their plans with others. You could provide the variables for less confident students so they can focus on planning the practical. You could also provide the plan and the results table.

More confident students could design their own ideal insulating material, incorporating factors that reduce thermal energy loss by conduction, convection, radiation and evaporation.

Extension ideas

Students could investigate how temperature, surface area and draught over a surface influence evaporation.

Homework suggestions

- Questions 12–13 in the Student's Book
- Give students an existing cooling device such as a heat sink. Students can work out how that device can cool down an item, such as a computer.
- Students can look for examples of insulation on their clothes or in their houses, and write a paragraph with a diagram to explain the purpose of the material or object they found.
- Workbook questions 7–8 on this topic

Chapter 8 Checking students' progress

End of chapter reflection

- To review the learning points of the topic
- To test understanding through answering questions

Resources

- Student's Book [End of chapter review questions]
- Workbook pages [Self-assessment grid, End-of-chapter questions]
- Worksheet 8.4

Approach

Use the Key facts and Check your skills progress boxes at the end of each topic in the Student's Book as a way of recapping on the main learning points and of identifying any areas students still find difficult.

Also ask students to complete the Self-assessment checklist at the end of Chapter 8 in the Workbook. Encourage students to discuss in small groups their self-assessment for each topic. Can they explain any difficult bits to one another? Are there questions that none of them feel confident to answer? If so, they should feed that back to the teacher. You may wish to revisit some Learning Objectives again if students find any areas particularly challenging. Use the Student's Book text and questions to revisit these aspects. Go through the answers orally in class. Take note of whether giving the answer appears to resolve students' uncertainty or not, and work through further examples.

The Teacher's comments in the Workbook can be completed to give students personalised feedback on next steps. Students could be asked to answer specific End of chapter review questions in their books. Alternatively go through some or all of the End of chapter review questions in the Student's Book orally in class.

Use the Context worksheet (Worksheet 8.4) as a framework for students to conduct their own research enquiries about thermal energy transfer. For example, students could focus on the design of yurts, the development of electric vehicles, thermographs, cooling by evaporation or the development of theories about thermal energy.

After any group revision sessions that you have decided to give, finish by asking students to complete the End-of-chapter questions at the end of Chapter 8 in the Workbook. Provide feedback on students' answers, and work with students individually to identify areas that still need more work.

Chapter 9 Forces

Chapter overview

In this chapter students will extend their knowledge of balanced and unbalanced forces, and pressure in liquids from Stage 8. The work also builds on ideas about how the mass and shape of an object can affect if it floats or sinks.

Students will investigate and explain which materials float and which sink. They will learn about the forces exerted when an object floats or sinks. They will explain how differences in pressure at different depths in a fluid cause the upthrust force. They will explain floating and sinking in water in relation to the density of the material and use this to make and test predictions about which solids or liquids will float on water.

Topic title	Number of 40-minute periods	Learning Objectives	Thinking and Working Scientifically Learning Objectives
9.1 Floating and sinking	3–4	9Pf.01 Use density to explain why objects float or sink in water	9TWSp.03 Make predictions of likely outcomes for a scientific enquiry based on scientific knowledge and understanding 9TWSp.04 Plan a range of investigations of different types to obtain appropriate evidence when testing hypotheses 9TWSc.07 Collect, record and summarise sufficient observations and measurements, in an appropriate form 9TWSp.01 Suggest a testable hypothesis based on scientific understanding 9TWSa.03 Make conclusions by interpreting results, explain the limitations of the conclusions and describe how the conclusions can be further investigated 9TWSc.01 Sort, group and classify phenomena, objects, materials and organisms through testing, observation, using secondary information, and making and using keys
End of chapter questions	1	Check student's progress	

Background science

Students should have a good understanding of balanced forces, and the cause of pressure in a liquid from Stage 8. They are also likely to have investigated floating and sinking in previous studies and this, combined with everyday observations should mean they have knowledge of what floats or sinks in water. This topic will develop these ideas to explain floating and sinking by considering the forces involved and the pressure exerted by water at different depths.

Key to this topic is the idea of an upthrust force on an object placed in a fluid such as water. Students may have experienced this as a 'buoyancy' effect when swimming but are unlikely to have considered where it comes from. As pressure in a liquid increases with depth, the pressure on the lower surface of an immersed object will be less than on the upper surface. This gives a resultant upward force – the upthrust.

Archimedes' principle states that when an object is partially or completely immersed in water, there is an upthrust equal to the weight of water displaced. In order to float, an object must displace its own weight of water. Clearly an object cannot displace more liquid than its own volume, so objects with a density higher than water cannot displace enough water to float. An object of very low density (such as polystyrene) has a very low weight so needs to displace very little water to float. This is why it floats high in the water.

Students do not need to know Archimedes' principle at this stage.

Students need to relate the density of water to the density of the floating or sinking object. Objects that float in water may not float in different liquids, because the different density of the fluid produces a different upthrust. The same ship will 'sit' lower in fresh water than in the sea because freshwater is less dense than saltwater and provides less upthrust. This is the principle behind the Plimsoll line, a system showing the safe loading level for ships in different types of water. This explanation is not required in Lower Secondary but could provide some interesting extension work.

Students need to know that shape is important as well as the density of the material. A boat made of steel can float because its overall density is less than that of water, due to it being largely full of air.

9.1 Floating and sinking

Topic overview

		Student's Book	Workbook	Downloadable material
9.1a	**Starting point** Students recap ideas of floating and sinking, observing that some things such as cork always float whereas others such as aluminium only float if the shape is right	Starting point table		
9.1b	**Floating and upthrust** Students investigate which materials float and sink and observe how high they float in the water Students identify the forces on objects immersed in water, draw force diagrams and recap the terms balanced and unbalanced forces Students suggest a testable hypothesis about which materials will float and describe how they would test their hypothesis Students measure the upthrust on various objects by finding the difference between their weight in air and their weight in water	Questions 1–4 Activity 9.1 Activity 9.2	Questions 1–4	Practical notes 9.1b(1) Practical notes 9.1b(2) Worksheet 9.1b(1) Worksheet 9.1b(2)
9.1c	**Floating and density** Students are presented with data on density. They link this to their observations to draw a conclusion about what makes objects sink in water, and predict what will happen to liquids of different densities poured into water	Questions 5–8 Activity 9.3	Questions 5–12	Practical notes 9.1c Worksheet 9.1c(1) Worksheet 9.1c(2)
9.1d	**Floating in air** Students apply their knowledge to predict what will happen to balloons of gases in air	Questions 9–13		

Learning Objectives

- 9Pf.01: Use density to explain why objects float or sink in water
- 9TWSc.01: Sort, group and classify phenomena, objects, materials and organisms through testing, observation, using secondary information, and making and using keys
- 9TWSc.07: Collect, record and summarise sufficient observations and measurements, in an appropriate form
- 9TWSp.01: Suggest a testable hypothesis based on scientific understanding
- 9TWSp.03: Make predictions of likely outcomes for a scientific enquiry based on scientific knowledge and understanding
- 9TWSp.04: Plan a range of investigations of different types to obtain appropriate evidence when testing hypotheses
- 9TWSa.03: Make conclusions by interpreting results, explain the limitations of the conclusions and describe how the conclusions can be further investigated

Links to other topics

- Students will apply their knowledge of balanced forces and pressure in liquids from Stage 8 Physics Chapter 8.
- Students need to relate floating or sinking in water to the density of water. Density is also studied in Stage 9 Chemistry Chapter 6, together with the differences in density between solids, liquids and gases and an explanation of density in terms of the particle model. If students have not yet covered density, an explanation and the equation for density is given in this chapter.
- Students will also explain in this chapter why hot air is less dense than cold, linking to convection, which is studied in Physics in Chapter 8 and in Earth and space (in the context of explaining the movement of tectonic plates) in Chapter 12.

Using models

- Students should be familiar with the particle model of matter and should be able to explain pressure in a liquid as due to particle motion.
- They should also be familiar with the use of force arrows, and the use of these to determine the size and direction of the resultant force on an object.

Science in context

Teaching ideas

Look out for opportunities to relate this content to local or everyday situations, such as the use of buoyancy aids, party balloons filled with helium gas, local woods that have a very high density and sink in water, or to news stories involving high-altitude balloon travel.

Common misunderstandings

- Students often think that it is weight that determines whether something floats or sinks rather than density. In order to remedy this, you could ask students why they think huge steel ships float on the sea – the answer is that they are hollow so even though they have a huge weight, they are still less dense than water. Emphasise that the density of the material and the density of an object made of that material are not necessarily the same. The shape of a steel ship (like a bowl, with air inside) means it has a much larger volume than the same mass of steel made into a solid ball. The larger volume makes the density of the ship less than the density of water so the ship floats. Note that the volume of the ship that is filled with air adds very little to the overall weight of the ship, the main effect is the increase in upthrust. This is addressed in the Starting Point activity where students see that a piece of foil will sink if screwed into a ball or float if shaped into a boat.
- Students may think that volume determines whether something floats or sinks, for example, they may think that small rocks are able to float but big ones are not. This can be dispelled by demonstration with appropriate small dense and larger less dense objects.
- Students may be surprised that ice floats as it is the same 'stuff' as water. Explain that water expands slightly when it freezes making ice slightly less dense than water. To explain why it expands involves knowledge of hydrogen bonding which is well above Stage 9.

9.1a Starting point

Resource preparation

For the demonstration: a large bowl or tank of water, a cork (or piece of polystyrene), two large pieces of aluminium foil and a block of aluminium or a piece of aluminium such as the ring pull from a drinks can.

Teaching and learning ideas

- Show the students the materials and ask which will float. They will probably know that cork (or polystyrene) will float. Demonstrate that they are correct.
- Ask about the aluminium block. They will predict sinking. Demonstrate this.
- Ask about the aluminium foil. This is more ambiguous. After letting students give their ideas, screw one piece tightly into a ball and make a boat shape from the other. Demonstrate that aluminium can either float or sink.

- Conclude that some materials always float. Others (such as metals) will only float if their shape is changed eg making into a boat shape. The reason why metals can float when their shape is changed will be explained in episode 9.1c.

9.1b Floating and upthrust

Students will learn:

- To explain how an object floats or sinks using density
- To use scientific knowledge to make predictions
- To collect and record observations and measurements appropriately
- To interpret results and form conclusions using scientific knowledge and understanding
- To use scientific knowledge and understanding to suggest a testable hypothesis
- To plan investigations used to test hypotheses

Glossary words

balanced forces, resultant force, unbalanced forces, upthrust

Resource preparation

For Activity 9.1: beaker, spill trays and objects made of a range of floating and non-floating materials

For Activity 9.2: newton meters, stands with bosses and clamps, blocks made of different metals and one of wood which floats, with a means of hanging, beakers, spill trays to catch overflowing water

Practical notes 9.1b(1) and 9.1b(2)

Teaching and learning ideas

- Discuss students' experience of floating and sinking. Establish the idea from Stage 8 that water pushes up and this can cause a 'buoyancy' effect called upthrust.
- Explain that some things float more easily than others (avoid discussing who floats easily as this is linked to body fat and could be a sensitive issue). Refer to the iceberg photo in the Student's Book – most of the iceberg is underwater.
- Students carry out Activity 9.1 *Investigating floating and sinking*. They are asked to draw the floating objects to draw attention to the fact that some things float higher in the water (details in Practical notes 9.1b(1) and Worksheet 9.1b(1).
- Students answer question 1 where they are asked to discuss a hypothesis relating sinking to 'heaviness' and how the data given supports this (or not). They then suggest an improved hypothesis and explain how they would test it. This is likely to involve using objects which are the same size and shape.
- Ask students where they think the force pushing up (upthrust) comes from. Remind them that water is made of particles which are in constant random motion. These particles hit the parts of the object in contact with the water, causing pressure. Remind them that in Stage 8 they learned that there is pressure on all sides because the particles in the water can move in any direction, and that pressure in a liquid increases with depth. Refer to the diagram of a submerged block in the Student's Book (figure 9.3) to explain how there is a resultant upward force on the bottom surface.
- Now consider all the forces on an immersed object: weight down and upthrust up. Remind students (from Stage 8 work) that balanced forces lead to no change in motion, hence if upthrust = weight, the object floats (it does not move up or down). If weight is larger than upthrust, it sinks.
- Remind students that we can measure weight using a newton meter and discuss how this allows us to calculate the upthrust on an immersed object.
- Refer students to Activity 9.2 *Investigating upthrust* (details in Practical notes 9.1b(2) and Worksheet 9.1b(2)). Show students how to suspend the newton meter from the clamp stand and use it to lower the block into the water until fully immersed (apart from the wooden block which will not become fully immersed). Students measure the weight of each block and the apparent weight in water and from this calculate the upthrust.
- Ask students to consider what else, other than the material might affect whether an object will float – this could be the shape of the object. Ask them to write a testable hypothesis and describe how they could test it.
- Students complete questions 2–4 in the Student's Book.

Differentiation

- Less confident students may need help formulating a hypothesis to test in question 1. Ask them to list all the variables in the experiment and discuss which it is they want to see the effect of changing.
- More confident students could add force arrows to each of their drawings of floating objects in Activity 9.1. All should show balanced forces (equal length arrows) but the lower the object sinks, the bigger both arrows should be.

Extension ideas

- Get students to investigate the amount of water displaced when an object is placed in water. Some students will be able to work out that the volume of the water displaced is same volume as the object, and that when an object floats, the volume of water displaced is equal to the volume of the object below the surface of the water. Students can be asked to design an experiment to demonstrate this.
- Students could investigate Plimsoll lines on ships. Students could investigate upthrust in beakers of salt water (upthrust increases as salinity increases) and follow this up by drawing force diagrams for the same ship floating in sea water and fresh water, and researching the purpose of Plimsoll lines on ships. Plimsoll lines are lines marked on the side of ships to indicate safe loading levels. Ships will be lower in the water if they are in warm or fresh water which has a lower density than cold or sea water. This could cause a ship to sink if it has been overloaded and then enters less dense water.

Homework suggestions

Students could investigate household items to see which float. Allow time for discussion of results next lesson in relation to average density, as some will be surprising e.g. cans of diet (no sugar) soft drinks will float but those with sugar have a higher average density and so sink. Sponges initially float but eventually fill with water and sink. An unpeeled orange floats because the peel contains tiny air pockets, but a peeled orange sinks.

Students complete Workbook questions 1–4.

9.1c Floating and density

Students will learn:

- To explain how an object floats or sinks using density
- To use observations, to organise and classify objects
- To use scientific knowledge to make predictions
- To collect and record observations and measurements appropriately
- To interpret results and form conclusions using scientific knowledge and understanding

Glossary words

Density, mass, volume

Resource preparation

For demonstration: bowl or bucket of water, an orange, a can of diet (sugar-free) fizzy drink and a can of full-sugar fizzy drink

For Activity 9.3 Beaker, cooking oil, coloured water, syrup

Practical notes 9.1c

Worksheets 9.1c(1) and 9.1c(2)

Teaching and learning ideas

- Introduce or revise the idea of density (depending whether students have done this in Chemistry Stage 9 Topic 6 already).
- Explain that density is a more correct way of expressing what we intuitively call heaviness. Illustrate this with questions such as 'which is heavier: a ton of feathers or a ton of coal?' for clarity we must compare the mass of equal volumes i.e. the density.
- Using Worksheet 9.1c(1) ask students to list the materials from Student's Book table 9.1 in order of density. Students colour those which float and those which sink. This will make it clear that those with lower densities than water float and those which have higher densities sink.
- Give students the densities of cooking oil and golden syrup. Ask them to predict what happens when each is added to water. Add the three liquids to the beaker (water first) and observe and record what happens.
- Follow up the Homework activity from the previous lesson by placing unpeeled and peeled oranges, or diet and regular soft drinks in a bowl of water. Ask students to explain what they observe, using what they have learned about density.
- Students answer questions 5-8. Worksheet 9.1c(2) can be used to record answers to question 5.
- Challenge students to explain why sea water is warmest at the top of the ocean.

Differentiation

- Less confident students may find it useful to pick up equal sized blocks of different materials to clarify what we mean by density.
- More confident students could be asked to calculate the density of blocks of materials they have tested.

Extension ideas

- Challenge students to investigate what happens when they drop different objects into the syrup/water/oil beaker. An iron nut will sink to the bottom, a grape will float just above the syrup, a plastic bottle top will float above the water and a cork floats above the oil. Challenge students to explain what is going on in terms of the different densities involved.

Homework suggestions

Students complete Workbook questions 5–12.

9.1d Floating in air

Students will learn:

- To explain how an object floats or sinks using density
- To use scientific knowledge to make predictions

Resource preparation

For demonstration: a helium balloon, an air-filled balloon and a carbon dioxide filled balloon

Teaching and learning ideas

- Tell students the densities of the three gases. Ask them to predict what will happen to each balloon when they are released.
- Demonstrate that a carbon dioxide balloon will sink and a helium balloon rises. Compare this to floating and sinking activities they have done.
- Discuss densities of gases given in Student's Book table 9.2. Point out the different units. Ask students to complete questions 9–13.
- An internet search will provide video clips of different gas-filled balloons and soap bubbles. These can lead to a discussion of the wider properties of gases and how they can be used. For example the doomed use of hydrogen in airships.
- Ask students to draw force diagrams to show why some balloons rise and others sink. They should be able to give an explanation such as 'if the gas is denser than air, upthrust is less than weight so the balloon will sink'.
- Students could explain how the Cartesian diver experiment works using the idea of particle theory and the idea that gases are compressible. The Cartesian diver experiment demonstrates Archimedes' principle of buoyancy. The experiment involves a large bottle filled with water and a small tube, which is open at one end, being placed inside the bottle. When the bottle is pressed inwards at the sides, the air inside the glass tube is compressed and water enters the tube. The tube sinks down through the water. When the pressure on the sides of the bottle is released, the air inside the glass tube expands, pushing out the water and causing the tube to float again.

Differentiation

Less confident students may need to recap force diagrams and may benefit from working through at least one example.

More confident students should be challenged to produce the force diagrams themselves.

Chapter 9 Checking students' progress

End of chapter reflection

- To review the learning points of the chapter
- To test understanding through answering questions

Resources

- Student's Book [End of chapter review questions]
- Workbook [Self-assessment grid, End-of-chapter questions]
- Worksheet 9.2

Approach

Use the Key facts and Check your skills progress boxes in the Student's Book for each topic in this chapter as a way of recapping on the main learning points.

Ask students to complete the Self-assessment checklist at the end of Chapter 9 in the Workbook. Encourage students to work together in small groups to discuss which aspects of the material are the most difficult to understand, and then have a class vote. You may wish to revisit some Learning Objectives again if students find any areas particularly challenging. Use the Student Book text and questions to revisit these aspects. Go through the answers orally in class. Take note of whether giving the answer appears to resolve students' uncertainty or not, and work through further examples.

The Teacher's comments in the Workbook can be completed to give students personalised feedback on next steps. Students could be asked to answer specific End of chapter review questions in their books. Alternatively go through some or all of the End of chapter review questions in the Student's Book orally in class.

Use the Context worksheet (Worksheet 9.2) as a framework for students to conduct their own research enquiries about the science ideas they have studied in this chapter. For example, students could investigate how ideas about density are used in designing bath toys for children or research different types of buoyancy aids and compare their pros and cons. Alternatively they can look into the environmental problems caused by helium and hot air party balloons.

Finish by asking students to complete the End-of-chapter questions at the end of Chapter 9 in the Workbook. Provide feedback on students' answers, and work with students individually to identify areas that still need more work.

Chapter 10 Electricity

Chapter overview

In this chapter students will extend their knowledge of circuits from Stage 7. They will use the terms voltage and resistance, along with the rope loop model and a central heating model to understand and explain circuits.

Students will measure current and voltage in series and parallel circuits and use their measurements to calculate resistance. They will also investigate the rules concerning current and voltage in series and parallel circuits.

Topic title	Number of 40-minute periods	Learning Objectives	Thinking and Working Scientifically Learning Objectives
10.1 Voltage and resistance	3	9Pe.02 Know how to measure current and voltage in series and parallel circuits, and describe the effect of adding cells and lamps 9Pe.03 Calculate resistance (resistance = voltage / current) and describe how resistance affects current	9TWSc.05 Carry out practical work safely, supported by risk assessments where appropriate 9TWSc.04 Take appropriately accurate and precise measurements, explaining why accuracy and precision are important 9TWSm.03 Use symbols and formulae to represent scientific ideas 9TWSa.03 Make conclusions by interpreting results, explain the limitations of the conclusions and describe how the conclusions can be further investigated 9TWSc.07 Collect, record and summarise sufficient observations and measurements, in an appropriate form 9TWSm.02 Describe some important models, including analogies, and discuss their strengths and limitations
10.2 Measuring current in series and parallel circuits	3	9Pe.02 Know how to measure current and voltage in series and parallel circuits, and describe the effect of adding cells and lamps 9Pe.01 Describe how current divides in parallel circuits 9Pe.04 Use diagrams and conventional symbols to represent, make and compare circuits that include cells, switches, resistors (fixed and variable), ammeters, voltmeters, lamps and buzzers	9TWSm.03 Use symbols and formulae to represent scientific ideas 9TWSp.03 Make predictions of likely outcomes for a scientific enquiry based on scientific knowledge and understanding 9TWSm.02 Describe some important models, including analogies, and discuss their strengths and limitations 9TWSc.07 Collect, record and summarise sufficient observations and measurements, in an appropriate form 9TWSc.05 Carry out practical work safely, supported by risk assessments where appropriate 9TWSc.04 Take appropriately accurate and precise measurements, explaining why accuracy and precision are important
End of chapter questions	1	Check students' progress	

Background science

Students should have a good understanding of current as the rate of flow of charge from Stage 7. In this topic they will expand this by considering voltage and resistance.

Voltage can be defined in different ways and is also referred to as potential difference.

A volt is defined as the potential difference between two points if one joule of energy is transferred in moving one coulomb of charge between the points. Charge is not quantified at Stage 9 and students do not need to know this definition, or what a coulomb of charge is.

In this text voltage is defined in terms of the amount of energy that charge (carried by electrons) has at different points in the circuit. Charge and electrons can be used interchangeably here, and some students find it easier to picture a flow of charge as a stream of electrons, each carrying their own little load of energy, which they can transfer to a lamp. In these terms, the voltage across a bulb is the difference between the electrons before a lamp with lots of energy, compared to those after the lamp which have little energy.

Resistance is defined in terms of current and resistance and can be calculated using the equation

$$\text{resistance} = \frac{\text{voltage}}{\text{current}}$$

This equation is the mathematical expression of Ohm's Law, but students do not need to know the term Ohm's Law at this stage.

Resistance can be measured directly using an ohmmeter but use of these here would be counterproductive. Measuring current and voltage then calculating resistance helps students understand resistance.

If students use symbols for quantities, ensure they use the correct conventions:

- V for voltage
- R for resistance
- I for current (from the French 'intensite du courant' meaning strength of current).

Do not let students abbreviate current to C because this may cause problems later when they meet coulombs.

10.1 Voltage and resistance

Topic overview

Learning episode		Student's Book	Workbook	Downloadable materials
10.1a	**Starting point** In this short activity students revisit the rope loop model first met in Stage 7 They use the model to plan how they would explain to younger students what happens when a cell is used to light a lamp	Starting point table		
10.1b	**Measuring voltage** Students are introduced to the idea of voltage and measure voltages using a voltmeter	Questions 1–3 Activity 10.1,	Questions 1–2	Practical notes 10.1b
10.1c	**Electrical resistance** Students investigate the effect of adding fixed and variable resistors in simple circuits and about some practical uses of resistors	Questions 4–6 Activity 10.2	Questions 3–10	Practical notes 10.1c
10.1d	**Resistance, voltage and current** Students use the equation Voltage = current × resistance and then investigate the resistance of different lengths of wire	Questions 7–8 Activity 10.3		Practical notes 10.1d(1) Worksheets 10.1d(1), 10.1d(2)

Learning Objectives

- 9Pe.02: Know how to measure current and voltage in series and parallel circuits, and describe the effect of adding cells and lamps
- 9Pe.03: Calculate resistance (resistance = voltage/current) and describe how resistance affects current
- 9TWSm.02: Describe some useful models including analogies and discuss their strengths and limitations
- 9TWSc.05: Carry out practical work safely supported by risk assessments where appropriate
- 9TWSm.03: Use symbols and formulae to represent scientific ideas
- 9TWSc.04: Take appropriately accurate and precise measurements, explaining why accuracy and precision are important
- 9TWSa.03: Make conclusions by interpreting results, explain the limitations of the conclusions and describe how the conclusions can be further investigated
- 9TWSc.07: Collect, record and summarise sufficient observations and measurements, in an appropriate form

Links to other topics

- Students should have met the rope loop model in Stage 7 Topic 9. It is useful to check this and to discuss any other circuit models they may have met.

Using models

The rope loop model is revisited from Stage 7, but it may be appropriate to look at other models, particularly the central heating model, which is introduced in this topic.

- In the rope loop model the teacher asks a group of students to gently hold a loop of rope in their hands. The loop of rope models an electrical circuit. The teacher or one of the students acts as the power supply and pulls the rope, supplying energy for the electrons to carry. When the teacher or student pulls the rope, the rope flows around through the students' hands. This models the flow of current through the circuit.
- In the central heating model the boiler supplies the energy, which is then carried as thermal energy in the water. The water represents the flow of electrons and flows through the pipes to the radiator. As the water flows through the radiator the thermal energy is transferred to the room. The water returning to the boiler has less energy, which is analogous to the electrons returning to the power supply to collect charge before circulating back around the circuit again.

Students could be given the opportunity to discuss the strengths and limitations of any models discussed to meet learning objective 9TWSm.02.

Discussion of which models they find most helpful can be very productive in helping students to formulate their ideas about circuits. Time spent on these discussions will help students as they move on to more complex circuits.

Science in context

Teaching ideas

Students should be encouraged to look at electrical appliances in the classroom and at home and ask questions about how they are powered and what materials are used in their construction. They should be reminded that mains electricity is much more powerful that the cells they are using and can therefore be hazardous.

Common misunderstandings

- Although the term voltage is used here it may be useful to introduce the term potential difference as this will be used at Cambridge IGCSE or equivalent level. The two terms are synonymous.
- Voltage is a widely misunderstood concept. It is sometimes described as a pushing force which makes current flow. It is more useful to describe voltage as supplying energy to make current flow. The cell supplies energy which is transported by the charges (electrons) around the circuit. A lamp converts energy to heat and light, so the charges return to the cell with less energy. A voltmeter effectively measures the drop is energy as the charge passes through the lamp.

10.1a Starting point

Resource preparation

Large and small rope loops as for Stage 7 Chapter 9

Have a simple series circuit set up with a cell and a lamp

Diagram or model of a simple central heating system

Teaching and learning ideas

- Show students the simple series circuit. Remind them of the rope loop model. If necessary have the whole class hold the rope for the demonstration and discuss what the circuit represents. Then challenge students to prepare a short demonstration for Stage 6 students to explain what an electric current is.
- Discuss any misunderstandings and remind the students that this model is just a way of looking at circuits and that there are other models. Show a picture or model of a central heating system and discuss what each feature of the central heating system represents.

10.1b Measuring voltage

Students will learn:

- To understand how to draw circuit diagrams using conventional symbols for the components
- To understand how current and voltage can be measured in series and parallel circuits
- To represent scientific ideas using recognised symbols or formulae

- To carry out practical work in a safe manner to minimise risks
- To collect accurate and precise measurements

Glossary words

voltage, voltmeter

Resource preparation

For the demonstration: a range of cells with different voltages, voltmeter

For Activity 10.1: each group will need a voltmeter, two cells, two lamps and 6 connecting wires

Practical notes 10.1b

Teaching and learning ideas

- Discuss the role of a cell in a circuit. Let students look at the cells and ask what differences they see. One difference is the number of volts. They will probably realise that this is a measure of how powerful the cell is, or how much energy it can supply.
- Explain that a voltmeter measures the drop in energy between two points in a circuit e.g. between the sides of a cell or across a lamp. Students will realise that to do this the voltmeter needs to be connected to both sides of the component to measure the change in energy.
- Demonstrate how to connect a voltmeter to a circuit. It is best to encourage students to make up the circuit without the voltmeter and then add the voltmeter to make it clear that the voltmeter is not part of the loop. You can reinforce this idea by making a circuit then placing your hands on the circuit to show where the voltmeter will go.
- Students complete Activity 10.1 *Using a voltmeter* to investigate the use of voltmeters in series circuits. Encourage discussion about the sums of voltages across the bulbs compared with the voltage across the cell. See Practical notes 10.1b for further information.
- Students answer questions 1–3 in the Student's Book.

Differentiation

Be aware of those students who find circuit construction challenging and ensure they take part in this activity. If a more confident student is helping, ensure it is 'hands off helping'.

Challenge faster workers to predict and check the effect of connecting more lamps in series.

Ask them what will happen if one of the lamps is the opposite way round.

Homework

Ask students to check what devices they have at home which use cells and to find out what voltages and how many cells are used.

10.1c Electrical resistance

Students will learn:

- To understand how to draw circuit diagrams using conventional symbols for the components
- To describe current in series and parallel circuits
- To understand how current and voltage can be measured in series and parallel circuits
- To understand how to calculate resistance and describe the effect of resistance on current
- To describe the uses, strengths and limitations of models and analogies
- To carry out practical work in a safe manner to minimise risks
- To collect accurate and precise measurements

Glossary words

resistance, resistor

Resource preparation

For the demonstrations: large and small rope loops; cell, buzzer, motor, lamp and connecting wires

For Activity 10.2: each group needs a cell, a lamp, two fixed resistors (10 ohm), a variable resistor and four connecting wires

Practical notes 10.1c

Have a range of resistors available for students to look at. Include colour-coded resistors, a rheostat, a light-dependent resistor (LDR), and a thermistor

For Extension: table of resistor colour codes

Teaching and learning ideas

- Use the rope loop model to define resistance. Resistance is modelled by students gripping the rope so that their hands become hot, and the person pulling the rope has to work harder. Note: the students should not grip the rope too tightly, or move the rope too quickly, as the movement of the rope through their hands could cause rope burns. It may be useful to have the students model this in groups to allow them to feel the effect of increasing resistance both as the cell and as the lamp. Ask them what would represent resistance in the central heating model – the radiators – a large radiator which emits a lot of heat would represent a large resistance.
- Explain that whilst lamps and buzzers have resistance, the purpose of a resistor is simply to resist and therefore control current.
- Show students some resistors. Discuss the colour code which indicates the value of a resistor. They may be surprised that resistors of the same size can have different values. Use a variable resistor to demonstrate dimming a light, and controlling the speed of a motor and the loudness of a buzzer.
- Explain that LDRs and thermistors are variable resistors whose resistance depends on the light/ temperature level.
- Students complete Activity 10.2 *Dimming the lights* to investigate how fixed and variable resistors can be used to control the current in a circuit. See Practical notes 10.1c for further information.
- Students answer questions 4–6 in the Student's Book.
- Challenge students to identify as many uses as they can for variable resistors (including LDRs and thermistors).

Extension

Some students may be interested in the colour code system. Have resistors available for them to look at and find their resistance, and a table similar to this for them to use (colourful versions are readily available online).

Colour	Digit	Multiplier	Tolerance (%)
Black	0	10^0 (1)	
Brown	1	10^1	1
Red	2	10^2	2
Orange	3	10^3	
Yellow	4	10^4	
Green	5	10^5	0.5
Blue	6	10^6	0.25
Violet	7	10^7	0.1
Grey	8	10^8	
White	9	10^9	
Gold		10^{-1}	5
Silver		10^{-2}	10
(none)			20

Fixed resistors have a series of colour bands instead of numbers. The colours of these bands can be used, along with the colour chart on the left to calculate the total resistance of a fixed resistor. The colours of the first and second bands give you the first and second digit value. The colour of the third band gives you the multiplier (the factor by which the first and second digits must be multiplied). The colour of the fourth band gives you the tolerance, which is the value by which the resistance can vary.

Homework

Workbook questions 1–2.

10.1d Resistance, voltage and current

Students will learn:

- To understand how to draw circuit diagrams using conventional symbols for the components
- To describe current in series and parallel circuits
- To understand how current and voltage can be measured in series and parallel circuits
- To understand how to calculate resistance and describe the effect of resistance on current
- To carry out practical work in a safe manner to minimise risks
- To collect accurate and precise measurements
- To record observations and measurements appropriately
- To interpret results and form conclusions using scientific knowledge and understanding

Glossary words

resistor, resistance, ohm

Resource preparation

For Activity 10.3: each group needs 1 m of resistance wire, a cell, a lamp, 5 connecting wires, 2 crocodile clips and a metre ruler. A set of fixed resistors may be needed for some students (differentiation). An ammeter, voltmeter and a set of cells and fixed resistors (10 Ω and 20 Ω) may be needed for some students (extension).

Practical notes 10.1d(1)

Teaching and learning ideas

- Challenge students to explain how current, voltage and resistance are linked. They will easily link any two. For example, students will understand that more cells means more voltage so more current flows but realise that it is more complicated when you try to include all three.
- Point out that one of the joys of physics is that we can use equations to express situations like this simply. Define resistance as the number of volts needed to make an amp flow and express this as

  $$\text{resistance} = \frac{\text{voltage}}{\text{current}}$$
- Encourage students to have a go at rearranging the equation to make voltage of current the subject.
- Work through some examples then get students to answer questions 7 and 8 in the Student's Book. Worksheet 10.1d(1) could be used to practise these calculations.
- Students complete Activity 10.3 *Calculating resistance*, to investigate how resistance varies with the length of resistance wire. For parts A8–A10 of Activity 10.3, provide the students with fixed resistors but do not tell them the value of these fixed resistors. They should use the equation shown above to calculate the value of the resistance. See Practical notes 10.1d(1) for further information.
- Worksheet 10.1d(2) can be used to record the results of Activity 10.3.

Differentiation

Activity 10.3 can be presented as an investigation into the effect of length of wire on resistance and students encouraged to plan the investigation and present their results, including a line graph. This will give more confident students the opportunity to work independently.

Students lacking confidence could be given fixed resistors to measure rather than wires. If the value of the resistor is covered up, students can measure resistance and then check the actual resistance. This will give immediate feedback and motivate students.

Extension

Challenge students to calculate what the voltage should be in a circuit then measure with a voltmeter to check their calculations. Supply 3 cells and resistors labelled 10 and 20 ohms. Students should start with one cell and a 10 ohm resistor. Connect these in series with an ammeter. They will know I and R and so can calculate V then check by adding a voltmeter. The students can try different combinations of cells and resistors.

Homework suggestions

Workbook questions 3–10

10.2 Measuring current and voltage in series and parallel circuits

Topic overview

Learning episode		Student's Book	Workbook	Downloadable materials
10.2a	**Starting point** This game allows students to check and consolidate their knowledge by describing electrical components and quantities	Starting point table		
10.2b	Series and parallel circuits	Questions 1–4 Activity 10.4		Practical notes 10.2b
10.2c	Making measurements in series and parallel circuits	Questions 5–7 Activity 10.5 Activity 10.6	Questions 1–8	Practical notes 10.2c(1), 10.2c(2) Worksheets 10.2c(1), 10.2c(2)

Learning Objectives

- 9Pe.01: Describe how current divides in parallel circuits
- 9Pe.02: Know how to measure current in series and parallel circuits and describe the effect of adding cells and lamps
- 9Pe.04: Use diagrams and conventional symbols to represent, make and draw circuits that include cells, switches, resistors ammeters, voltmeters, lamps and buzzers
- 9TWSc.04: Take appropriately accurate and precise measurements, explaining why accuracy and precision are important
- 9TWSc.05: Carry out practical work safely supported by risk assessments where appropriate
- 9TWSp.03: Make predictions of likely outcomes for a scientific enquiry based on scientific knowledge and understanding
- 9TWSm.03: Use symbols and formulae to represent scientific ideas
- 9TWSm.02: Describe some important models, including analogies, and discuss their strengths and limitations
- 9TWSc.07: Collect, record and summarise sufficient observations and measurements, in an appropriate form

Using models

- The central heating model is very good for modelling parallel circuits. Students readily grasp the idea that in series arrangement all the water flows through the whole system, but in a parallel circuit some will flow through one loop and some through the other. It will make sense to them that the currents in the separate loops of a parallel circuit add up to the total current before it splits because with this model it is clear that there is nowhere else for water to go.

Common misunderstandings

- Students often assume that current will always split equally if there are two loops in a parallel circuit. This is countered in Activity 10.5 by having unequal branches so that one branch has less resistance and therefore receives more current. Students could be challenged to predict how the current would split for different circuits with different numbers of lamps in the branches.
- Voltage in a parallel circuit can lead to the impression of getting something for nothing as two lamps can be lit just as brightly as one. Asking the students to think about what is happening with the energy stored in the cell – a finite store – will help them realise that the energy will be used up more quickly and the cell will go flat sooner. Consider setting up circuits and leaving them until the bulb goes flat to demonstrate this, though thought needs to be given to monitoring this.

10.2a Starting point

Resource preparation

Prepare a set of cards for each group with one of the following words on each card:

current	amp	ammeter	voltage	volt	voltmeter	charge
energy	electron	lamp	buzzer	resistor	resistance	ohms

Teaching and learning ideas

- The starting activity can be done with the class as a whole or in small groups. A student takes a card at random and must tell the others what it is without using the word on the card. It can be made into a competition by timing how many cards each group can get through in a minute. It can also be made harder by adding a list of words which cannot be used in the explanation, for example explain amps without being able to use the words current or ammeter.

10.2b Series and parallel circuits

Students will learn:

- To understand how to draw circuit diagrams using conventional symbols for the components
- To describe current in series and parallel circuits
- To represent scientific ideas using recognised symbols or formulae
- To describe the uses, strengths and limitations of models and analogies

Glossary words

series, parallel

Resource preparation

For the demonstrations: large and small rope loops; cells, lamp and connecting wires

For Activity 10.4: each group needs 1 cell, 2 lamps, 1 switch, 6 wires

Practical notes 10.2b

Teaching and learning ideas

- Before setting up parallel circuits it is worth spending some time considering models so students can make predictions about current flow, effect of switches etc.
- The rope loop model can be extended by adding a second loop. This demonstrates that the cell can power both loops. As the rope is double just before and after the cell, then single loops going past each lamp, this can be used to model what happens to the current at junctions in a parallel circuit. However, it is worth trying this beforehand as moving the two different sized loops simultaneously can be awkward.
- The central heating model is much better for explaining current in parallel circuits as students can envisage the flow of water. This is a good opportunity to discuss the fact that a model is just a representation and that all models have strengths and limitations.
- It may also help to explain that in science, a model can be a simple pictorial representation (e.g. diagrams of a central heating system), and does not need to be a physical representation (e.g. the rope loop model). This can help lead into the idea of mathematical models in later study.
- Students should be shown how to set up a parallel circuit. It is best to show them this step by step, talking through what you are doing:
  1. Set out the cells and lamps as in your circuit diagram.
  2. Set up the first loop of circuit just as you would a series circuit.
  3. Add on the second loop. Point out that although the diagram makes it look like you are connecting to the middle of a wire, the connection can be made to either end. Demonstrate that it makes no difference which end.
- Remind students to take apart a circuit and start from scratch for the next one. Rearranging parallel circuits easily leads to confusion.
- Students complete Activity 10.4 *Connecting lamps in series and parallel circuits* to investigate different arrangements of lamps and switches in series and parallel circuits. Encourage discussion of the advantages and disadvantages of the different arrangements and get them to suggest real life situations which use different switching options. See Practical notes 10.2b for further information.
- Students should then answer questions 1–4 in the Student's Book.

Extension

Ask students to produce a poster describing and explaining their preferred circuit model. Challenge them to include as much as possible eg ammeters, switches, variable resistors, what happens at junctions etc.

10.2c Making measurements in series and parallel circuits

Students will learn:

- To understand how to draw circuit diagrams using conventional symbols for the components
- To describe current in series and parallel circuits
- To understand how current and voltage can be measured in series and parallel circuits
- To describe how the addition of cells and lamps affects the measurement of voltage
- To collect accurate and precise measurements
- To carry out practical work in a safe manner to minimise risks
- To use scientific knowledge to make predictions
- To collect and record observations and measurements appropriately

Glossary words

ammeter, voltmeter

Resource preparation

For the demonstration (Differentiation): cells, ammeters, connecting wires, lamps

For Activities 10.5 and 10.6: each group needs an ammeter, a voltmeter, 2 cells, 3 lamps and 7 connecting wires

Practical notes 10.2c(1), 10.2c(2)

Teaching and learning ideas

- Use models to get students to make predictions about what will happen to current at different points in the circuit.
- Students complete Activity 10.5 *Measuring current in series and parallel circuits* to investigate current in series and parallel circuits. Note that the activity puts two lamps in one branch and one in the other. This is to avoid the misunderstanding that current will always split equally. See Practical notes 10.2c(1) for further information. Worksheet 10.2c(1) can be used to record the results of Activity 10.5.
- Students should then answer question 5 in the Student's Book.
- Remind students that voltmeters go across a component and as such should be the last thing added to a circuit. There is no need to remake a circuit when measuring voltage across different point as the voltmeter is just added on to the circuit.
- Students complete Activity 10.6 *Measuring voltage in series and parallel circuits* to investigate voltage in series and parallel circuits. See Practical notes 10.2c(2) for further information. Worksheet 10.2c(2) can be used to record the results of Activity 10.6.
- Students should then answer questions 6–7 in the Student's Book.
- The parallel circuit results can lead students to query how a cell can light two lamps as brightly as one. This seems wrong as we seem to be getting something for nothing. The explanation is that the energy of the cell is being use up twice as quickly so the cell will run out of energy (become flat) twice as quickly.

Differentiation

The circuit which causes most difficulty here is the current in a parallel circuit as it is easy to get the ammeters in the wrong place. For more able students, predicting the split in current and then doing the experiment to confirm their predictions is a useful exercise as students will realise when they have made a mistake in the circuit and self-check.

For less able students it is worth setting this up as a demonstration, with an ammeter at each point (4 ammeters). The ammeters could all be switched off to start with and then switched on one at a time, asking students to predict what the current will be for the second, third and fourth ammeters. Ask them to justify their predictions (e.g. it will be less than ammeter one because some current has gone the other way) – this can lead to useful discussion and dispel misunderstandings.

Extension

Students can research different types of bulbs and produce a poster or leaflet explaining why moving to low energy bulbs is good both for individuals and the planet.

Homework suggestions

Poster/leaflet as above.

Workbook questions 1–8

Chapter 10 Checking students' progress

End of chapter reflection

- To review the learning points of the chapter
- To test understanding through answering questions

Resources

- Student's Book [End of chapter review questions]
- Workbook [Self-assessment grid, End-of-chapter questions]
- Worksheet 10.3

Approach

Use the Key facts and Check your skills progress boxes in the Student's Book for each topic in this chapter as a way of recapping on the main learning points.

Ask students to complete the Self-assessment checklist at the end of Chapter 10 in the Workbook. Encourage students to work together in small groups to discuss which aspects of the material are the most difficult to understand, and then have a class vote. You may wish to revisit some Learning Objectives again if students find any areas particularly challenging. Use the Student's Book text and questions to revisit these aspects. Go through the answers orally in class. Take note of whether giving the answer appears to resolve students' uncertainty or not, and work through further examples.

The Teacher's comments in the Workbook can be completed to give students personalised feedback on next steps. Students could be asked to answer specific End of chapter review questions in their books. Alternatively go through some or all of the End of chapter review questions in the Student's Book orally in class.

Use the Context worksheet (Worksheet 10.3) as a framework for students to conduct their own research enquiries about the development of cells and batteries. For example, students could focus on how ideas about rechargeable cells were developed or on the global environmental impact of mining for the materials needed for these cells.

Finish by asking students to complete the End-of-chapter questions at the end of Chapter 10 in the Workbook. Provide feedback on students' answers, and work with students individually to identify areas that still need more work.

Chapter 11 Sound

Chapter overview

In this chapter students will learn about the factors that affect the loudness and pitch of sounds. This builds on ideas about the vibrations of particles in a sound wave from Stage 7 Chapter 8, and on description of sounds in terms of high or low pitch and loud or quiet volume. Students will also learn about what happens when two sound waves meet and interfere with each other.

Note: many of the activities associated with this chapter depend on the use of an oscilloscope. If you do not have an oscilloscope available, you can access a virtual oscilloscope online at one of the following websites:

- https://academo.org/demos/virtual-oscilloscope/
- https://www.oszilloskope.net/en/oscilloscope/
- http://eleceng.dit.ie/dsp/elab/

Topic title	Number of 40-minute periods	Learning Objectives	Thinking and Working Scientifically Learning Objectives
11.1 Loudness and pitch	2–3	9Ps.01 Draw and interpret waveforms, and recognise the link between loudness and amplitude, pitch and frequency	9TWSa.02 Describe trends and patterns in results, identifying any anomalous results and suggesting why results are anomalous 9TWSp.01 Suggest a testable hypothesis based on scientific understanding 9TWSp.03 Make predictions of likely outcomes for a scientific enquiry based on scientific knowledge and understanding 9TWSp.04 Plan a range of investigations of different types to obtain appropriate evidence when testing hypotheses 9TWSa.01 Evaluate the strength of the evidence collected and how it supports, or refutes, the prediction 9TWSa.03 Make conclusions by interpreting results, explain the limitations of the conclusions and describe how the conclusions can be further investigated 9TWSc.07 Collect, record and summarise sufficient observations and measurements, in an appropriate form
11.2 Interference	2	9Ps.02 Use waveforms to show how sound waves interact to reinforce or cancel each other	9TWSp.03 Make predictions of likely outcomes for a scientific enquiry based on scientific knowledge and understanding 9TWSm.02 Describe some important models, including analogies, and discuss their strengths and limitations
End of chapter questions	1	Check students' progress	

Background science

Sound is created by vibrations. Sound travels as a longitudinal wave that creates a series of compressions and rarefactions. Compressions are areas of the longitudinal wave where the particles are close together and rarefactions are areas where the particles are further apart.

In longitudinal waves, the particles vibrate in the same direction as the movement that created the wave. The wave travels parallel to the direction of vibration. In transverse waves, the wave travels at 90° to the direction of vibration. The particles vibrate about a fixed position. Students do not need to know details about longitudinal and transverse waves in Lower Secondary Science.

The volume of a sound is determined by the size of the vibrations and hence the amplitude of the sound wave. The amplitude is the maximum displacement of the particle from its resting position. The pitch of a sound is determined by the frequency of the vibrations producing the sound wave. The more frequent the vibrations, the higher the frequency and the higher the pitch. The frequency of a sound source is the number of vibrations each second, this equals the frequency of the sound wave produced. The frequency of a wave is a measure of the number of complete waves passing a point in one second. The higher the frequency of the wave, the shorter its wavelength will be. Note that the relationship between wavelength and frequency of a wave is not required until IGCSE or equivalent level. A simple idea that the higher the frequency, the closer together the waves are, is all that is needed here.

The human ear can normally detect sounds with a frequency of between 20 Hz and 20 000 Hz. Sounds with frequencies higher than this are called ultrasound. Those with a frequency below 20 Hz are called infrasound. Ultrasound has many uses, including finding distances to surfaces from reflected pulses of ultrasound (in medical imaging, for internal structures in the body and in sonar, for underwater exploration). Animals such as elephants can hear infrasound. Seismic P waves are longitudinal infrasound waves and reflection (and refraction) of seismic waves can be used to find out more about the internal structure of the Earth. Students do not need to learn these contexts.

The lowest pitch (and therefore the lowest frequency) of a musical instrument is determined by the length of the instrument, as this determines the maximum possible wavelength for the sound waves produced. This lowest possible frequency is called the fundamental frequency. Different sources of sound also produce differently shaped waveforms; for example two notes of the same pitch and loudness, played on different instruments do not sound the same because the waveforms are different shapes. This is a result of each instrument producing additional sound waves with higher frequencies than the fundamental frequency. These 'overtones' are superimposed onto the fundamental wave, changing the overall shape of the final waveform produced. A wave is the signal produced and the waveform is the shape or form of the wave signal. This produces sounds of different quality. Students do not need to know about fundamental frequencies, overtones or quality of notes in Lower Secondary.

Sound travels in the form of a wave. When two sound waves meet the waves combine. In some cases they will combine constructively, forming a wave with an amplitude equal to the sum of the amplitudes of the original waves. In other cases they will combine destructively, with the 'crest' of one wave combining with the 'trough' of the other. This results in a wave with a lower amplitude than either of the original waves. In this case the amplitude of the resultant wave will equal the difference between the amplitudes of the two original waves. If the original waves had equal amplitudes, destructive interference results in zero amplitude, that is, there would be no wave. In practice, this means if two sound waves of the same frequency meet, they will interfere constructively to produce areas where the sound is loud and interfere destructively to produce other areas where the sound is quiet or even disappears. The overall pattern formed is called an interference pattern.

Although to explain interference we use transverse waves (the shape of waveforms on an oscilloscope screen, or ripples on the surface of water) as an analogy or visual representation, remember that sound waves are longitudinal (made up of a series of compressions and rarefactions) and it is the compressions and rarefactions that combine. Transverse waves have crests (peaks) and troughs, and it is easy to visualise what happens when two waves meet each other (crest meeting crest, trough meeting trough and crest meeting trough).

When we look at sound waveforms on an oscilloscope screen we are also equating the crests (peaks) and troughs of the waveform with positions of maximum particle displacement (compressions and rarefactions) in the sound wave.

When the compressions and the rarefactions of two sound waves line up, they strengthen each other and create a wave with larger maximum particle displacement, producing sounds with more energy (louder). When the compressions of one sound wave line up with the rarefactions from another wave they cancel each other producing quieter sounds.

11.1 Loudness and pitch

Topic overview

		Student's Book	Workbook	Downloadable materials
11.1a	**Starting point** This short activity provides an opportunity to establish the extent of your students' prior knowledge on pitch and volume	Starting point table		Practical notes 11.1a (1) 11.1a(2)
11.1b	**Why do sounds have different pitches and loudness?** Students learn about the link between the volume and amplitude of a sound wave, and between pitch and frequency and compare the amplitude and frequency of different waveforms	Questions 1–7	Questions 1–3	Practical notes 11.1b
11.1c	**The frequency range of human hearing** Students learn about the range of human hearing. They are introduced to ultrasound and infrasound. They explore their own hearing range	Questions 8–9 Activity 11.1	Questions 4–10	Practical notes 11.1c
11.1d	**Musical instruments** Students explore how the pitch of a sound can be changed. They make hypotheses, predict outcomes, look for patterns and consider the strength of their evidence	Activity 11.2 Activity 11.3		Practical notes 11.1d(1), 11.1d(2), 11.1d(3) Worksheets 11.d(1), 11.d(2)

Learning Objectives

- 9Ps.01: Draw and interpret waveforms, and recognise the link between loudness and amplitude, pitch and frequency
- 9TWSa.02: Describe trends and patterns in results, identifying any anomalous results and suggesting why results are anomalous
- 9SIC.02: Describe how science is applied across societies and industries, and in research
- 9TWSp.03: Make predictions of likely outcomes for a scientific enquiry based on scientific knowledge and understanding
- 9TWSp.01: Suggest a testable hypothesis based on scientific understanding
- 9TWSp.04: Plan a range of investigations of different types to obtain appropriate evidence when testing hypotheses
- 9TWSa.01: Evaluate the strength of the evidence collected and how it supports, or refutes, the prediction
- 9TWSa.03: Make conclusions by interpreting results, explain the limitations of the conclusions and describe how the conclusions can be further investigated
- 9TWSc.07: Collect, record and summarise sufficient observations and measurements, in an appropriate form

Links to other topics

- This topic builds on the Sound chapter in Stage 7 and on ideas about energy transfer from Stage 7 Chapter 6.

Using models

- As in Stage 7 Chapter 8 *Sound*, the use of a slinky spring as a model or analogy for how sound travels through a medium as a wave is helpful, and can be used to visualise the amplitude and frequency of the movement of particles in the medium the wave passes through. To demonstrate amplitude, you can use bigger or smaller movements of the slinky spring up and down. To demonstrate frequency, move the slinky spring up and down faster or slower.

Science in context

The question in the Science in context: Your unique voiceprint box in Topic 11.1 of the Student's Book covers the following learning objective:

- 9SIC.02 Describe how science is applied across societies and industries, and in research

Teaching ideas

Look out for opportunities to discuss with students issues about the use of scientific technology where they can apply their scientific understanding to evaluate the advantages and disadvantages of the technology, for example, security concerns about voice identification technology.

Common misunderstandings

- When showing waveforms representing waves on an oscilloscope screen, students often think that the amplitude of a wave is the height of the whole wave, from the bottom of a trough to the top of a crest. When you first introduce the term *amplitude*, it is worth stressing that it is a measure of the maximum height of a wave measured from its normal (undisturbed) position. For a sound wave, this is the maximum change in position of a single particle from its undisturbed position as the sound wave passes through. The slinky spring model should help explain the amplitude of particle motion in a sound wave, although students should be aware that in a real sound wave, the amplitude of particle motion is tiny, of the order of nanometres rather than centimetres or millimetres as in the slinky spring model.
- Students sometimes think that both amplitude and frequency change when the volume of a sound changes. It is worth stressing that the effects on pitch and volume are independent of each other.

11.1a Starting point

Resource preparation

For the student practical work, one 30 cm plastic ruler per student/group

For the demonstration: signal generator connected to a loudspeaker, microphone. If you do not have a signal generator you can use a musical instrument which can produce notes of different pitches and volumes.

Practical notes 11.1a(1), 11.1a(2)

Teaching and learning ideas

- This activity reviews ideas about pitch as well as touching on their work on how sounds are made in the Sound topic at Stage 7.
- In the student practical, students vibrate plastic rulers, changing the length of the part of the ruler which is vibrating and note the effect on the pitch of the sound produced. Full instructions are given in Practical notes 11.1a(1).
- In the demonstration (see Practical notes 11.1a(2)) a signal generator and loudspeaker are used to produce a range of different sounds. First, demonstrate two sounds that are at the same pitch, but which have different volumes. Second, demonstrate two sounds that have the same volume but different pitches. In each case ask students what is the same and what is different for each pair of sounds they hear. Ask students to suggest in each case what might be the difference in the sound waves that causes the different sounds they hear.

  If you do not have a signal generator available you can make the different volume and pitched sounds by other means, for example, speaking or playing a musical instrument into a microphone.

11.1b Why do sounds have different pitches and loudness?

Students will learn:

- To understand how to draw and label waveforms and describe the links between loudness and amplitude and pitch and frequency
- To describe trends and patterns shown in a set of results, identifying, and explaining, any anomalous results present
- To interpret results and form conclusions using scientific knowledge and understanding
- Describe how the movement of coils in a slinky spring is analogous to the movement of particles in a sound wave

Glossary words

amplitude, frequency, hertz, pitch, wavelength, volume, waveform

Resource preparation

For the demonstration: signal generator connected to a loudspeaker and an oscilloscope, microphone. If you do not have a signal generator then a microphone can be used in its place, possibly with a musical instrument which can produce notes of different pitches and volumes.

Slinky spring

Teaching and learning ideas

- Use the signal generator to produce sounds that have the same pitch but different volumes. Play these through the loudspeaker and also use the microphone to pick up these sounds. Show the sound waves generated in each case on the oscilloscope screen. Ensure the students can see that as the volume increases, the amplitude of the wave increases, but that no other property of the sound wave changes.
- If you do not have a signal generator then you can speak into the microphone, keeping your pitch the same but altering the volume of your speech. Alternatively play the same note from a musical instrument at different volumes into the microphone.
- At this point, it would be useful to introduce the term *amplitude*. A large diagram of a waveform as shown on an oscilloscope screen would be useful here, such as the one below. Explain that the 'wave' seen on the oscilloscope screen is a graph with time on the horizontal axis, and that the ups and downs represent the position (displacement) of an air molecule as the sound wave passes through air in the room and makes the particles in air vibrate. At any time, some particles have moved far away from their undisturbed position, some are back in their undisturbed position and some have moved far away from their undisturbed position in the opposite direction (shown as negative on the graph).

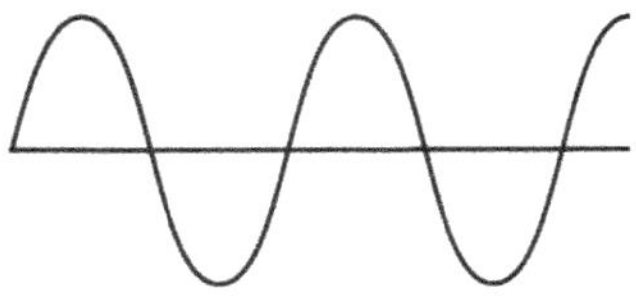

- Ask students to suggest where on the wave you would show the wave's amplitude. The completed version (shown below) could then be shown.

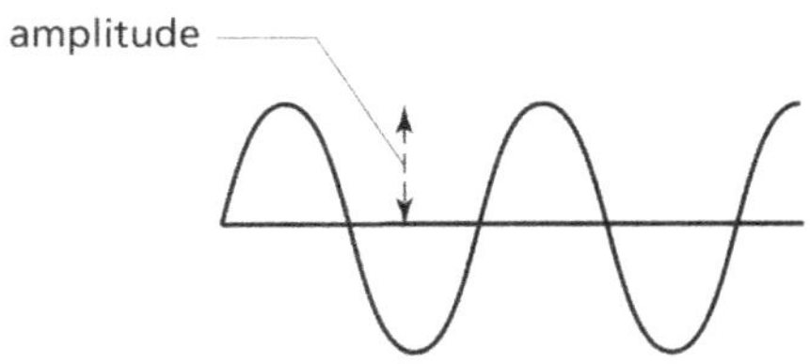

- The signal generator could be replaced with a microphone and students could be asked, individually, to make a sound into the microphone (such as by singing or whistling), making the sound louder or quieter while the class can then see how the amplitude of the waves on the oscilloscope screen changes.
- It might be worth using a slinky spring to recap the positions of vibrating particles in a medium in a sound wave and then demonstrate the amplitude of motion of particles in a sound wave. The size of a vibration is its amplitude. Mark one coil with a piece of coloured string to make it easier to see how one particle in the medium moves back and forth.
- Use the signal generator to produce sounds that have the same volume but different pitches. Show the waveforms generated in each case on the oscilloscope screen. Ensure the students can see that as the pitch increases, the peaks get closer together or more frequent, but the amplitude of the wave stays the same. Make the point that the width of the screen shows the same time interval, so if a greater number of complete waves are shown in a given time, the waves must be being made more frequently.
- If students are uncertain about the term 'frequently' then it might be worth using a slinky spring to demonstrate what it means. Move the end of the slinky backwards and forwards to make a wave with a low frequency. Then push the end of the slinky back and forth more quickly to demonstrate a wave with a higher frequency.
- It is worth pointing out to students that sound waves have higher frequencies than we can show with a slinky spring. (The range of human hearing is from about 20 Hz to 20 kHz and 'middle C' has a frequency of about 200 Hz).
- At this point, it would be useful to introduce the term *wavelength*. A large diagram of a waveform as shown on an oscilloscope screen would be useful here. An unlabelled diagram could be introduced first (as before, noting that this is a snapshot of the oscilloscope trace at an instant in time) and students asked to suggest where on the wave you would show the wave's wavelength. The completed version (see below) could then be shown.

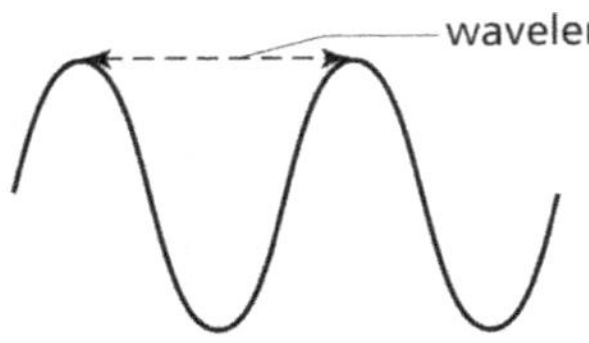

- The link between wavelength and frequency will now need to be shown. Using diagrams showing two waves that have different frequencies (but are identical in all other ways), will allow you to introduce the term *frequency* and show that, when the wavelength is shorter, more waves have been made in a given time, so the frequency of the waves must be higher. This can be seen by comparing Sounds D & E in figure 11.4 in the Student's Book.
- As it is actually the frequency of a sound wave that determines its pitch, it is important that frequency, rather than wavelength, is used when discussing pitch. (If the sound wave travels through a different medium, the change of speed means that its wavelength will change, but its frequency and pitch will not.)
- The signal generator could be replaced with a microphone and students could be asked, individually, to make a sound into the microphone and then change the pitch of the sound, while the class can then see how the wavelength and frequency of the waves on the oscilloscope screen changes.
- Questions 1–7 in the Student's Book will help students consolidate their understanding of the links between loudness and amplitude and between pitch and frequency.
- Challenge students to draw some waveforms to represent different sounds they are familiar with. This works best if they draw waveforms for two or three different sources, to the same scale, so they can compare them. Examples could be a cow mooing and a bat squeaking or a small dog yapping and a large dog growling.

Differentiation

In the challenge activity above, less confident students could be given the waveform for one sound and then only have to draw the second sound, being guided by questions such as Would it be louder or quieter? Would it be higher or lower pitch? More confident students could work with those that need more support by asking them the types of questions outlined above. This will help the more confident consolidate their understanding even further.

Extension ideas

Students could research why the same pitched sounds (the same note) sound different when played by different musical instruments and how this related to the waveforrn of the sound wave produced.

Homework suggestions

Students answer questions 1–3 in Topic 11.1 Loudness and pitch in the Workbook.

11.1c The frequency range of human hearing (extension work)

Students will learn:

- To understand how to draw and label waveforms and describe the links between loudness and amplitude and pitch and frequency
- To interpret results and form conclusions using scientific knowledge and understanding

Glossary words

infrasound, ultrasound

Resource preparation

For Activity 11.1: signal generator connected to a loudspeaker, an oscilloscope (optional)

Practical notes 11.1c

Note: Please be aware of any students with existing hearing loss, as they may not be able to participate in some activities or may be uncomfortable in taking part.

If any of your students have an existing hearing problem, it may be better to omit this material as it is extension and not part of the LSS curriculum framework.

Also, be alert for any students who may discover that they could have hearing issues during these practicals and advise them to seek further medical assistance rather than self-diagnosing.

Teaching and learning ideas

- Introduce the frequency range of human hearing. Discuss ultrasound and infrasound using the Student's Book as a guide. Students can do questions 8–9.
- Activity 11.1 *Testing your hearing range* in the Student's Book can be done as a whole-class demonstration. Use the signal generator to produce sounds that have different pitches. Gather the class round the loudspeaker and ensure they can all also see the oscilloscope screen. Set the signal generator at just over 20 Hz. Ask the students to raise their hands if they can hear the sound. Gradually increase the frequency and ask the students to lower their hands when they can no longer hear any sound. This is likely to happen when the frequency is approximately 20 000 Hz. To give students more confidence in lowering their hands, raise your hand when you first hear the sound and lower it when you can no longer hear it. This is likely to happen at a frequency that the students can still hear, as the older you are, the lower the maximum frequency you can hear.
- Using the oscilloscope will allow you to demonstrate to students that sound is still present even when the frequency is too high for humans to hear. From this you can start to discuss ultrasound.
- It is possible to demonstrate that sounds with a frequency of less than about 20 Hz cannot be heard. There is sometimes a problem with this as sounds may be heard because of the characteristics of the loudspeaker used. It is worth trying this out first before trying to demonstrate it to a class.

Differentiation

Less confident students may need to be supported by specifically pointing out to them, through looking at the oscilloscope screen, that sound waves are still being produced even when we can't hear them. This is more obvious at the higher end of the audible frequency range. Decreasing the frequency gradually from just above the audible range to sounds students can hear and asking them to put their hands up when they can start to hear again is also an effective way to reinforce the idea that there are some sound waves which have too high a frequency for us to hear. More confident students could try and explain why we cannot hear the sound even though the sound waves are still being produced.

Extension ideas

Ask students to find out some of the uses of ultrasound and some of the problems that can be caused by infrasound.

Homework suggestions

Students answer questions 4–10 in topic 11.1 Loudness and pitch in the Workbook.

Students could research the uses of ultrasound further.

11.1d Musical instruments

Students will learn:

- To understand how to draw and label waveforms and describe the links between loudness and amplitude and pitch and frequency
- To use scientific knowledge and understanding to suggest a testable hypothesis
- To plan investigations used to test hypotheses
- To use scientific knowledge to make predictions
- To collect and record observations and measurements appropriately
- To interpret results and form conclusions using scientific knowledge and understanding
- To evaluate how the prediction is supported by the evidence collected

Resource preparation

For the demonstration: signal generator capable of producing more than one type of wave (e.g. sine waves, square waves, triangular waves) connected to a loudspeaker and an oscilloscope, microphone, selection of musical instruments or other means of making different types of sounds.

For Activity 11.2: beaker, pencil, water

For Activity 11.3: everyday materials to make a 'bottle based' musical instrument (e.g. yoghurt pots, glass and plastic bottles of different heights and shapes, beakers, mugs), water

Practical notes 11.1d(1), 11.1d(2), 11.1d(3)

Worksheet 11.1d(1), 11.1d(2)

Teaching and learning ideas

- By using different instruments to make sounds that are picked up by the microphone you can demonstrate which sounds made by different instruments make differently shaped waveforms. Practical notes 11.1d(1) gives more details.
- The different waveforms produced by different sources can also be demonstrated by asking students to make different types of sounds into the microphone, for example, whispering, whistling or singing.
- Students can then do Activity 11.2 *Changing the pitch*, as described in the Student's Book. Worksheet 11.1d(1) and Practical notes 11.1d(2) support this.
- Following on from this, students develop and test a hypothesis for the pitch of notes produced by musical bottles as described in Activity 11.3 in the Student's Book. Worksheet 11.1d(2) and Practical notes 11.1d(3) support this.
- Challenge students to suggest ways of comparing the amplitudes and frequencies of the sounds produced, and/or investigate quantitative relationships between their variables (using a microphone connected to an oscilloscope to view the waveforms produced by their instrument, and relate the pitch/frequency and volume/amplitude of the sound to what is seen on the oscilloscope trace).

Differentiation

Less confident students could be provided with support for making a hypothesis. You could provide some possible variables or prompt students to suggest the significant variables and help them select a suitable one to investigate (ask, for example, If you fill two different bottles with the same level of water, do both bottles make the same note? If the bottles are different, which bottle will produce the highest pitch, and why? You could help them frame this as a hypothesis that clearly links the dependent variable to an identified independent variable, and to use scientific ideas to say what they think will happen and why. You could also help students identify which variables to keep the same in the same way. More confident students should be encouraged to make their own hypothesis and decide on the variable to investigate and those to control.

Extension ideas

Ask students to carry out research to find out why strings of the same length in instruments such as a guitar, produce different pitched notes.

Homework suggestions

Students select a musical instrument and consider how changes in pitch and volume are produced.

11.2 Interference

Topic overview

		Student's Book	Workbook	Downloadable materials
11.2a	**Starting point** This is an opportunity to review the meaning of key terms needed to understand interference. It also reviews how frequency and amplitude are represented on diagrams and displays of waveforms	Starting point table		
11.2b	**What is interference?** Students look at examples of the effects of two waves meeting and combining to form a larger wave or cancelling each other, with an emphasis on interference of sound waves This section also identifies the terms crest and trough which students will have to know before being able to describe and explain interference	Questions 1–7 Activity 11.4	Questions 1–5	Practical notes 11.2b(1) and 11.2b(2)

Learning Objectives

- 9Ps.02: Use waveforms to show how sound waves interact to reinforce or cancel each other
- 9SIC.02: Describe how science is applied across societies and industries, and in research
- 9SIC.03: Evaluate issues which involve and/or require scientific understanding
- 9TWSp.03: Make predictions of likely outcomes for a scientific enquiry based on scientific knowledge and understanding
- 9TWSm.02: Describe some important models, including analogies, and discuss their strengths and limitations

Links to other topics

- Students studied energy transfer in Stage 7. Energy is needed to make waves, like sound waves. Waves transfer energy from one place to another.

Using models

- Students use the analogy of the water levels when two sets of water ripples combine to explain how two sets of sound waves interact to reinforce or cancel each other. Students can be asked to discuss the strengths and limitations of this analogy to meet the learning objective 9TWSm.02.

Science in context

Questions 9 and 10 in the Science in context: Shutting out the noise box in Topic 11.2 of the Student's Book covers the following learning objectives:

- 9SIC.02 Describe how science is applied across societies and industries, and in research
- 9SIC.03 Evaluate issues which involve and/or require scientific understanding.

Teaching ideas

Look out for opportunities to discuss with students examples of interference, for example the noise-cancelling technology discussed in the Science in context: Shutting out the noice box in the Student's Book, what happens to the water on the surface of a lake when two power boats pass each other and why there are weak areas for TV, and radio reception in areas between two transmitters broadcasting at the same frequency. (TV and radio signals travel through the air in the form of waves and are transmitted at a certain frequency, such as 210.25 MHz)

Common misunderstandings

- Students will struggle unless they are secure in their understanding of the difference between amplitude and frequency and how each of these is shown on a diagram or an oscilloscope display of a waveform.
- Students may believe that the amplitude of a wave is its maximum height from the bottom of a trough to the top of a crest. Watch out for this and reinforce the correct meaning. Reminding students that it shows the maximum displacement of particles from their original position will help them understand that amplitude has to be measured from the centre of the wave. They may often also forget that amplitude can be measured from the bottom of a trough to the centre and not just from the top of a crest to the centre.
- There are not likely to be many misconceptions about interference when students first meet it. They do, however, often find the idea of destructive interference difficult to grasp. The more examples they meet, the more likely they are to accept and understand that two waves can cancel each other out (so that two sound waves interfering can produce silence).
- Students are likely to think that combining two or more waves will always make a bigger wave. They find it difficult to imagine two waves cancelling each other out. A demonstration using water waves (using a source from the internet if you do not have a ripple tank) shows what happens quite well. Overlaying two acetate sheets, each with the same sinusoidally shaped wave (same amplitude and wavelength) drawn on them can help too.
- To convince students that two sound waves can combine to produce a quieter sound, detect an area of quiet in Activity 11.4 then cover up one loudspeaker with an absorbent material (e.g. foam). The sound will immediately get louder, showing that one loudspeaker produces a louder sound than two loudspeakers combined in that particular spot.
- Students may have met the term interference in a different context because when we talk about interference in everyday language it usually means a 'noise' or 'crackle' in a radio receiver caused by radio frequency emission from other radio or TV signals, or from electrical signals (or even solar flares or lighting). In Physics, interference means the effect produced when two or more sets of any waves overlap – not just radio waves.

11.2a Starting point

Teaching and learning ideas

- Draw a large diagram of a waveform on the board and ask students to identify its amplitude. You may draw on more than one option for them to choose from. Now ask a student to draw (on the board) a waveform with a larger amplitude than the one you drew. Ask the class whether the waveform drawn by the student has the same frequency as your original wave. (It may or may not have unless you specified this and at this stage it is probably better not to and just look at one variable at a time). Now ask a student to draw a waveform with a higher frequency than the one the first student drew. You can then ask students which of the waveforms on the board represents the sound with the highest/lowest pitch and which represents the sound with the highest/lowest volume.

11.2b What is interference?

Students will learn:

- To understand how to represent the interaction or reinforcement of sound waves using waveforms
- To use scientific knowledge to make predictions
- To describe the uses, strengths and limitations of models and analogies

Glossary words

constructive interference, crest, destructive interference, interference, trough

Resource preparation

For demonstration 1 (Practical sheet 11.2b(1)): signal generator, two identical loudspeakers, microphone, oscilloscope. If no signal generator is available you could use some other means of playing a single note through two loudspeakers.

For demonstration 2: (Practical sheet 11.2b(2)) ripple tank and a means of making two waves (e.g. two pipettes, a double dipper or a single dipper plus a barrier with two gaps for the wave produced to be divided in two). Note: the two pipette or double dipper method matches the diagram in the practical notes better than the two gaps method.

Practical notes 11.2b(1), 11.2b(2)

Teaching and learning ideas

- Start by asking student what they think will happen if two sound waves meet up at a point and combine. This then leads into demonstration 1.
- Use demonstration 1 (refer to Practical notes 11.2b(1)) to show how combining two sound waves of the same frequency can produce both constructive and destructive interference. Talk through what is happening with the students. Figure 11.5 in the Student's Book provides good support for this explanation. Full instructions are given for this activity in Activity 11.4 *Interference of sound waves* in the Student's Book.
- Use demonstration 2 (refer to Practical notes 11.2b(2)) to show interference between two, ideally identical, water waves. This helps students to see areas where the waves cancel to produce flat water and giving a practical example of two waves interfering destructively. Doing this demonstration at this point helps reinforce all the ideas about interference that have been met so far in this topic.
- If a ripple tank is not available, use a suitable animation instead.
- Explicitly make the analogy/comparison between water waves and sound waves and discuss the strengths and weaknesses of the analogy. Strength: water waves can be simply visualised; 'flat water' is an analogy for 'no sound'; in both cases you have the idea of energy adding together to produce a bigger wave when the waves reinforce each other; in both cases you have the idea of no energy being transferred by wave means no movement (water doesn't move, air molecules don't move). Weaknesses: ripples cause water particles to vibrate up and down but sound waves cause air particles to vibrate back and forth.
- Challenge students to answer question 5 in the Student's Book, on the interference of radio waves broadcast simultaneously from two transmitters. Ensure students understand that the waves that interfere are the radio waves (of the same frequency) and not the sound waves of the music produced by the electronics in the radio.

Extension ideas

You could ask students to find out how destructive interference proves that sound travels in the form of a wave.

- Provide, or ask students to find out about, other examples or real-life situations where waves interfere destructively and ask students to discuss what is happening. In later work in Physics students will discover that all types of wave, not just sound waves and water waves, can interfere.
- In demonstration 1, students could work together in small groups to map out the positions of the areas of constructive and destructive interference (no detailed measurements are required). You could also repeat the demonstration using a sound of different frequency and they could map out the new positions of the areas of constructive and destructive interference. They could then suggest a hypothesis relating how close together these areas are to the frequency of the sound used. (The higher the frequency, the closer together they are.)

Homework suggestions

Students answer questions 1–5 in topic 11.2 Interference in the Workbook

Ask students to find out more about noise-cancelling technology and how it has been developed.

Students could also be asked to research other applications of or problems caused by interference of sound waves, such as in some concert halls, where there may be areas where the audience sometimes cannot hear the musicians properly.

Chapter 11 Checking students' progress

End of chapter reflection

- To review the learning points of the chapter
- To test understanding through answering questions

Resources

- Student's Book [End of chapter review questions]
- Workbook [Self-assessment grid, End-of-chapter questions]
- Worksheet 11.3

Approach

Use the Key facts and Check your skills progress boxes in the Student's Book for each topic in this chapter as a way of recapping on the main learning points.

Ask students to complete the Self-assessment checklist at the end of Chapter 11 in the Workbook. Encourage students to work together in small groups to discuss which aspects of the material are the most difficult to understand, and then have a class vote. You may wish to revisit some Learning Objectives again if students find any areas particularly challenging. Use the Student's Book text and questions to revisit these aspects. Go through the answers orally in class. Take note of whether giving the answer appears to resolve students' uncertainty or not, and work through further examples.

The teacher response in the Workbook can be completed to give students personalised feedback on next steps. Students could be asked to answer specific End of chapter review questions in their books. Alternatively go through some or all of the End of chapter review questions in the Student's Book orally in class.

Use the Context worksheet (Worksheet 11.3) as a framework for students to conduct their own research enquiries about sound. For example, students could focus on voice recognition systems and how they work.

Finish by asking students to complete the End-of-chapter questions at the end of Chapter 11 in the Workbook. Provide feedback on students' answers, and work with students individually to identify areas that still need more work.

Chapter 12 Plate tectonics

Chapter overview

This chapter provides opportunities to describe and use models and explain how they can be supported or changed based on new evidence.

Students will first examine in detail the different types of evidence for plate tectonics, including the matching shapes of continental coastlines, matching coastal rock layers, volcano and earthquake zones, fossils of the same extinct organism found on different continents and magnetic field patterns in the ocean floor. In the second topic, students investigate convection currents and learn how these currents in the Earth's mantle provide an explanation for the movement of tectonic plates.

Topic title	Number of 40-minute periods	Learning Objectives	Thinking and Working Scientifically Learning Objectives
12.1 Evidence for plate tectonics	4	9ESp.02 Explain why the jigsaw appearance of continental coasts, location of volcanoes and earthquakes, fossil record and alignment of magnetic materials in the Earth's crust are all evidence for tectonic plates	9TWSm.01 Understand that models and analogies reflect current scientific evidence and understanding and can change 9TWSm.02 Describe some important models, including analogies, and discuss their strengths and limitations 9TWSa.01 Evaluate the strength of the evidence collected and how it supports, or refutes, the prediction 9TWSa.02 Describe trends and patterns in results, identifying any anomalous results and suggesting why results are anomalous 9TWSp.02 Describe examples where scientists' unexpected results from enquiries have led to improved scientific understanding 9TWSc.05 Carry out practical work safely, supported by risk assessments where appropriate 9TWSp.03 Make predictions of likely outcomes for a scientific enquiry based on scientific knowledge and understanding
12.2 Explaining plate movement	2	9ESp.01 Explain the movement of tectonic plates in terms of convection currents	9TWSm.02 Describe some important models, including analogies, and discuss their strengths and limitations
End of chapter questions	1	Check students' progress	

Background science

The structure of the Earth can be described as a four-layer system. The inner core is the solid layer at the centre of the Earth and is made up of iron and nickel. The outer core is a liquid layer of iron and nickel surrounding the solid inner layer. The mantle surrounds the outer core and is made up of semi-molten rock known as magma. The outer layer of the Earth is known as the crust. The crust is a relatively thin solid rocky layer whose depth varies between 5 and 70 km depending on the location.

The crust is divided up into sections called tectonic plates. These plates can move around due to the convection currents occurring in the magma inside the mantle. The plates can move in different directions depending on the strengths and directions of the convection currents, and this movement is known as plate tectonics.

Some plates will move apart, causing magma to move up from the mantle until it reaches the surface of the Earth, where it then cools and solidifies to form igneous rock. If this pulling apart of plates and consequential movement of magma continues for a prolonged amount of time, the formation of the igneous rock becomes so great that a volcano is formed.

In other places the tectonic plates may move towards each other. Sometimes one plate is denser than the other and will move under the other in a process known as subduction. Earthquakes can occur because of this subduction. As the denser plate is pushed down under the other, the magma below the crust gathers into any gaps left. This causes a build-up of pressure, which is released through volcanic eruptions, where the magma erupts through the surface of the crust.

If the plates are both made up of sedimentary rocks and have the same density, when they move towards each other the edges of the plates meet and crumple up against each other. This creates mountains of sedimentary rock known as fold mountains.

In the early 20th century, Alfred Wegener proposed the theory of Continental Drift, which suggested that all the continents had once been joined to form one giant landmass that later became named Pangaea. He suggested that the continents on their tectonic plates moved apart through the movement of convection currents in the magma inside the mantle.

Fossil records provided one piece of evidence for this theory. Fossils of plants and animals which were of the same types and ages were found at the edges of continents otherwise separated by oceans. Magnetic materials in the Earth's crust located at the edges of these plates have been found to be aligned with each other. The edges of these continents also appeared to match each other like the pieces of a jigsaw puzzle. These three pieces of evidence all suggest that the continents were once joined.

The formation of volcanoes and mountains, along with the location of earthquakes, seems to be focused along the edges of tectonic plates. Volcanic eruptions and earthquakes are thought to be triggered by the movement of the tectonic plates over time, so this also provides evidence for the tectonic plate theory.

This chapter covers the evidence for the theory of plate tectonics and the reason why tectonic plates move.

A more detailed investigation of the evidence supporting plate tectonics reinforces the ideas from Stage 7 about the structure of the Earth. It presents opportunities to discuss matched coastlines of continents, rock layers and fossil evidence. At each stage, students should be encouraged to consider why the evidence presented supports the theory of plate tectonics, and examine whether there are limitations to that evidence.

The next section discusses convection currents in the Earth's mantle. The hands-on practical activity to demonstrate a convection current is also designed to focus on the assessment of risk and safety precautions by the students; if they have seen or carried out a similar convection currents activity before, then the emphasis on risks and safety provide a different aspect of practical work for students to consider.

12.1 Evidence for plate tectonics

Topic overview

		Student's Book	Workbook	Downloadable materials
12.1a	**Starting point** This short discussion provides an opportunity for students to recall their understanding of plate tectonics and a brief summary of evidence for the theory	Starting point table		
12.1b	**Continental coasts** Students learn about the 'jigsaw' model of continents and coastal rock layers, showing how the coasts of some continents appear to match up and the idea of a supercontinent, Pangaea, that broke up to form the continents	Questions 1–4 Activity 12.1	Questions 1–2	Practical notes 12.1b Worksheet 12.1b
12.1c	**Volcanoes and earthquakes** Students learn how volcanoes and earthquakes occur much more frequently along plate boundaries	Questions 5–7 Activity 12.2	Question 3	Practical notes 12.1c
12.1d	**Fossils** Students learn how fossils of extinct species that lived on land have been found in separate continents, and how this provides evidence for the existence of Pangaea	Questions 8–11 Activity 12.3	Questions 4–5	Practical notes 12.1d Worksheet 12.1d(2)
12.1e	**Magnetic materials** Students learn how the Earth's magnetic field reverses direction from time to time and how new rocks produced at a plate boundary beneath an ocean show alternating magnetic field patterns as a result	Questions 12–15 Activity 12.4	Questions 6–10	Practical notes 12.1e

Learning Objectives

- 9ESp.02: Explain why the jigsaw appearance of continental coasts, location of volcanoes and earthquakes, fossil record and alignment of magnetic materials in the Earth's crust are all evidence for tectonic plates
- 9TWSm.01: Understand that models and analogies reflect current scientific evidence and understanding, and can change
- 9TWSm.02: Describe some important models, including analogies, and discuss their strengths and limitations
- 9TWSp.02: Describe examples where scientists' unexpected results from enquiries have led to improved scientific understanding
- 9TWSc.05: Carry out practical work safely, supported by risk assessments where appropriate
- 9TWSa.01: Evaluate the strength of the evidence collected and how it supports, or refutes, the prediction
- 9TWSa.02: Describe trends and patterns in results, identifying any anomalous results and suggesting why results are anomalous
- 9TWSp.03: Make predictions of likely outcomes for a scientific enquiry based on scientific knowledge and understanding
- 9SIC.03: Evaluate issues which involve and/or require scientific understanding
- 9SIC.05: Discuss how the uses of science can have a global environmental impact

Links to other topics

- Earth and Space Stage 6 Formation of sedimentary rocks and fossils
- Earth and Space Stage 7 Chapter 10 Topic 10.1 The Earth's crust
- Earth and Space Stage 7 Chapter 10 Topic 10.2 Earthquakes, volcanoes and mountains
- Physics Stage 8 Chapter 10 Topic 10.1 Magnets and magnetic materials

Using models

- Discussing models of the effects of plate tectonics assists in relating the theory to the evidence that supports it.
- There are opportunities to discuss some strengths and limitations of the model to meet learning objective 9TWSm.02. Most evidence matches with the strengths of the model, but two limitations can be explored as time permits: (1) although we can show general effects of plate movement over time, the theory does not provide specific, measurable predictions, such as when exactly earthquakes will occur or volcanoes erupt, and (2) where a plate is pushed downwards by another plate, the crust enters the mantle and becomes molten – all evidence on this side of a plate is destroyed.

Science in context

Questions 1 and 2 in the Science in context: Making use of plate tectonics box in Topic 12.1 of the Student's Book cover the following learning objectives:

- 9SIC.03 Evaluate issues which involve and/or require scientific understanding
- 9SIC.05 Discuss how the uses of science can have a global environmental impact.

Teaching ideas

Look out for the opportunities and questions that encourage students to discuss the risks and modern practical uses that the movements of tectonic plates provide. Countries formed on top of a plate boundary where the plates are moving apart, such as Iceland, can make use of the thinner crust and the areas of heated rock that exist close to the surface. Here, renewable energy provided by hot rocks (geothermal energy) is used so that fossil fuels are not needed. The position of Iceland means there are risks (in the form of volcanoes, in particular), but also benefits.

Common misunderstandings

- Some students may think and state that fossils in general provide evidence for plate tectonics. It is important to highlight that only *some* fossils do so. The fossilised organisms must have lived at a time when continents were joined together and have become extinct after that time. They must also have been land-based organisms or organisms that lived in or near freshwater (as marine or flying animals could have moved between continents).
- Sometimes students confuse continents with tectonic plates. Continental coastlines and their matching shapes are evidence for plate tectonics, but they do not define all plate boundaries.
- Linked to this is the idea that earthquakes occur because 'continents bump together'. It is important to compare the very slow movement of plates (over millions of years) with the momentary movements due to earthquakes. Many, but not all, earthquakes are the result of plate movement, but the sudden effects are due to the build-up of forces over very long periods of time.
- Sometimes students connect the theory of plate tectonics and in particular the fossil evidence with the asteroid collision that is thought to have resulted in the extinction of the dinosaurs. It is important to emphasise that plate movements are due to the internal structure of the Earth, not the impact of external objects. Remind students that dinosaurs and many other species became extinct around 65 million years ago – long after Pangaea is believed to have separated.

12.1a Starting point

Review students' understanding of plate movements and the evidence they studied briefly in Stage 7. Check that they remember the timescales (hundreds of millions of years) and that the slow movement of tectonic plates causes events such as volcanoes and earthquakes at plate boundaries.

Teaching and learning ideas

Encourage students to discuss between themselves the following tasks/questions:

- Why are earthquakes and volcanoes found mostly along particular lines across the Earth's surface?
- How do fossils form?
- How long does the process of plate movement take?

- What were the weaknesses of Wegener's model of continental drift?
- What further evidence helped scientists to modify the model so that it addressed those weaknesses, forming the theory of plate tectonics?

12.1b Continental coasts

Students will learn:

- To explain the evidence for tectonic plate movement including coastal shapes
- To understand that models and analogies can change according to scientific evidence
- To describe the uses, strengths and limitations of models and analogies
- To evaluate how the prediction is supported by the evidence collected
- To describe trends and patterns shown in a set of results, identifying, and explaining, any anomalous results present
- To describe examples where scientific understanding has been improved by the unexpected outcomes of scientific enquiries

Glossary words

coastline, Pangaea, supercontinent

Resource preparation

Online or printed maps showing Pangaea and how today's continents formed

Figure 12.2 in the Student's Book

Small jigsaw puzzle – does not have to have a particular subject

A biscuit or chocolate bar that contains layered biscuit (examples: Kit Kat, Oreo, etc.)

Worksheet 12.1b for investigating how the shape of today's continents fit together

Scissors for cutting out continent shapes from the map in Worksheet 12.1b

Sheets of blank paper (A4 or A3) and coloured pens/pencils for the Extension activity

Practical notes 12.1b

Teaching and learning ideas

- Display a diagram (e.g. from Worksheet 12.1b or from a reliable online source) that shows the continents of the world.
- Explain that this was not how the land areas of the world looked about 300 million years ago.
- Discuss how figure 12.2 in the Student's book shows two continents that match up in shape, if arranged in the correct way.
- Have a small (12 pieces or fewer) jigsaw puzzle that students can solve while discussing how the shapes of some continents fit together.
- Demonstrate how breaking apart a multi-layered biscuit reveals two parts, and the layers of both parts of the biscuit match up. Encourage students to discuss how this could be a model for two coastlines that are being pulled apart.
- Ask what evidence students can describe that shows how two continents have formed from one (shape of coastline and exposed layers of rock on coastline).
- Ask students to complete Activity 12.1, which involves them cutting out the shapes of the main continents and investigating how some coastlines fit together. (Use Worksheet 12.1b, a simple map of the main continents of the world.)

Differentiation

- Suggest to students they imagine they are explorers travelling from the west coast of Africa across the ocean to South America. Ask students to describe the evidence they might encounter which would help to show the continents were once joined together. Encourage a group role play.
- With Activity 12.1, if less confident students are finding it difficult to find good matches between shapes, pair them up with more confident students who have found examples and encourage discussion and peer learning.

Extension ideas

- Have students prepare a simple model of two layered rock 'coastlines' (this can be as simple as two sheets of paper, each with a set of identical layers coloured in).
- Have the students present their model to others in the class, and encourage other students/groups to exchange ideas on the weaknesses of the model.

Homework suggestions

Students answer questions 1–2 in Topic 12.1 in the Workbook.

12.1c Volcanoes and earthquakes

Students will learn:

- To explain the evidence for tectonic plate movement including the locations of volcanoes and earthquakes
- To understand that models and analogies can change according to scientific evidence
- To describe the uses, strengths and limitations of models and analogies
- To evaluate how the prediction is supported by the evidence collected
- To describe trends and patterns shown in a set of results, identifying, and explaining, any anomalous results present
- To describe examples where scientific understanding has been improved by the unexpected outcomes of scientific enquiries

Resource preparation

On-screen map of the world, ideally one plain and one showing the 'hotspots' of areas with large numbers of earthquakes and volcanoes (such as that in the Student's Book figure 12.4).

Teaching and learning ideas

- Display a plain map of the world and have students volunteer to suggest where some plate boundaries occur.
- Now display a map of the world showing the 'hotspots' of areas containing frequent earthquakes and many volcanoes. Ask students to volunteer to come up to the front and, using an interactive whiteboard or a piece of A3 tracing paper and a marker, suggest where a plate boundary can be found.
- After five or six volunteers, a number of boundaries should be marked. Students should discuss the accuracy of the original predictions.
- Encourage students to explain how volcanoes and earthquakes can be caused by the forces that build up over time due to plate movements.
- Discuss the material in the Student's Book that describes how detectors developed to monitor the testing of nuclear weapons had another unexpected benefit: they produced detailed maps of earthquakes and volcanic eruptions.
- In Stage 7, students will have looked at fold mountains where two plates are pushing together and upwards, and volcanoes and earthquakes where two plates are pushing together and one plate dives below the other. Revisit this learning if you think a refresher is needed.
- Move on to discuss plates that are moving apart. Ask students to complete Activity 12.2, which involves them examining the Great Rift Valley on the eastern side of Africa.
- Emphasise the timescale of these changes – over many millions of years.

Differentiation

- With Activity 12.2, if less confident students are finding it difficult to interpret the diagram in the Student's Book (figure 12.5), sketch a diagram of Africa on the whiteboard and mark the approximate line of the plate boundary. Ask students to add arrows showing the movement of the two plates (away from each other).

Extension ideas

- Ask students to suggest how a (3D) model could be made to show two plates moving apart and forming a rift valley. An example would be a cake or loaf of bread with visible layers of 'rocks', which when pulled apart slowly from the top would break leaving exposed, matching layers.

Homework suggestions

Students answer question 3 in Topic 12.1 in the Workbook.

12.1d Fossils

Students will learn:

- To explain the evidence for tectonic plate movement including fossil records
- To understand that models and analogies can change according to scientific evidence
- To describe the uses, strengths and limitations of models and analogies
- To evaluate how the prediction is supported by the evidence collected
- To describe trends and patterns shown in a set of results, identifying, and explaining, any anomalous results present
- To describe examples where scientific understanding has been improved by the unexpected outcomes of scientific enquiries

Glossary words

Cynognathus, *Glossopteris*, *Lystrosaurus*, *Mesosaurus*

Resource preparation

Worksheet 12.1d(2) for analysing whether some fossils can or cannot produce evidence for plate tectonics

Online access to sites with information about the organisms in Activity 12.3 (or a reference book or books if students can access these in school)

Practical notes 12.1c

Teaching and learning ideas

- Use the Making links question in the Student's Book as a starter activity. As a class, have students suggest or write up on the whiteboard the step-by-step process that explains how fossils form.
- Use the example of the *Glossopteris* fern to ask questions of the students such as:
  - What kind of organism was it? (plant)
  - Could it move on its own? What is the evidence for this? (No, because it was fixed in its place by its roots.)
  - How would *Glossopteris* have spread more widely? (Reproduction by spores – although it is not essential to discuss the method.)
  - Fossils of *Glossopteris* have been found on several continents. How could this have happened, if the plant could not move and the land is now separated by thousands of miles? (Supercontinent Pangaea breaking apart and continents moving due to plate tectonics.)
- Ask students to carry out Activity 12.3 in small groups. For a larger version of the table in the Student's Book, use Worksheet 12.1d(2). Students should look for information about the organisms listed in the table, either online or in a reference book if a suitable one is available.
- Check that groups have covered at least three of the listed organisms before asking them to research another of their own choice (if time permits).
- Gather the class back together and go through the answers together, with students volunteering their group's information.
- If time permits, discuss question 10 in the Student's Book in class. Ensure that all students are confident in recalling and using the information.

Differentiation

- With Activity 12.3, encourage more confident groups to work together with groups that are finding it challenging to locate the information in time. A combined group could assign tasks to pairs of students, either by giving one organism or one column of information to each pair to research.

Extension ideas

- Challenge students to think about why some extinct organisms leave no fossils (one reason is: animals with only soft tissue and no bone structure/hard materials to survive the squeezing of rock layers).
- Ask students in pairs to discuss where in the world they would go looking for fossil evidence of plate tectonics today, and justify their agreed answer.

Homework suggestions

Students answer questions 4–5 in Topic 12.1 in the Workbook.

12.1e Magnetic materials

Students will learn:

- To explain the evidence for tectonic plate movement including the position of magnetic materials
- To understand that models and analogies can change according to scientific evidence
- To describe the uses, strengths and limitations of models and analogies
- To evaluate how the prediction is supported by the evidence collected
- To describe trends and patterns shown in a set of results, identifying, and explaining, any anomalous results present
- To describe examples where scientific understanding has been improved by the unexpected outcomes of scientific enquiries
- To use scientific knowledge to make predictions

Glossary words

magnetic alignment, magnetic field reversal

Resource preparation

For initial review work: A bar magnet, a sheet of stiff paper (A4 is large enough), iron filings and a non-magnetic (plastic or cardboard) tray to catch any dropped filings

For Activity 12.4, for each group: a bar magnet, a compass, paper and pencils

Practical notes 12.1e

Teaching and learning ideas

- Use the Making links section in the Student's Book to review students' recollection and understanding of magnetic fields.
- As a demonstration, ask students to guide you in setting up an experiment using iron filings to model magnetic field lines. They should guide you to place the tray to catch iron filings, then place the magnet flat in the centre of the tray. Then place the stiff paper on top and finally gently scatter iron filings on the paper above the magnet.
- Ask students what device can be used to show the direction of the Earth's magnetic field (a compass, or a bar magnet suspended in a thread cradle so it is free to rotate and align itself to Earth's magnetic field).
- Explain how the direction of the Earth's magnetic field reverses roughly once every 450 000 years. Encourage students to suggest the effects of a field reversal, particularly how it affects magnetic rocks.
- Get students to describe how fresh molten rocks rises to fill the gaps left as two tectonic plates move apart. Encourage them to decide what will happen to hot rocks produced under the ocean in this way (they cool very quickly to form solid rock).
- Explain how this rapid solidification means magnetic rocks are 'trapped' with their magnetic fields aligned with that of the Earth.
- Ask students to carry out Activity 12.4 about field reversals. Encourage students in groups to think of ways they can build their own model to show what happens when Earth's magnetic field reverses, using a bar magnet.
- Students should record a prediction based on the method they choose, then test that prediction and make changes to the method as needed to improve the results.

Differentiation

- With Activity 12.4, if less confident students are finding it difficult to construct a model, have them start with two bar magnets and ask them to demonstrate how we can observe magnetic attraction and repulsion. Build up to the idea that if Earth is one of the bar magnets, we can cause the 'Earth' magnet to move if we bring another magnet close to it in the right position. More confident students could explain their models to the less confident students.

Extension ideas

- Students could list ways in which our lives would be affected if the Earth's magnetic field direction reversed suddenly. How many devices that we depend upon would be confused or made useless by the change?
- Students could use the ideas that (i) tectonic plates move 2–4 cm a year and (ii) the Earth's magnetic field reverses every 450 000 years to predict how wide a strip of newly formed magnetised rock would be before the field reversed. (The answer is not vital for this exercise, it is more important that students present a logical method for arriving at a value.)

Homework suggestions

Students answer questions 6–10 in Topic 12.1 in the Workbook.

12.2 Explaining plate movement

Topic overview

		Student's Book	Workbook	Downloadable materials
12.2a	**Starting point** Students review what they have learnt about convection currents	Starting point table		
12.2b	**Convection currents** Students plan an investigation designed to demonstrate convection currents in water, including a risk assessment	Questions 1–4 Activity 12.5	Questions 1–8	
12.2c	**The effect on tectonic plates** Students apply their knowledge of convection currents to explain why tectonic plates move	Questions 5–8	Questions 9–10	Worksheet 12.2c

Learning Objectives

- 9ESp.01: Explain the movement of tectonic plates in terms of convection currents
- 9TWSm.02: Describe some important models, including analogies, and discuss their strengths and limitations

Links to other topics

- Physics Stage 9 Chapter 8 Topic 8.3 Conduction, convection, radiation and evaporation

Using models

- Students can use their previous studies of convection to create a visual model of convection currents in liquids. They can explore some limitations of this model including that the rock in the mantle is a very slow-moving, hot, sticky substance, so a liquid like water moves much more quickly.
- Students can explore some limitations of the convection current model for plate tectonics, such as when a plate is pushed downwards by another plate, all evidence is lost because the rock becomes molten, so although the model could be correct it is very difficult to prove.

Science in context

Teaching ideas

There is an opportunity to discuss how scientific knowledge is developed through collective understanding and scrutiny over time. Refer back to the Stage 7 material on Wegener's hypothesis about continental drift, then in this chapter the development of the modern theory of plate tectonics with its accompanying evidence and the realisation that plate movement could be explained by convection currents in the Earth's mantle.

Common misunderstandings

- Many students may assume the convection current in the Earth's mantle is as rapid as that demonstrated in heated water. It is important to stress that the Earth's mantle is not a liquid in the sense that students can observe in everyday life; it is a thick, slow-moving, sticky substance.
- Some students may think that the Earth's tectonic plates float freely on the mantle, so that e.g. earthquakes are plates 'bumping together' and plates moving apart leave a large gap that is rapidly filled with molten rock, but the reality is much more slow-moving. Encourage students to think about whether a movement of 1 cm in a year is visible, and scale their ideas of time accordingly.

12.2a Starting point

Resource preparation

Diagram of the structure of Earth, such as the one below.

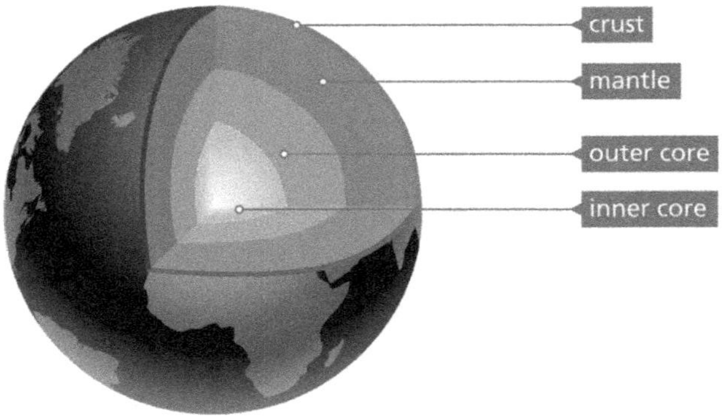

Teaching and learning ideas

- Display an image of the structure of the Earth to stimulate discussion.
- Ask students to say where tectonic plates are found and how deep they go into the Earth's structure (within the crust, and just on the surface so a few km thick under the oceans, thicker than this under land masses).
- Encourage students to say how hot the mantle is (very hot, so that rock is molten).
- Point out that tectonic plates of solid, cool rock are therefore resting on top of a very hot, deep layer of molten rock in the mantle.

12.2b Convection currents

Students will learn:

- To explain how convection currents cause tectonic plate movement

Glossary words

convection current

Teaching and learning ideas

- Check with students what they can recall about the transfer of heat, in particular convection and convection currents.
- Recap the convection demonstration using the potassium manganate(VII) from Chapter 8 if students have already done this.
- Alternatively, the concept of convection currents can be introduced using syrup or molasses to represent the magma and pieces of biscuit to represent the tectonic plates. The syrup/molasses are heated up and the convection currents cause the pieces of biscuits to move around on the surface like the tectonic plates on top of the magma.
- Conclude by explaining that although the mantle of Earth is very different to liquid water, there is a source of heat (from the core) and this sets up convection currents under the crust.

Differentiation

- Consider having a short whole-class recap of the three types of heat transfer: conduction, convection and radiation, and guide students to focus on just what is happening in the water in the beaker, not in the surroundings.

Extension ideas

- Students could consider what would happen if two crystals were placed in different parts of the beaker, but with the same single heat source in the same place. If time permits, they could predict the effects and then test their predictions.

Homework suggestions

Students answer questions 1–8 in Topic 12.2 in the Workbook.

12.2c The effect on tectonic plates

Students will learn:

- To explain how convection currents cause tectonic plate movement
- To make predictions based on scientific knowledge and understanding
- To carry out practical work safely
- To discuss strengths and weaknesses of the plate tectonics model

Resource preparation

- Optional: jar of very thick treacle or molasses (or search on the internet for 'pitch drop experiment' for a University of Queensland experiment)
- Display of figure 12.11 in the Student's Book, ideally on an interactive whiteboard where different labels and arrows can be hidden for students to explain
- Worksheet 12.2c

Teaching and learning ideas

- Return to the concept that the Earth's mantle is not a runny liquid, but is more like a sticky, thick substance that flows very slowly.
- Optional: demonstrate a typical runny liquid like water and compare how it moves freely with a substance like very thick treacle or molasses. Ask students to comment on the speed at which each liquid moves. (The 'Pitch drop experiment' at the University of Queensland in Australia shows an extremely slow-moving substance called pitch, where one drop falls roughly every 14 years!)
- Show the mantle convection current diagram (figure 12.11 in the Student's Book). Cover up sections of the diagram to show a single plate at a time and discuss each current and direction of plate movement in turn, before then discussing the full diagram including plate boundaries. Hide direction arrows and ask students to volunteer to mark them back on the diagram.
- Students complete Worksheet 12.2c, where they have to mark the circulation of convection currents in the mantle so as to explain the observed plate boundaries.

Differentiation

- Less confident students may need support in considering one convection circulation and the effect it has on the plate above, before moving on to consider two convection currents as found interacting at plate boundaries. Sketch or have the students sketch diagrams for each type of plate boundary separately. More confident students should be able to move straight on to the interaction of the two convection currents.

Extension ideas

- Students draw an extension of figure 12.11 in the Student's Book, where the mantle and a number of plates form a circle (to better model a cross-section through Earth). They should draw convection current circulation arrows in the mantle, and plate movement arrows to match. Students should observe it is impossible to have an arrangement of only plates pushing together or only plates moving apart – to have one type of boundary requires there to be one of the other type as well.

Homework suggestions

Students answer questions 9–10 in Topic 12.2 in the Workbook.

Chapter 12 Checking students' progress

End of chapter reflection

- To review the learning points of the chapter
- To test understanding through answering questions

Resources

- Student's Book [End of chapter review questions]
- Workbook [Self-assessment grid, End-of-chapter questions]
- Worksheet 12.3

Approach

Use the Key facts and Check your skills progress boxes in the Student's Book for each topic in this chapter as a way of recapping on the main learning points.

Ask students to complete the Self-assessment checklist at the end of Chapter 12 in the Workbook. Encourage students to work together in small groups to discuss which aspects of the material are the most difficult to understand, and then have a class vote. You may wish to revisit some Learning Objectives again if students find any areas particularly challenging. Use the Student's Book text and questions to revisit these aspects. Go through the answers orally in class. Take note of whether giving the answer appears to resolve students' uncertainty or not, and work through further examples.

The Teacher's comments in the Workbook can be completed to give students personalised feedback on next steps. Students could be asked to answer specific End of chapter review questions in their books. Alternatively go through some or all of the End of chapter review questions in the Student's Book orally in class.

Use the Context worksheet (Worksheet 12.3) as a framework for students to conduct their own research enquiries about plate tectonics and some of the effects of plate movement. Students should comment on the variety of sources of information they found and how good those sources were.

Finish by asking students to complete the End-of-chapter questions at the end of Chapter 12 in the Workbook. Provide feedback on students' answers, and work with students individually to identify areas that still need more work.

Chapter 13 Climate change

Chapter overview

This chapter provides opportunities for students to plan investigations, make risk assessments and decide what equipment to use for investigations.

Students will first explore the different parts of the carbon cycle: photosynthesis and feeding, respiration, decomposition and combustion. Students will also use word equations to represent chemical reactions for photosynthesis and respiration, linking to both chemistry and biology.

To reinforce and develop the concept of the overall cycle, a diagram is built up at each stage until the full carbon cycle is revealed and explained. The chapter concludes by describing the impacts of climate change including the increased frequency and severity of extreme weather events.

Topic title	Number of 40-minute periods	Learning Objectives	Thinking and Working Scientifically Learning Objectives
13.1 The carbon cycle	5	9ESc.01 Describe the carbon cycle (limited to photosynthesis, respiration, feeding, decomposition and combustion)	
13.2 Impacts of climate change	2	9ESc.02 Describe the historical and predicted future impacts of climate change, including sea level change, flooding, drought and extreme weather events	9TWSa.02 Describe trends and patterns in results, identifying any anomalous results and suggesting why results are anomalous 9TWSp.03 Make predictions of likely outcomes for a scientific enquiry based on scientific knowledge and understanding 9TWSp.05 Make risk assessments for practical work to identify and control risks 9TWSc.02 Decide what equipment is required to carry out an investigation or experiment and use it appropriately
End of chapter questions	1	Check students' progress	

Background science

Carbon is an essential element that is found inside the cells of all living organisms. The carbon cycle describes the stages through which carbon can be stored and transferred between different living organisms.

Carbon is contained in the atmosphere in the form of carbon dioxide gas. This carbon dioxide gas is taken in by plants during the process of photosynthesis and converted into glucose, which is further converted into starch and stored. When consumers eat the green plants, they take in the starch and increase the amount of carbon in their own bodies.

Respiration occurs in the cells of living things to release energy. The process of respiration produces carbon dioxide as a waste product, which is then released by the organism into the atmosphere. The use of fossil fuels through the combustion process also releases carbon dioxide into the atmosphere as a pollutant. The decomposition of living organisms involves the breakdown of glucose and other organic matter to produce carbon dioxide gas, which is also released into the atmosphere.

Climate change describes the changes in atmospheric conditions over long periods of time and over large areas. The Earth's climate has undergone many changes through time. The levels of carbon dioxide and other greenhouse gases, such as methane and water, in the atmosphere have increased due to human activities such as burning fossil fuels, deforestation and farming, as well as natural emissions such as volcanic eruptions, decomposition of marshlands and emissions from farm animals (note that deforestation is not covered in Stage 9).

These greenhouse gases absorb the infrared radiation that bounces off the surface of the Earth, stopping it from leaving the atmosphere. The higher the concentration of greenhouses gases in the atmosphere, the higher the amount of infrared radiation absorbed and the warmer the Earth's climate becomes.

The warming of the Earth's climate is known as global warming and has led to changes such as increased sea levels, flooding, droughts and extreme weather events such as severe storms.

This chapter covers the main processes of the carbon cycle and some of the impacts of climate change.

Students should already have encountered photosynthesis, decomposition and respiration as chemical reactions and as vital processes for life on Earth. Here, these reactions are set in context as parts of the carbon cycle with global significance. The human contributions to the carbon cycle are explored, both as part of agriculture and via the combustion of fossil fuels.

The next section builds upon Stage 8 Chapter 11 The Earth and its Resources, to describe some of the current and likely future impacts of climate change. The emphasis is on long-term changes such as rising sea levels, and the increase in frequency and strength of extreme weather events.

13.1 The carbon cycle

Topic overview

		Student's Book	Workbook	Downloadable materials
13.1a	**Starting point** This short discussion provides an opportunity for students to review their knowledge of climate change and its causes	Starting point table		
13.1b	**Photosynthesis and feeding** Students learn about the photosynthesis reaction in plants and its significance globally, and also feeding/the uptake of nutrients by all organisms	Questions 1–4	Questions 1 and 6	
13.1c	**Respiration** Students learn about aerobic respiration in plants and animals, and its importance within the carbon cycle	Questions 5–7	Questions 2 and 7	
13.1d	**Combustion** Students learn the significance of combustion within the carbon cycle and in particular how human activities involving combustion are returning larger amounts of carbon dioxide to the atmosphere	Questions 8–10	Question 5	
13.1e	**The complete carbon cycle** Students learn about the full carbon cycle and how the individual processes link together	Questions 11–14	Questions 3, 4, 8, 9 and 10	Worksheets 13.1e(1), 13.1e(2)

Learning Objective

- 9ESc.01: Describe the carbon cycle (limited to photosynthesis, respiration, feeding, decomposition and combustion)

Links to other topics

- Biology Stage 7 Chapter 2 Topic 2.2 Microorganisms and decay
- Biology Stage 9 Chapter 1 Topic 1.1 Photosynthesis and plant growth
- Biology Stage 8 Chapter 1 Topic 1.2 Human respiratory system
- Chemistry Stage 8 Chapter 6 Topic 6.1 Using word equations
- Chemistry Stage 9 Chapter 7 Topic 7.2 Word and symbol equations
- Earth and Space Stage 8 Chapter 11 Topic 11.2 Climate change

Using models

- Describe the carbon cycle as a model for understanding how carbon compounds are exchanged between the atmosphere, organisms and materials found in the Earth's crust.

Science in context

Teaching ideas

Look out for opportunities to consider the dependence of some countries on fossil fuels for their income and the potential impact on some less industrialised nations of a rapid switch away from those fuels to reduce the production of carbon dioxide.

Common misunderstandings

- Many students recognise some of the processes in the carbon cycle but either do not make all the links in the full cycle or do not appreciate how carbon is transferred and transformed in different processes.
- Some students think that plants obtain most of their mass from nutrients in the soil and fail to appreciate that carbon dioxide from the atmosphere actually provides much of the mass of plants via photosynthesis, which is why plants (and trees in particular) are important carbon dioxide sinks.
- Some students think that carbon as an element is what circulates in the carbon cycle, for understandable reasons given the cycle's name. However, it is important to emphasise at each stage of the cycle that carbon in its elemental form is generally only present in coal as a fossil fuel, and that carbon elsewhere in the cycle is transferred between different compounds (e.g. carbon dioxide, hydrocarbons, carbohydrates including glucose).

13.1a Starting point

Review students' understanding of photosynthesis and feeding by displaying an image of a tropical rainforest. Discuss the types of plants present and check that students remember the purpose of photosynthesis and the conditions under which it occurs (presence of light and carbon dioxide). Briefly review food chains by discussing what animals eat the plants in the rainforest, and what types of predators may eat those animals.

Teaching and learning ideas

Encourage students to discuss between themselves the following tasks/questions:

- What organisms can carry out photosynthesis and why is it important to all organisms?
- What happens to the remains of organisms after they die?
- Where do we get fossil fuels from and how does this relate to organisms that lived a long time ago?
- Why is it important that we understand the processes that produce or use up carbon dioxide?

13.1b Photosynthesis and feeding

Students will learn:

- To describe how photosynthesis, respiration, feeding, decomposition and combustion make up the carbon cycle

Glossary words

chloroplasts, feeding, photosynthesis, sink

Resource preparation

Optional: Figure 13.6 from the Student's Book

Teaching and learning ideas

- Optional: display on a whiteboard figure 13.6 from the Student's Book, which shows the full carbon cycle, but cover up all arrows and arrow labels. As each process of the cycle is discussed, ask the class to suggest where on the cycle the process is found, the direction of the arrow showing the movement of carbon compounds and how to label it.
- Ask the students to write out the word equation for photosynthesis, either each individually or have volunteers come to draw it on the whiteboard.
- Note the conditions required: light (sunlight in a natural setting) and the presence of chlorophyll.
- Discuss the role of chloroplasts that are found in plant cells but not animal cells, and why these organelles are required for photosynthesis to occur.
- Have students in groups construct a short food chain to show how animals and the predators that hunt those animals are all dependent on finding food from plants.

- Encourage students in groups to discuss the importance of plants and photosynthesis in helping to limit the amount of carbon dioxide in the atmosphere.

Homework suggestions

Students answer questions 1 and 6 in Topic 13.1 in the Workbook.

13.1c Respiration

Students will learn:

- To describe how photosynthesis, respiration, feeding, decomposition and combustion make up the carbon cycle

Glossary words

aerobic respiration, decomposition, mitochondria, respiration, source

Resource preparation

Optional: Figure 13.6 in the Student's Book

Teaching and learning ideas

- Optional: display on a whiteboard figure 13.6 from the Student's Book, which shows the full carbon cycle, but cover up all arrows and arrow labels. As each process of the cycle is discussed, ask the class to suggest where on the cycle the process is found, the direction of the arrow showing the movement of carbon compounds and how to label it.
- Review the products of photosynthesis, drawing attention to glucose as a store of energy.
- Ask students to suggest what this energy is needed for in plants (staying alive, growth and reproduction).
- Ask students to suggest what the energy animals take from food is needed for (staying alive, movement, growth and reproduction).
- Recap aerobic respiration as the process by which organisms release energy through the conversion of glucose and oxygen to carbon dioxide and water. Write the word equation on the whiteboard.
- Challenge students to suggest where they have seen the products and reactants before (photosynthesis, but the reverse of that reaction).
- Ask students to describe how aerobic respiration is connected to climate change (production of carbon dioxide).
- Note the key reactant oxygen that is needed for respiration to be aerobic. Ask students to suggest whether respiration can occur if limited or no oxygen is present.
- Point out the importance of respiration in bacteria and other organisms as part of the process of decomposition.

Extension ideas

- Ask students to draw up a table comparing photosynthesis and aerobic respiration. They should include: products, reactants, conditions, at what time of day they occur, which organisms carry out both processes and in which organelles in cells the reactions take place.

Homework suggestions

Students answer questions 2 and 7 in Topic 13.1 in the Workbook.

13.1d Combustion

Students will learn:

- To describe how photosynthesis, respiration, feeding, decomposition and combustion make up the carbon cycle

Glossary words

combustion, fossil fuels, fossils, hydrocarbons

Teaching and learning ideas

- Optional: display on a whiteboard figure 13.6 from the Student's Book, which shows the full carbon cycle, but cover up all arrows and arrow labels. As each process of the cycle is discussed, ask the class to suggest where on the cycle the process is found, the direction of the arrow showing the movement of carbon compounds and how to label it.
- As a class, agree and write up the word equation for combustion of wood. Emphasise the essential reactant (oxygen) and the need for an initial input of energy (e.g. a spark or lit splint).
- Check students' understanding of fossil fuels as hydrocarbons (see also Stage 8 Chapter 11 Topic 11.3 Renewable and non-renewable resources).
- Use question 8 in the Student's Book as a class discussion. Students discuss and agree the similarities and differences between combustion, photosynthesis and respiration.
- Emphasise the importance of combustion and respiration as reactions that release energy that can be used for other purposes, in both natural and human-produced processes.
- Write the word equations for photosynthesis, respiration and combustion on the whiteboard. Ask students to draw linking lines or shade in colour where each of oxygen, carbon dioxide and water appear, either as reactants or products.
- Return to the idea of photosynthesis as a means of using up carbon dioxide from the atmosphere and storing carbon in plants. Encourage students to think of combustion and respiration as means of releasing the stored carbon back into the atmosphere as carbon dioxide.

Differentiation

- Less confident students may need help to link the role of carbon with carbon dioxide in the processes of combustion, respiration and photosynthesis.
- More confident students could be challenged to identify the ways in which carbon is taken in and given out from these processes in the form of carbon compounds.

Extension ideas

- Students could suggest how to investigate the different amounts of energy produced by the combustion of different fossil fuels and wood. (This should only be discussed as a concept, not set up as an actual experiment.)

Homework suggestions

Students answer question 5 in Topic 13.1 in the Workbook.

13.1e The complete carbon cycle

Students will learn:

- To describe how photosynthesis, respiration, feeding, decomposition and combustion make up the carbon cycle
- Understand that the model of the carbon cycle is simplified and more evidence is being used to improve the model

Resource preparation

Optional: Figure 13.6 in the Student's Book

Worksheets 13.1e(1) and 13.1e(2)

Teaching and learning ideas

- Optional: display on a whiteboard figure 13.6 from the Student's Book, which shows the full carbon cycle, but cover up all arrows and arrow labels. Now ask the class to volunteer individually and add all the process arrows and labels, one at a time until the cycle is complete.
- Review the reactants and products in the different processes. Either by using figure 13.6, or by drawing a similar diagram of the full cycle on the whiteboard, ask students to suggest 'carbon loops' by drawing different coloured arrows that form circles or ovals. For example, one such circle starts at the top of the diagram, follows the carbon dioxide that is taken in by plants during photosynthesis, and its return to the atmosphere in the respiration of plants.
- Emphasise how the 'carbon' in the carbon cycle changes form, for example as carbon dioxide molecules in the atmosphere but as carbon in coal.
- Ask students to identify the process that is largely dependent on human activities (combustion).
- Ask students to complete Worksheet 13.1e(1), which takes them step by step through how the use of fossil fuels increases the amount of carbon dioxide put into the atmosphere by combustion. Students complete the extended answer questions as time permits.

- Conduct a vote by show of hands on what overall effect humans are having on the amount of carbon dioxide in the atmosphere: are they increasing or decreasing it?
- Review students' understanding of the effects of increased carbon dioxide on climate (increasing temperature due to enhanced greenhouse effect).
- Emphasise that there are more processes involved in the carbon cycle, such as the absorption of carbon dioxide into the oceans.
- Explain that the model of the carbon cycle is simplified to make it easier to understand; climate scientists use supercomputers to model the effects of many more processes on the Earth's atmosphere and are adding more evidence to improve the model.

Differentiation

- Ask students to complete Worksheet 13.1e(2), which provides a scaffolded route from a skeleton carbon cycle to a fully completed cycle.
- Draw a circle made from an anticlockwise arrow on the whiteboard or on A3 sheets of paper. Students use the names of objects or place cut-out diagrams of them to show where they are involved in the cycle. For example, a tree could go on the left-hand side of the diagram (as something that takes carbon dioxide out of the atmosphere, on balance). Fossil fuels could be placed at the bottom. A factory or power station could be placed on the right (as something that produces carbon dioxide that is given out to the atmosphere). Point out that plants also have their own cycle with photosynthesis and respiration.

Extension ideas

- Students could discuss and speculate how changes on Earth would affect the carbon cycle: which processes would increase and which would decrease? What would this mean for the amount of carbon dioxide in the atmosphere? Examples of events include: an asteroid colliding with Earth; a change in energy output from the Sun; all humans changing to a vegan/meat-free diet.

Homework suggestions

Students answer questions 3, 4, 8, 9 and 10 in Topic 13.1 in the Workbook.

13.2 Impacts of climate change

Topic overview

		Student Book	Workbook	Downloadable materials
13.2a	**Starting point** Students discuss extreme weather events they can remember or have read about	Starting point table		
13.2b	**Droughts** Students learn about the causes and effects of droughts	Questions 1–3 Activity 13.1	Questions 1, 2, 6 and 7	Practical notes 13.2b Worksheet 13.2b
13.2c	**Floods** Students learn about the causes and effects of floods from rain water that falls on land	Questions 4–5		Worksheets 13.2c(1) 13.2c(2)
13.2d	**Sea level change and sea flooding** Students learn about the causes and effects of sea level change and sea flooding	Questions 6–8 Activity 13.2	Question 5	Practical notes 13.2d Worksheets 13.2d(1) 13.2d(2)
13.2e	**Other extreme weather events** Students describe the different types of extreme weather events and learn how increasing average global temperatures are affecting their frequency and severity	Questions 9–12 Activity 13.3	Questions 3, 4, 8, 9 and 10	Practical notes 13.2e

Learning Objectives

- 9ESc.02: Describe the historical and predicted future impacts of climate change, including sea level change, flooding, drought and extreme weather events
- 9TWSp.03: Make predictions of likely outcomes for a scientific enquiry based on scientific knowledge and understanding
- 9TWSa.02: Describe trends and patterns in results, identifying any anomalous results and suggesting why results are anomalous
- 9TWSp.05 Make risk assessments for practical work to identify and control risks
- 9TWSc.02 Decide what equipment is required to carry out an investigation or experiment and use it appropriately

Links to other topics

- Biology Stage 9 Chapter 1 Topic 1.2 Water and mineral salts
- Earth and Space Stage 8 Chapter 11 Topic 11.1 Climate and weather
- Earth and Space Stage 8 Chapter 11 Topic 11.2 Climate change

Using models

- Students recognise that models of the atmosphere are complex and require supercomputers for scientists to analyse the effects of changes to the atmosphere. As more evidence is gathered, these computer-based models are improved to make more reliable predictions.

Science in context

Worksheet 13.2b and Questions 1, 2 and 3 below the Science in context: Droughts in Malawi box in Topic 13.2 of the Student's Book cover the following learning objectives:

- 9SIC.02 Describe how science is applied across societies and industries, and in research
- 9SIC.03 Evaluate issues which involve and/or require scientific understanding
- 9SIC.05 Discuss how the uses of science can have a global environmental impact.

Teaching ideas

Look out for opportunities to discuss the effects of climate change and extreme weather events on people and wildlife. A case study of Malawi, which is affected regularly by drought, provides the opportunity to understand some of the reasons why climate change has a direct but different impact on people in different countries. It also discusses how scientists are producing improved methods of forecasting drought, developing faster-growing crop seeds that are more tolerant of drought and providing watering systems for the worst-affected areas.

Common misunderstandings

- Some people think particular weather events occur only because of climate change. This is not a scientific approach. In science, we cannot be certain why a particular extreme weather event occurred at a particular time; what we can be certain about is that climate change is helping these extreme events to happen more often and with increased effects.

13.2a Starting point

Resource preparation

Optional: Pictures of climate effects and extreme weather events and their effects such as drought, thunderstorms, tornadoes, hurricanes, flooding.

Teaching and learning ideas

- Optional: Display up to four pictures of climate effects and extreme weather events and their effects to prompt discussion.
- Prompt students to talk about weather events they have experienced or read about. Write a list of the effects that students recall.
- After sufficient discussion, go through the list and have the students suggest causes (e.g. unusually heavy rainfall may cause flooding).

13.2b Droughts

Students will learn:

- To describe droughts as impacts of climate change
- To predict and test the outcome of a scientific enquiry
- To plan investigations
- To decide what equipment is needed for an investigation and use it correctly
- To evaluate the reliability of secondary information
- To make predictions of likely outcomes for a scientific enquiry
- To collect, record and summarise sufficient observations and measurements

Glossary words

drought, wild fires

Resource preparation

Worksheet 13.2b

For Activity 13.1: Four recently germinated seedlings of easy-to-grow plant species, such as beans, prepared in individual small pots with compost, space on a sunny windowsill or tabletop, white card or stiff paper for students to label individual pots, water supply.

Practical notes 13.2b

Teaching and learning ideas

- Review students' understanding of why organisms (plants and animals) need water, including its importance in maintaining cell structure, as part of many organs and systems, and in particular water's role in transport in plants.
- Explain that drought is defined as an unusually long period of time with very low or no rainfall. Ask students to describe the possible effects they can think of.
- Use the Making links questions from the Student's Book to prompt discussion: Why does drought affect some plants and not others? (Adaptations such as fleshy stems/trunks that store water to reduce the effects of drought.) How might animals cope with a drought? (Adaptations such as storing water; only being active at night and finding cool places to rest during the day to reduce water loss; moving out of drought-affected areas in search of food and shelter.)
- Discuss the example of a country affected regularly by drought, Malawi. Have students read the short case study in the Student's Book and then in small groups discuss the answers to the questions on Worksheet 13.2b.
- Long-term project: Ask students to complete Activity 13.1 in groups. Students plan how they will demonstrate the effects of different levels of watering on four seedlings, predict the likely outcomes and then evaluate their predictions against the real results.

Differentiation

- Less confident students may need help to describe the effects of droughts on plants and animals, and may need support on the geographical areas in which droughts are more common.
- More confident students should be able to explain why droughts are more likely in countries such as Malawi and describe the short and long-term impacts of these droughts.

Extension ideas

- For Activity 13.1, have students list as many control variables as they can and explain how each can be (or was) controlled. Students then evaluate their results and highlight anything they did not expect, and include in their conclusion a statement about what other variables they might have controlled.

Homework suggestions

Students answer questions 1, 2, 6 and 7 in Topic 13.2 in the Workbook.

13.2c Floods

Students will learn:

- To describe flooding as an impact of climate change
- To evaluate the reliability of secondary information
- To describe trends and patterns in results, identifying and explaining any anomalous results
- To make conclusions by interpreting results

Glossary words

flood defences

Resource preparation

- Figure 13.10 from the Student's Book
- Worksheets 13.2c(1) and 13.2c(2)

Teaching and learning ideas

- Use figure 13.10 from the Student's Book as a starting point for a discussion about flooding. Students consider what the picture shows and what has caused this to happen.
- Explain how flooding from rivers is usually caused by extremely heavy rainfall, which can be over a short period of time or a longer period, and this increases the volume of water in a river until it rises above its banks.
- Explain that flooding can also occur if there is insufficient drainage, for example if the drains in a town or city are overwhelmed by the amount of water flowing through them.
- Ask students to consider that weather is in part determined by the amount of energy in the atmosphere, and that a rising global average temperature causes there to be more energy stored in the atmosphere.

- As a class, ask students to suggest what might happen to weather patterns and extreme weather events if the atmosphere stores more energy (more frequent and more severe).
- Ask students to complete Worksheet 13.2c(1) on rainfall and flooding. This shows rainfall by month in a particular area. Students are asked to identify whether there are anomalous readings and whether or not high readings should be included or excluded from calculations of average rainfall. It also asks students to comment on possible flood defences and their effectiveness.

Differentiation

- For Worksheet 13.2c(1), before answering the questions explain to students that the city was flooded in March 2019. Ask students to identify the amount of rainfall in March (171 mm) and explain that although the rainfall value is unusually high, it is not an anomalous reading. Less confident students may also need assistance in calculating the average: write the calculation on the whiteboard so students can follow along with calculators. More confident students may just be able to work through Worksheet 13.2c(1) without support.

Extension ideas

- Students could discuss the case study in Worksheet 13.2c(2), where the effects of floods and the construction of flood defences have had some intended and some unintended consequences.

13.2d Sea level change and sea flooding

Students will learn:

- To describe sea level changes as impacts of climate change
- To evaluate the reliability of secondary information
- To present and interpret observations and measurements appropriately

Resource preparation

For Activity 13.2: Student's Book and online access to maps of Shanghai, China

For Activity 13.2: Worksheet 13.2d(2) for differentiation

Practical notes 13.2d

Worksheet 13.2d(1)

Teaching and learning ideas

- Explain how rising sea levels and sea flooding may occur at different times to periods of extreme rainfall.
- Ask students to suggest why the risk of sea flooding in many places around the Earth is increasing by completing Worksheet 13.2d(1). (Human activities are introducing more carbon dioxide into the atmosphere, causing the average global temperature to rise, which in turn causes ice at the poles and in glaciers to melt more rapidly, and this excess water increases sea levels globally.)
- Students complete Activity 13.2 using the Student's Book, which involves students suggesting why Shanghai is at increasing risk of flooding and how the city can work towards reducing the impact of sea flooding.
- Bring the class back together to discuss the answers to Activity 13.2, in particular to share the range of ideas suggested to help defend Shanghai from the effects of sea flooding.
- Challenge students to answer the question: why is a coastal city at more risk of flooding at certain times of each month? (Spring tides.)

Differentiation

- For Activity 13.2, less confident students use Worksheet 13.2d(2) to provide scaffolding for their research and answers. More confident students could just use Activity 13.2 from the Student's Book.

Extension ideas

- Students suggest all the possible causes of flooding in Shanghai, by considering not just rising sea levels and spring high tides, but also river flooding. Students predict the combination of events that put a city most at risk of flooding.
- Students in pairs prepare and give a short presentation in which they explain why countries working together to reduce carbon dioxide emissions will reduce the effects of sea flooding in the long term.

Homework suggestions

Students answer question 5 in Topic 13.2 in the Workbook.

13.2e Other extreme weather events

Students will learn:

- To describe extreme weather events as impacts of climate change
- To sort, group and classify events as changes in climate or as extreme weather events
- To evaluate the reliability of secondary information sources

Resource preparation

For Activity 13.3: Student's Book and online resources

Practical notes 13.2e

Teaching and learning ideas

- Start with a class brainstorming session, where students suggest examples of extreme weather events other than the ones already discussed in this section and comment on what features of these events are most likely to cause damage/affect people and natural habitats. Examples include hurricanes and typhoons (very high winds, intense rainfall, storm surges); heat waves (very high temperatures that threaten life); cold spells (very low temperatures that threaten life); tornadoes (very high winds); wild fires (linked to drought and heat waves, may threaten life and destroy habitats).
- Ask students to answer the Making links question to classify events or patterns as extreme weather events or as changes in climate more generally.
- Students complete Activity 13.3 as a case study and research investigation into Hurricane Katrina in 2005, using online resources as sources of information.
- Students comment on the reliability of the information they have used in Activity 13.3.

Differentiation

- For Activity 13.3, pre-select data and direct less confident students to use them as their main reference (such as the National Geographic magazine website and the 'Hurricanes in History' section of the US National Oceanic and Atmospheric Administration). More confident students could be challenged to find their own resources.

Extension ideas

- Students could research why a hurricane or typhoon may also produce a 'storm surge', where sea levels rise far above normal high tide or spring tide levels.
- Students could identify another extreme weather event to research instead of or as well as Hurricane Katrina.

Homework suggestions

Students answer questions 3, 4, 8, 9 and 10 in Topic 13.2 in the Workbook.

Chapter 13 Checking students' progress

End of chapter reflection

- To review the learning points of the chapter
- To test understanding through answering questions

Resources

- Student's Book [End of chapter review questions]
- Workbook [Self-assessment grid, End-of-chapter questions]
- Worksheet 13.3

Approach

Use the Key facts and Check your skills progress boxes in the Student's Book for each topic in this chapter as a way of recapping on the main learning points.

Ask students to complete the Self-assessment checklist at the end of Chapter 13 in the Workbook. Encourage students to work together in small groups to discuss which aspects of the material are the most difficult to understand, and then have a class vote. You may wish to revisit some Learning Objectives again if students find any areas particularly challenging. Use the Student's Book text and questions to revisit these aspects. Go through the answers orally in class. Take note of whether giving the answer appears to resolve students' uncertainty or not, and work through further examples.

The Teacher's comments in the Workbook can be completed to give students personalised feedback on next steps. Students could be asked to answer specific End of chapter review questions in their books. Alternatively go through some or all of the End of chapter review questions in the Student's Book orally in class.

Use the Context worksheet (Worksheet 13.3) as a framework for students to conduct their own research enquiries about extreme weather events in their country or local area, and to assess what effects of climate change are likely to affect their country or local area most.

Finish by asking students to complete the End-of-chapter questions at the end of Chapter 13 in the Workbook. Provide feedback on students' answers, and work with students individually to identify areas that still need more work.

Chapter 14 Astronomy

Chapter overview

This chapter provides opportunities for students to sort, group and classify information, describe how models reflect scientific evidence and can change, and evaluate the strength of evidence and how it supports or contradicts a prediction.

Students will first explore three different models of how the Moon formed and discuss evidence for and against each, before going on to investigate in more detail the collision hypothesis, which is currently the most widely accepted model. They then learn about other asteroid collisions with Earth and their effects and consider the evidence for an asteroid collision causing the mass extinction of 75% of the Earth's living species, including the dinosaurs.

Students then learn about nebulae (clouds of interstellar dust and gas) and how they arise from massive star explosions, and how new stars may form from some nebulae. The chapter concludes with a brief activity in which students build up a life cycle for a very large star.

Topic title	Number of 40-minute periods	Learning Objectives	Thinking and Working Scientifically Learning Objectives
14.1 Collisions	3	9ESs.02 Describe the evidence for the collision theory for the formation of the Moon 9ESs.01 Describe the consequences of asteroid collision with the Earth, including climate change and mass extinctions	9TWSm.02 Describe some important models, including analogies, and discuss their strengths and limitations 9TWSm.01 Understand that models and analogies reflect current scientific evidence and understanding and can change 9TWSa.01: Evaluate the strength of the evidence collected and how it supports, or refutes, the prediction 9TWSa.03 Make conclusions by interpreting results, explain the limitations of the conclusions and describe how the conclusions can be further investigated 9TWSc.01 Sort, group and classify phenomena, objects, materials and organisms through testing, observation, using secondary information, and making and using keys
14.2 Observing the Universe	2	9ESs.03 Know that nebulae are clouds of dust and gas, and can act as stellar nurseries	9TWSm.01: Understand that models and analogies reflect current scientific evidence and understanding and can change 9TWSp.01 Suggest a testable hypothesis based on scientific understanding 9TWSp.04 Plan a range of investigations of different types to obtain appropriate evidence when testing hypotheses 9TWSa.01 Evaluate the strength of the evidence collected and how it supports, or refutes, the prediction 9TWSa.03 Make conclusions by interpreting results, explain the limitations of the conclusions and describe how the conclusions can be further investigated 9TWSc.01 Sort, group and classify phenomena, objects, materials and organisms through testing, observation, using secondary information, and making and using keys 9TWSp.03 Make predictions of likely outcomes for a scientific enquiry based on scientific knowledge and understanding 9TWSp.05 Make risk assessments for practical work to identify and control risks 9TWSc.05 Carry out practical work safely, supported by risk assessments where appropriate 9TWSa.04 Evaluate experiments and investigations, including those by others, and suggest improvements, explaining any proposed changes
End of chapter questions	1	Check students' progress	

Background science

This chapter covers different hypotheses for the formation of the Moon, including the currently accepted model in which an object roughly the same size and mass as the planet Mars collided with the Earth soon after the Earth had formed, and large pieces of rock from the asteroid and from the Earth were pulled together due to the effects of gravity to form the Moon we know today. The evidence for this collision theory comes in part from the Apollo missions which brought samples of rock and soil from the Moon back to Earth. Analysis of these samples showed similarities in the composition of the Moon rocks and those found on Earth.

Analysis also showed that the minerals contained in the Moon rock were likely to have been formed at high temperatures during the collision of the two objects. Students should be encouraged to consider why the evidence given supports or contradicts one or more of the hypotheses and to apply their understanding of asteroids gained in Stage 8 Chapter 12.

An asteroid collision with Earth is considered, one that is thought to have caused mass extinctions around 65 million years ago. Evidence for this comes mainly from the fossil record, which shows that many species became extinct over a relatively short time period, and the discovery of a partly hidden massive crater near Chicxulub in Mexico.

Evidence for how stars are formed and how very massive stars end can be found by observing clouds of interstellar dust and gas called nebulae. Most of the gas found in nebulae is hydrogen and most of the rest is helium. These are essential fuels for nuclear fusion, the reaction that takes place inside stars and releases large amounts of energy.

14.1 Collisions

Topic overview

		Student's Book	Workbook	Downloadable materials
14.1a	**Starting point** This short discussion provides an opportunity for students to recall their understanding of asteroids, where they are found and how they formed	Starting point table		
14.1b	**Ideas about how the Moon formed** Students learn about three different models for how the Moon formed and how evidence suggests one model, the collision theory, is more likely than the others	Questions 1–2 Activity 14.1	Questions 1, 3 and 10	Practical notes 14.1b Worksheet 14.1b
14.1c	**The Moon and its evidence** Students learn more about evidence supporting the collision theory and produce a sequence of diagrams to describe it	Questions 3–6 Activity 14.2	Questions 2, 5 and 8	Practical notes 14.1c
14.1d	**Asteroid collisions** Students learn about asteroid collisions with the Earth and evidence in the form of meteorites	Questions 7–9	Question 4	Worksheet 14.1d
14.1e	**The effects of asteroid collisions** Students learn about the possible effects over large areas of the Earth after an asteroid collision, including mass extinctions of organisms	Questions 10–13 Activity 14.3	Questions 6, 7 and 9	Practical notes 14.1e Worksheet 14.1e

Learning Objectives

- 9ESs.01: Describe the consequences of asteroid collision with the Earth, including climate change and mass extinctions
- 9ESs.02: Describe the evidence for the collision theory for the formation of the Moon
- 9TWSc.01: Sort, group and classify phenomena, objects, materials and organisms through testing, observation, using secondary information, and making and using keys
- 9TWSa.01: Evaluate the strength of the evidence collected and how it supports, or refutes, the prediction
- 9TWSm.01: Understand that models and analogies reflect current scientific evidence and understanding and can change
- 9TWSm.02: Describe some important models, including analogies, and discuss their strengths and limitations
- 9TWSa.03: Make conclusions by interpreting results, explain the limitations of the conclusions and describe how the conclusions can be further investigated

Links to other topics

- Earth and Space Stage 7 Chapter 10 Topic 10.1 The Earth's crust

Using models

- This part of the topic provides the opportunity to address learning objectives 9TWSm.01 and 9TWSm.02.
- Discussing three different models put forward as explanations of how the Moon formed and considering the evidence for and against them. These three models are discussed in the Student's Book. The first idea is known as the **capture hypothesis**, which suggests that a rocky object such as an asteroid was pulled into orbit around the Earth due to the gravitational force of the Earth. The second idea is the **co-formation hypothesis**, which suggests that the Moon formed at the same time as the Earth. The third idea is the **collision hypothesis**, which suggests that a large asteroid crashed into the Earth and split the mass of the rock into two objects - the Earth and the Moon.
- Investigating in more detail the collision model for the formation of the Moon.
- Discussing how scientists collect data about previous asteroid collisions and use this to inform and improve mathematical models that predict which asteroids may one day collide with Earth, when this may happen and what the effects could be. This activity allows students to meet learning objective 9TWSm.01 where they can discuss the way in which the change in the nature of the data collected allows the models to be altered.

Science in context

Teaching ideas

Look out for the opportunities to discuss how teams of scientists gather evidence from past asteroid collisions and build computer simulations for predicting the outcome of future collisions. There are also international teams dedicated to finding and tracking all asteroids that may approach Earth in future.

Common misunderstandings

- Some students may think the reason we can only ever see one side of the Moon from Earth is because the Moon does not spin on its axis. Actually the Moon does spin, but the time it takes to spin one complete rotation is exactly matched to the time it takes to orbit the Earth once. This is due to the stability of the orbit over a very long period of time.
- Some students may think that we can only observe the Moon at night, but the Moon is often visible during the day. It can be a useful exercise to investigate this by going outside during a science lesson at an appropriate time to see the Moon.
- Some students may think that the side of the Moon we cannot see from Earth is always dark, but the Sun always illuminates one half of the Moon, and as the Moon spins on its axis and orbits the Earth, so the illuminated half changes. The best way to explain this is to consider when we see a 'new' Moon, and the side facing Earth is only very dimly lit, then the far side of the Moon is fully lit by the Sun.

14.1a Starting point

Review students' understanding of the Solar System and how it formed. Include a discussion of asteroids and how they formed at the same time as the planets in the Solar System, and explore with the students what asteroids are made of and the evidence we have for these things.

Teaching and learning ideas

Encourage students to discuss between themselves the following tasks/questions:

- What caused the planets and the Sun to form from the solar nebula?
- When did asteroids form?
- What are asteroids formed from?
- How have scientists found evidence for these things?

14.1b Ideas about how the Moon formed

Students will learn:

- To describe the evidence for the collision theory for the formation of the Moon
- To describe how there are different models for the formation of the Moon
- To understand that models and analogies can change according to scientific evidence
- To describe the uses, strengths and limitations of models and analogies

- To evaluate how well the prediction is supported by the evidence collected
- To interpret results and form conclusions using scientific knowledge and understanding

Glossary words

capture hypothesis, co-formation hypothesis, collision hypothesis

Resource preparation

For Activity 14.1: Worksheet 14.1b for evaluating evidence about the formation of the Moon, Student's Book, scissors, large sheet of paper

Practical notes 14.1b

Teaching and learning ideas

- Explain to students that analysis of rocks from the Moon has told us that the Moon formed at least 100 million years after the Sun and planets in the Solar System formed.
- Use the statements in the Student's Book of the three hypotheses for the formation of the Moon to conduct a quick vote by showing hands. Students predict which hypothesis they think is most likely to be correct – based only on what they know about the Moon already.
- Select or have three volunteer pairs of students. Give each pair one of the hypotheses for the formation of the Moon. Have the class conduct a debate, where each pair uses ideas from the Student's Book, plus their own ideas, to argue the case for 'their' hypothesis.
- Use this as an opportunity to remind students of the difference between scientific evidence (facts that are measurable) and opinions (statements of personal preference). Encourage students to make notes during the debate to decide which statements made by the debating pairs of students are evidence and which are opinions.
- Following the debate, ask students to vote on the hypothesis that is most likely to be correct, this time based only on the evidence they have heard.
- Ask students to complete Activity 14.1 using Worksheet 14.1b. Students cut out a set of statements that are either hypotheses or evidence; students should sort the statements (i) into the hypotheses and evidence then (ii) decide which pieces of evidence support or contradict each hypothesis.
- Conclude by summarising that the evidence students have considered, plus many other pieces of evidence, have led scientists to believe that the collision hypothesis is the most likely process by which the Moon formed. (Students may reach the same or a different conclusion – reassure students that their conclusions, if based on evidence, are valid, even if they do not reach the same conclusion as is currently accepted; there is much more evidence that there is not time to discuss in the lesson.)

Differentiation

- You could draw a timeline to assist less confident students in explaining how and when the Moon formed compared to the rest of the Solar System, so that they see how asteroids and planets formed before the Moon.
- For Activity 14.1, identify the three hypotheses from among the different statements as a class. At the front of the class, set up a version of the table the students should use in the activity, including the three hypotheses pre-installed in the correct positions, as scaffolding for less confident students.
- For Activity 14.1, you could also have Worksheet 14.1b printed out on a large sheet of paper, and complete the whole table as a class with volunteer students being given a piece of evidence and deciding where to place it.
- More confident students can complete Activity 14.1 alone.

Extension ideas

- Have students discuss moons of the other planets, such as of Jupiter. Most moons of Jupiter are thought to have been captured and pulled into an orbit by Jupiter. Students should suggest reasons why this might be, and/or conduct online research into the shapes of Jupiter's moons. (Many are irregularly shaped, suggesting they were once asteroids but have been pulled into orbit around Jupiter by the gravitational effects of Jupiter's very large mass.)

Homework suggestions

Students answer questions 1, 3 and 10 in Topic 14.1 in the Workbook.

14.1c The Moon and its evidence

Students will learn:

- To describe the theory for the Moon formation known as collision theory and its evidence
- To understand that models and analogies can change according to scientific evidence

- To describe the uses, strengths and limitations of models and analogies
- To evaluate how well the prediction is supported by the evidence collected
- To interpret results and form conclusions using scientific knowledge and understanding

Glossary words

collision theory, crater, robotic, theory

Resource preparation

Displayed image of the Moon such as figure 14.1 in the Student's Book

For Activity 14.2, for each pair of students: a large sheet of plain paper, a ruler, coloured pens and pencils

Practical notes 14.1c

Teaching and learning ideas

- Display an image of the Moon (such as figure 14.1 in the Student's Book). Ask students to point out features they can see and identify them.
- Ask students to suggest what has caused the circular, or nearly circular, depressions in the surface (collisions with small asteroids).
- Also comment on how the Moon is almost exactly spherical (round), unlike many asteroids.
- Ask students to form small groups in which they conduct a brainstorming session. Students list the features they have observed from the Moon image, along with other evidence discussed in section 14.1b, and suggest reasons for this evidence. Prompt groups with questions they need to answer, such as 'is there an atmosphere on the Moon?', 'are there volcanoes on the Moon?', 'is there water on the Moon?'.
- Gather the class and note down their ideas. Include short discussions of the questions posed and what reasons have been suggested.
- Ask the class how they suggest evidence can be collected to support or contradict these ideas. (Space probes, robotic landers, people landing on the Moon and bringing back rock samples, etc.)
- Optional: if time permits, set students a mini-project to collect information about the Apollo 11 Moon landing. Set the questions: what were the risks to the astronauts and what precautions were taken to reduce those risks?
- Ask students to complete Activity 14.2, in which students working in pairs produce a poster that shows the stages in the formation of the Moon according to the collision theory and includes reference to evidence that supports it.

Differentiation

- For less confident students, provide additional support by showing on the whiteboard a labelled map of the Moon's surface as seen from Earth. This will highlight certain features, particularly the collision craters.
- For the brainstorming session, produce prompt cards with key pieces of evidence written on them; each group of students selects and displays the card they think matches the prompt questions.
- For less confident students during Activity 14.2, include items of evidence written on prompt cards that the students can match to the relevant parts of their poster. More confident students should be encouraged to find their own pieces of evidence in addition to any provided.

Extension ideas

- Explain to students that the darker grey, smoother areas of the Moon's surface were thought at one time to be oceans. Ask students to suggest how this was proved to be wrong (better telescopes, and of course landing probes and people on the Moon). Challenge students to suggest how these areas of darker, smooth rock may have formed. (Early in its history, the Moon had molten rock on or just beneath its surface that flowed to fill large impact craters, then cooled gradually to form solid rock.)

Homework suggestions

Students answer questions 2, 5 and 8 in Topic 14.1 in the Workbook.

14.1d Asteroid collisions

Students will learn:

- To describe how an asteroid colliding with the Earth could cause climate change and mass extinction events
- To understand that models and analogies can change according to scientific evidence
- To describe the uses, strengths and limitations of models and analogies
- To evaluate how well the prediction is supported by the evidence collected

- To interpret results and form conclusions using scientific knowledge and understanding
- To identify hazards and identify ways to minimise risks
- To carry out practical work in a safe manner to minimise risks
- To evaluate experimental methods, explaining any improvements suggested

Glossary words

meteor, meteorite

Resource preparation

Displayed image of Meteor Crater, Arizona, United States, such as figure 14.3 in the Student's Book

Worksheet 14.1d for describing an experiment to model asteroid collisions

Teaching and learning ideas

- Show an image of Meteor Crater, Arizona, United States, such as figure 14.3 in the Student's Book. Ask students to describe its shape. Explain that this is where the remains of an asteroid hit the ground and exploded.
- Describe the dimensions (see the Student's Book) to give students a sense of scale.
- Describe the difference between an asteroid, a meteor and a meteorite. Ask students to draw a diagram to show how an asteroid enters the Earth's atmosphere before colliding with the surface.
- Ask students to suggest an explanation why only small parts of meteors end up reaching the Earth's surface.
- Ask students to complete Worksheet 14.1d, which describes a basic method for an experiment to model the impact of an asteroid. Students evaluate potential risks, predict likely outcomes and use their predictions to suggest improvements to the experiment. *Note: students should not carry out the experiment.*

Differentiation

- For Worksheet 14.1d, pre-prepare some multiple-choice answers for less confident students to the likely risks, safety precautions and improvements to the experiment. More confident students should be encouraged to think of these for themselves.

Extension ideas

- For Worksheet 14.1d, challenge students to suggest different materials to model the Earth's surface and describe up to six limitations of the model.

Homework suggestions

Students answer question 4 in Topic 14.1 in the Workbook.

14.1e The effects of asteroid collisions

Students will learn:

- To describe how an asteroid colliding with the Earth could cause climate change and mass extinction events
- To understand that models and analogies can change according to scientific evidence
- To describe the uses, strengths and limitations of models and analogies
- To use observations, measurements, secondary sources of information and keys, to organise and classify organisms, objects, materials, or phenomena
- To use scientific knowledge to make predictions
- To classify asteroid collisions according to their effects using secondary information

Glossary words

ecosystem, mass extinction, shock wave

Resource preparation

For Activity 14.3: Worksheet 14.1e for differentiation

Practical notes 14.1e

Teaching and learning ideas

- Ask students to describe what they think would be the effects of an asteroid colliding with the Earth's surface. Suggest that they should think about the effects on the Earth's crust and compare them to the effects of earthquakes and volcanoes.
- Explain how very large amounts of dust are thrown up into the atmosphere after an asteroid collision. Ask students to suggest why this can cause climate change.
- Ask the students to form groups of three or four and discuss ideas why there are no longer any dinosaurs on Earth. If necessary, prompt groups with statements such as 'dinosaurs were reptiles which can only survive in warmer climates' and 'nearly all the energy that living things use comes originally from sunlight'.
- Describe the evidence that has been found for the large meteor crater near Chicxulub in Mexico; display a map or image of the area from space if available. Encourage students to suggest how large the asteroid that caused the impact might have been and its effects. *(Note: it is not certain that an asteroid caused this impact, there are also ideas that this might have been the nucleus of a comet; however, as students will not have covered comets it is best to refer only to the idea that the impact was of an asteroid.)*
- Ask students to suggest what they think the term 'mass extinction' means. Try to convey a sense of scale: there are currently estimated to be 8.7 million species of living things on Earth. A mass extinction on the scale of the Chicxulub asteroid impact would mean only just over 2.2 million species left.
- Ask students to complete Activity 14.3, in which they use secondary information to predict the effects of collisions with different sizes of asteroids.

Differentiation

- For Activity 14.3, use the scaffolded Worksheet 14.1e for less confident students. More confident students do not need to use Worksheet 14.1e.

Extension ideas

- For Activity 14.3, give students the statement 'the asteroids that enter the Earth's atmosphere most often are the asteroids that are most difficult to observe and predict their movements'. Challenge students to say whether they agree with this statement and explain their answer. (Table 14.2 shows that the smaller the asteroid, the larger the chance there is of one entering the Earth's atmosphere/colliding with Earth. However, the smallest asteroids are very difficult to observe and predict their movements because they only reflect a little light.)

Homework suggestions

Students answer questions 6, 7 and 9 in Topic 14.1 in the Workbook.

14.2 Observing the Universe

Topic overview

		Student's Book	Workbook	Downloadable materials
14.2a	**Starting point** Students describe what they remember of how the Solar System formed from a cloud of interstellar dust and gas	Starting point table		Practical notes 14.2b Worksheet 14.2b
14.2b	**Nebulae** Students learn that a large percentage of the matter in the Universe consists of dust and gas, and that this matter is often gathered in nebulae	Activity 14.4	Questions 1, 4, 5, 6, 9 and 10	Practical notes 14.2c Worksheet 14.2c
14.2c	**Star formation** Students learn how new stars are formed from nebulae	Activity 14.5	Questions 2, 3, 7 and 8	

Learning Objectives

- 9ESs.03: Know that nebulae are clouds of dust and gas, and can act as stellar nurseries
- 9TWSm.01: Understand that models and analogies reflect current scientific evidence and understanding, and can change
- 9TWSp.01: Suggest a testable hypothesis based on scientific understanding
- 9TWSp.04: Plan a range of investigations of different types to obtain appropriate evidence when testing hypotheses
- 9TWSa.01: Evaluate the strength of the evidence collected and how it supports, or refutes, the prediction
- 9TWSa.03: Make conclusions by interpreting results, explain the limitations of the conclusions and describe how the conclusions can be further investigated
- 9TWSc.01: Sort, group and classify phenomena, objects, materials and organisms through testing, observation, using secondary information, and making and using keys
- 9TWSp.03: Make predictions of likely outcomes for a scientific enquiry based on scientific knowledge and understanding
- 9TWSp.05: Make risk assessments for practical work to identify and control risks
- 9TWSc.05: Carry out practical work safely, supported by risk assessments where appropriate
- 9TWSa.04: Evaluate experiments and investigations, including those by others, and suggest improvements, explaining any proposed changes

Links to other topics

- Earth and Space Stage 7 Chapter 11 Topic 11.1 The planets and the Solar System

Using models

- Computer modelling is used in all areas of astronomy to understand how stars, galaxies, nebulae and other phenomena are formed and evolve. In Activity 14.4, students are asked to suggest how such computer modelling could be used to make predictions.
- Models of the life cycles of different sizes of stars have been made and improved by astronomers analysing the evidence from images, radio signals and other measurements taken by telescopes. Evidence supporting or modifying these models has improved over time as the technology of telescopes has been improved. Although the life cycles last billions of years, we are able to observe every stage in a star's life because there are so many stars we can observe.

Science in context

Teaching ideas

Look out for opportunities to discuss how new astronomical telescopes are being built around the world on high mountains (to avoid atmospheric distortion) and with very large, perfect mirrors so as to observe very dim and distant objects including nebulae.

Common misunderstandings

- Many people think that all nebulae are the same, but there are many types. Even the name is occasionally misleading: before telescope technology had developed sufficiently to produce clear images of distant objects, anything that looked diffuse or cloud-like was called a nebula. We now know that many of these objects called 'nebulae' are in fact galaxies containing hundreds of millions of stars. 'True' nebulae are collections of interstellar dust and gas, but there are several types including:
  - diffuse nebulae that are spread over very large areas
  - emission nebulae containing gases ionised by a nearby star, which emit light in several colours
  - planetary nebulae, which again have a misleading name, as they are rarely connected with planets in any way; they are formed from stars not much larger than our Sun
- Some students may think that all stars are the same and follow identical life cycles. In fact, stars vary in several ways:
  - in size, from dwarf stars that are not much larger than the planet Jupiter, up to so-called 'hypergiant stars' such as VY Canis Majoris, which has a diameter 2600 times that of the Sun
  - in colour from red through yellow, white and blue, depending on age, size and the elements they contain
  - in brightness
  - in age – some are recently formed, many are several billions of years old; the shortest lifespan stars exist for about 50 million years.
- Some students may think that all stars end as supernovae. In fact, only a small percentage of stars, which are massive (many times the size of our Sun) will form supernovae. Most stars will eventually run out of nuclear fuel (hydrogen and helium) and expand, but then gradually shrink to form white dwarf stars.

14.2a Starting point

Resource preparation

Displayed image of a nebula such as in figure 14.6b in the Student's Book (the Omega Nebula)

Teaching and learning ideas

- Display an image of a nebula and ask students to describe what they observe (stars, bright and dark clouds of dust).
- Students discuss some prompt questions to link the image with a review of what they have learnt about the formation of the Solar System including the Sun, such as:
  - What kind of matter do they think is shown as a whitish-pink cloud? (Interstellar gas reflecting light from stars – specifically, hydrogen gas)
  - What kind of matter shows as darker clouds? (Interstellar dust blocking light from reaching us)
  - How did our Sun form and what did it form from? (a cloud of interstellar gas drawn together due to the effects of gravity)

14.2b Nebulae

Students will learn:

- To understand how stars can form from the clouds of dust and gas known as nebulae
- To understand that models and analogies can change according to scientific evidence
- To use scientific knowledge and understanding to suggest a testable hypothesis
- To plan investigations used to test hypotheses

Glossary words

matter, nebula

Resource preparation

Image of a nebula that mostly appears pink or red (e.g. figures 14.5a or 14.5b in the Student's Book)

For Activity 14.4: Image of two nebulae, one that is clearly expanding outwards from a central point and one that contains a number of young, bright blue or white stars (for example, figures 14.6a and b in the Student's Book); Worksheet 14.2b

Practical notes 14.2b

Teaching and learning ideas

- Display an image of a nebula that appears mostly pink or red.
- Explain that the pink colour is due to hydrogen gas reflecting light from stars.
- Ask students to form groups of three or four and agree how a star forms from interstellar gas. Use prompt questions to assist groups, such as: 'What is the force that pulls interstellar gas together?', 'How did the Sun form?'.
- Explain that some massive stars explode and produce nebulae.
- Ask students to complete Activity 14.4, where they examine two different types of nebulae and classify them as either recently formed from an exploding massive star or areas where new stars are forming.
- Ask students to research more different types of nebulae using books and online sources. Worksheet 14.2b provides instructions and guidance. Students should prepare a short presentation on their findings and evaluate the reliability and accuracy of the information sources they have used.

Differentiation

- For Activity 14.4, research and pre-select up to eight images of different nebulae for students to classify as either having been formed from an exploding massive star or be in the process of forming new stars. This should provide further support for less confident students in understanding the differences between the two types of nebulae. More confident students could be encouraged to find their own examples.
- Less confident students could be provided with examples of possible sources of information such as books or websites. More confident students should be encouraged to find their own.

Extension ideas

- For Activity 14.4, Worksheet 14.2b, challenge students to find out: (1) where the name 'nebula' originates (2) whether all astronomical objects with 'nebula' in their names are in fact nebulae and (3) whether a 'planetary nebula' contains planets.

Homework suggestions

Students answer questions 1, 4, 5, 6, 9 and 10 in Topic 14.2 in the Workbook.

14.2c Star formation

Students will learn:

- To evaluate how well the prediction is supported by the evidence collected
- To understand that models and analogies can change according to scientific evidence
- To understand how stars can form from the clouds of dust and gas known as nebulae

Glossary words

nuclear fusion, red supergiant, supernova

Resource preparation

Display image of star-forming nebula (for example figure 14.7 in the Student's Book)

Two foam or sponge balls to represent hydrogen nuclei

For Activity 14.5: Worksheet 14.2c for differentiation

Practical notes 14.2c

Teaching and learning ideas

- Display an image of a star-forming nebula such as figure 14.7 of the Eagle Nebula in the Student's Book. Explain that this contains interstellar gas and dust that is being pulled together by the force due to gravity, to form new stars.
- Explain that eventually hydrogen atoms are pulled so closely together they overlap. Use two foam balls to demonstrate the two nuclei of the hydrogen atoms pulling closer together, then eventually squashing together.
- Describe how the two nuclei merge to form a nucleus of helium, and in the process they release a very large amount of energy. Explain that this is how a star produces light.
- Ask students to work in pairs to complete Activity 14.5, where they prepare a model of a life cycle of a massive star.
- Emphasise the cyclical nature of massive stars: they form from nebulae, carry out the process of nuclear fusion until all the hydrogen and helium has been used up, expand to a red supergiant and then collapse to form an explosive supernova, after which an expanding cloud of gas and dust is left behind that eventually forms more stars.

Differentiation

- For Activity 14.5, use Worksheet 14.2c where less confident students sort images of the life cycle of a star into the correct order, then label each stage (nebula, massive star, red supergiant, supernova). More confident students could be encouraged to produce their own timelines.

Extension ideas

- For Activity 14.5, as a follow-up, students could research and prepare a life cycle for the Sun, where the later stages are the Sun expanding to form a red/orange star, then shrinking to form a white dwarf.

Homework suggestions

Students answer questions 2, 3, 7 and 8 in Topic 14.2 in the Workbook.

Chapter 14 Checking students' progress

End of chapter reflection

- To review the learning points of the chapter
- To test understanding through answering questions

Resources

- Student's Book [End of chapter review questions]
- Workbook [Self-assessment grid, End-of-chapter questions]
- Worksheet 14.3

Approach

Use the Key facts and Check your skills progress boxes in the Student's Book for each topic in this chapter as a way of recapping on the main learning points.

Ask students to complete the Self-assessment checklist at the end of Chapter 14 in the Workbook. Encourage students to work together in small groups to discuss which aspects of the material are the most difficult to understand, and then have a class vote. You may wish to revisit some Learning Objectives again if students find any areas particularly challenging. Use the Student's Book text and questions to revisit these aspects. Go through the answers orally in class. Take note of whether giving the answer appears to resolve students' uncertainty or not, and work through further examples.

The Teacher's comments in the Workbook can be completed to give students personalised feedback on next steps. Students could be asked to answer specific End of chapter review questions in their books. Alternatively go through some or all of the End of chapter review questions in the Student's Book orally in class.

Use the Context worksheet (Worksheet 14.3) as a framework for students to conduct their own research enquiries about past or possible future asteroid collisions with Earth and the Moon. Students should comment on the variety of sources of information they found and how good those sources were.

Finish by asking students to complete the End-of-chapter questions at the end of Chapter 14 in the Workbook. Provide feedback on students' answers, and work with students individually to identify areas that still need more work.

Planning guide

Number of 40-minute periods	Student's Book topic number	Learning episode	Learning and thinking and working scientifically focus	Downloadable lesson resources
Biology				
3–4	1.1 Photosynthesis	1.1a Starting point		
		1.1b Leaf structure and photosynthesis	• To describe how plants make their own food, using photosynthesis • To explain why plants need a continuous supply of water • To explain how leaves are adapted for photosynthesis • Make careful observations • Decide what equipment is required to carry out an investigation or experiment and use it appropriately.	Practical notes 1.1b Worksheet 1.1b(1), Worksheet 1.1b(2)
		1.1c Investigating the formation of oxygen	• To describe how plants make their own food, using photosynthesis • To explain why plants need a continuous supply of water • To explain how leaves are adapted for photosynthesis • Make careful observations	Practical notes 1.1c Worksheet 1.1c
		1.1d Investigating the need for chlorophyll or light	• To describe how plants make their own food, using photosynthesis • To explain why plants need a continuous supply of water • To explain how leaves are adapted for photosynthesis • Plan investigations to test ideas • Make predictions using scientific knowledge and understanding • Identify hazards and design ways to control the risk of harm to themselves and others • Identify important variables; choose which variables to change, control and measure • Use a range of equipment correctly • Discuss explanations for the results using scientific knowledge and understanding; communicate these clearly to others • Compare results with predictions	Practical notes 1.1d, 1.1e Worksheet 1.1d
	1.2 Transport of water and mineral salts in plants	1.2a Starting point		
		1.2b Absorbing water and mineral salts	• To explain what a plant needs water and mineral salts for • To describe how water and mineral salts enter a plant • Use a range of equipment correctly	Practical notes 1.2b(1), 1.2b(2) Worksheets 1.1b, 1.2b

Number of 40-minute periods	Student's Book topic number	Learning episode	Learning and thinking and working scientifically focus	Downloadable lesson resources
		1.2c Transport of substances in plants	• To describe the route that water takes through a plant • Select ideas and turn them into a form that can be tested • Plan investigations to test ideas • Use a range of equipment correctly • Present results as appropriate in tables and graphs • Take appropriately accurate measurements • Discuss explanations for results using scientific knowledge and understanding; communicate these clearly to others	Practical notes 1.2c(1), 1.2c(2) Worksheets 1.2c(1), 1.2c(2)
		1.2d The importance of mineral salts	• To explain what a plant needs mineral salts for • Use a range of equipment correctly • Suggest a testable hypothesis • Make predictions • Evaluate experiments	Practical notes 1.2d Worksheet 1.2d
	End of chapter questions		• Review the learning points of the topic • Test understanding through answering questions	Worksheet 1.3
2	2.1 Producing Urine	2.1a Starting point		
		2.1b The excretory system	• Describe the structure of the human excretory (renal) system • Describe how the kidneys filter blood to remove urea, which is excreted in urine. • Identify and develop analogies and comment on their strengths and weaknesses	Practical notes 2.1b Worksheet 2.1b
		2.1c Kidney problems	• Describe the structure of the human excretory (renal) system • Describe how the kidneys filter blood to remove urea, which is excreted in urine. • Describe examples where scientists' unexpected results from enquiries have led to improved scientific understanding. • Plan a range of investigations of different types to obtain appropriate evidence when testing hypotheses.	Practical notes 2.1c Worksheet 2.1c

Number of 40-minute periods	Student's Book topic number	Learning episode	Learning and thinking and working scientifically focus	Downloadable lesson resources
			• Make risk assessments for practical work to identify and control risks. • Carry out practical work safely, supported by risk assessments where appropriate. • Make conclusions and describe how the conclusions can be further investigated.	
1	End of chapter questions		• To review the learning points of the chapter • To test understanding through answering questions	Worksheet 2.2
2–3	3.1 Variation within a species	3.1a Starting point		
		3.1b Discontinuous variation	• To identify discontinuous variation • To choose charts and graphs to display different sorts of information • Make careful observations including measurements • Present results in the form of tables, bar charts and line graphs • Make conclusions from collected data, including those presented in a graph, chart or spreadsheet • Present conclusions using different methods	
		3.1c Continuous variation	• To identify discontinuous and continuous variation, data, range • To choose charts and graphs to display different sorts of information • Make careful observations including measurements • Present results in the form of tables, bar charts and line graphs • Make conclusions from collected data, including those presented in a graph, chart or spreadsheet • Present conclusions using different methods	Practical notes 3.1c(1), 3.1c(2) Worksheets 3.1c(1), 3.1c(2)
		3.1d Causes of variation	• To identify examples of environmental and inherited variation • Present results in the form of tables	

Number of 40-minute periods	Student's Book topic number	Learning episode	Learning and thinking and working scientifically focus	Downloadable lesson resources
1	3.2 Chromosomes, genes and DNA	3.2a Starting point		Worksheet 3.2a
		3.2b Inheritance and variation	• To describe the nature of chromosomes, DNA and genes • To outline how genes control an organism's characteristics • Recall some different types of model and discuss their importance in the development of scientific ideas	
2	3.3 Fertilisation and inheritance	3.3a Starting point		
		3.3b Sexual reproduction	• To outline the role of specialised cells (gametes) in sexual reproduction • To explain how human gametes are adapted to their functions • To identify the parts and functions of the male and female reproductive systems • To explain how sex is inherited in humans (in terms of X and Y chromosomes) • Use and develop different sorts of models	Worksheet 3.3b
		3.3c Fertilisation	• To describe the parts and functions of the human reproductive system • To describe what happens during fertilisation and how a fertilised egg cell develops into an embryo • To interpret information from diagrams and secondary sources	Worksheets 3.3c(1), 3.3c(2)
		3.3d Sex determination	• To explain how sex is inherited in humans (in terms of the X and Y chromosomes) • To develop and use physical models to aid explanations of scientific ideas	
2	3.4 Fetal growth and development	3.4a Starting point		
		3.4b The developing fetus	• To describe the development of the fetus • To explain how the fetus is protected and nourished • Interpret data from secondary sources • Present and interpret results, and predict results between the data points collected.	Practical notes 3.4b Worksheets 3.4b(1), 3.4b(2)

Number of 40-minute periods	Student's Book topic number	Learning episode	Learning and thinking and working scientifically focus	Downloadable lesson resources
		3.4c The effect of diet, drugs and disease on the developing fetus	• To explain how fetal development can be affected by the health and actions of the mother • To interpret data from secondary sources • To present conclusions to others in appropriate ways	
1	3.5 Natural selection	3.5a Starting point		Worksheet 3.5a
		3.5b Adaptation	• To describe how organisms are adapted to where they live	
		3.5c Natural selection	• To state what is meant by 'evolution' • To explain Darwin's idea about how species evolve by natural selection • Discuss and explain the importance of questions, evidence and explanations, using historical and contemporary examples • Discuss the way that scientists work today and how they worked in the past, including reference to experimentation, evidence and creative thought	Practical notes 3.5c Worksheet 3.5c
1	End of chapter questions		• To review the learning points of the topic • To test understanding through answering questions	Worksheets 3.2a, 3.3b, 3.3c(1), 3.3c(2), 3.4b(1), 3.4b(2), 3.5a, 3.6
2	4.1 Populations and extinction	4.1a Starting point		Worksheet 4.1a
		4.1b Resources and adaptation	• To recall the resources that plants and animals need • To explain ways in which organisms are adapted to their ecosystems	
		4.1c Factors affecting population size	• To explain reasons why populations change • Look critically at sources of secondary data • Interpret and explain results using scientific knowledge and understanding • Present and interpret results, and predict results between the data points collected.	Worksheets 4.1c(1), 4.1c(2)

Number of 40-minute periods	Student's Book topic number	Learning episode	Learning and thinking and working scientifically focus	Downloadable lesson resources
		4.1d Extinction	• To explain reasons why species may become extinct • Look critically at secondary sources of data • Evaluate experiments and investigations, including those by others, and suggest improvements, explaining any proposed changes	Worksheet 4.1d
1	End of chapter questions		• To review the learning points of the topic • To test understanding through answering questions	Worksheets 4.1a, 4.1c(1), 4.1c(2), 4.1d, 4.2

Number of 40-minute periods	Student's Book topic number	Learning episode	Learning and thinking and working scientifically focus	Downloadable lesson resources
Chemistry				
2-3	5.1 Atomic structure and the Periodic Table	5.1a Starting point		
		5.1b Atomic structure	• To describe the structure of an atom • To compare the structures of some simple atoms • Explain why scientific models change over time	Practical notes 5.1b Worksheet 5.1b
		5.1c The Periodic Table	• To understand that the structure of the Periodic Table is related to the atomic structure of the elements. • To understand that the Periodic Table can be used to predict an element's structure.	Worksheet 5.1c
2-3	5.2 Trends in the Periodic Table	5.2a Starting point		Worksheet 5.2a
		5.2b Trends in the Periodic Table	• To understand that groups within the Periodic Table have trends in physical and chemical properties • To be able to describe the trends within group 1 as an example • Identify trends and patterns in data • Present data as a bar chart • Predict a result between the data points	Practical notes 5.2b
		5.2c Developing the Periodic Table	• To explain how the Periodic Table was developed • Understand how models reflect current scientific evidence, and can change • Sort, group and classify elements based on their properties	Worksheet 5.2c
1	End of chapter questions		• To review the learning points of the chapter • To test understanding through answering questions	Worksheet 5.3
2-3	6.1 Chemical Bonds	6.1a Starting point		Worksheet 6.1a
		6.1b Covalent bonding	• Understand that molecules are formed by covalent bonds • Describe how a covalent bond is formed • Discuss the strengths and weaknesses of a dot-and-cross diagram as a model of a molecule	

Number of 40-minute periods	Student's Book topic number	Learning episode	Learning and thinking and working scientifically focus	Downloadable lesson resources
		6.1c Ionic bonding	• Describe an ion as an atom which has lost or gained one or more electrons • Describe how an ionic bond is formed • Use models to show the structure of ions and ionic bonds • Use symbols and formulae	
2–3	6.2 Simple and giant structures	6.2a Starting point		
		6.2b Giant structures	• Describe how some elements and compounds exist in giant structures • Explain how the structure of a giant structure influences its physical properties • Use models to show the structure of giant structures	
		6.2c Simple structures	• Describe how some elements and compounds exist as simple structures • Explain how the structure of a simple structure influences its physical properties • Use models to show the structure of simple structures • Classify substances based on their structure using secondary information	
3	6.3 Density	6.3a Starting point		
		6.3b Determining the density of a regular solid	• Define and calculate density • Calculate the density of regular solids • Make observations and measurements • Decide which measurements and observations are necessary and what equipment to use	Practical notes 6.3b Worksheets 6.3b(1), 6.3b(2), 6.3b(3)
		6.3c Determining the density of an irregular solid	• Calculate the density of an irregular solid • Select ideas and produce plans for testing based upon previous knowledge, understanding and research • Decide which measurements and observations are necessary and what equipment to use	Practical notes 6.3c Worksheet 6.3c

Number of 40-minute periods	Student's Book topic number	Learning episode	Learning and thinking and working scientifically focus	Downloadable lesson resources
		6.3d Determining the density of a liquid and a gas	• Calculate and compare densities of solids, liquids and gases • Use particle model to explain the difference in densities between solids, liquids and gases • Select ideas and produce plans for testing based upon previous knowledge, understanding and research • Make observations and measurements • Draw conclusions	Practical notes 6.3d(1), 6.3d(2) Worksheet 6.3d
1	End of chapter questions		• To review the learning points of the chapter • To test understanding through answering questions	Worksheet 6.4
2-3	7.1 Changes in chemical reactions	7.1a Starting point		
		7.1b Conservation of mass and energy	• To understand that in chemical reactions mass and energy are conserved • Make predictions based on scientific understanding	Practical notes 7.1b
		7.1c Mass change when metals react with oxygen	• Understand that in chemical reactions mass and energy are conserved • Take accurate measurements, explaining why this is important • Carry out practical work safely	Practical notes 7.1c Worksheet 7.1c
1-2	7.2 Word and symbol equations	7.2a Starting point		
		7.2b Word and symbol equations	• To use word and symbol equations to describe reactions • Use symbols and formulae to represent scientific ideas	Worksheet 7.2b
4-5	7.3 Methods for making salts	7.3a Starting point		
		7.3b Reactions between metals and acids	• To describe and write word equations for the formation of chloride, nitrate and sulfate salts using reactions with acids • Use a range of equipment appropriately • Carry out practical work safely	Practical notes 7.3b Worksheets 7.3b(1), 7.3b(2)

Number of 40-minute periods	Student's Book topic number	Learning episode	Learning and thinking and working scientifically focus	Downloadable lesson resources
		7.3c Reactions between acids and carbonates	• To describe and write word equations for the formation of chloride, nitrate and sulfate salts using reactions with acids • Use a range of equipment appropriately • Carry out practical work safely • Present results as appropriate in tables	Practical notes 7.3c Worksheet 7.3c
		7.3d Making a soluble salt	• To describe how to prepare some common salts by the reactions of metals with acids and metals with carbonates • To understand how to purify salts using filtration, evaporation and crystallisation • Decide what equipment is required to carry out an investigation or experiment and use it appropriately. • Carry out practical work safely	Practical notes 7.3d Worksheet 7.3d
2–3	7.4 Displacement reactions	7.4a Starting point		
		7.4b Displacement reactions	• To give examples of displacement reactions • To use the reactivity series to predict whether a displacement reaction will occur • Describe some important models, including analogies, and discuss their strengths and weaknesses • Make predictions of likely outcomes for a scientific enquiry based on scientific knowledge and understanding • Carry out practical work safely • Describe trends and patterns in results	Practical notes 7.4b(1), 7.4b(2) Worksheets 7.4b(1), 7.4b(2), 7.4b(3)
3–4	7.5 Rates of reaction	7.5a Starting point		
		7.5b What is the rate of a reaction?	• To describe three factors that affect the rates of chemical reactions • Describe some important models, including analogies • Carry out practical work safely • Make conclusions by interpreting results	Practical notes 7.5b Worksheet 7.5b

Number of 40-minute periods	Student's Book topic number	Learning episode	Learning and thinking and working scientifically focus	Downloadable lesson resources
		7.5c Investigating the effect of temperature on rate of reaction	• To explain how temperature affects the rate of a reaction • Describe some important models, including analogies • Make predictions of likely outcomes for a scientific enquiry based on scientific knowledge and understanding • Carry out practical work safely • Describe trends and patterns in results, identifying any anomalous results and suggesting why results are anomalous • Make conclusions by interpreting results	Practical notes 7.5c Worksheet 7.5c
		7.5d Investigating the effect of concentration on rate of reaction	• To explain how concentration affects the rate of a reaction • To plan and carry out investigations into the factors that affect the rate of reaction • Describe some important models, including analogies • Make predictions of likely outcomes for a scientific enquiry based on scientific knowledge and understanding • Carry out practical work safely • Describe trends and patterns in results • Make conclusions by interpreting results	Practical notes 7.5d Worksheet 7.5d
		7.5e Investigating the effect of surface area on rate of reaction	• To explain how surface area affects the rate of a reaction • To plan and carry out investigations into the factors that affect the rate of reaction • Describe some important models, including analogies • Make predictions of likely outcomes for a scientific enquiry based on scientific knowledge and understanding • Carry out practical work safely • Describe trends and patterns in results • Make conclusions by interpreting results	Practical notes 7.5e Worksheet 7.5e
1	End of chapter questions		• To review the learning points of the topic • To test understanding through answering questions	Worksheet 7.6

<table>
<tr><th>Number of 40-minute periods</th><th>Student's Book topic number</th><th>Learning episode</th><th>Learning and thinking and working scientifically focus</th><th>Downloadable lesson resources</th></tr>
<tr><td colspan="5">Physics</td></tr>
<tr><td rowspan="3">1-2</td><td rowspan="3">8.1 Energy conservation</td><td>8.1a Starting point</td><td></td><td>Worksheets 8.1a(1), 8.1a(2)</td></tr>
<tr><td>8.1b Pendulum</td><td><ul><li>Understand that energy is always conserved: it cannot be created or destroyed</li><li>Recognise that not all energy transfers are useful</li><li>Solve problems using the principle of conservation of energy</li><li>Collect, record and summarise observations</li><li>Present and interpret results</li></ul></td><td>Practical notes 8.1b</td></tr>
<tr><td>8.1c Measuring energy changes</td><td><ul><li>Understand that energy is always conserved: it cannot be created or destroyed</li><li>Recognise that not all energy transfers are useful</li><li>Recall that the unit of energy is the joule</li><li>Solve quantitative problems using the principle of conservation of energy</li><li>Make an informed decision whether to use evidence from first-hand experience or secondary sources</li></ul></td><td></td></tr>
<tr><td rowspan="2">2</td><td rowspan="2">8.2 Heating and cooling</td><td>8.2a Starting point</td><td></td><td></td></tr>
<tr><td>8.2b Heating and cooling</td><td><ul><li>State that heating is an energy transfer and causes an increase in temperature</li><li>State that energy is transferred from a warm object to a cooler object</li><li>Describe that objects at a higher temperature than their surroundings transfer energy to their surroundings</li><li>Describe that transferring energy away from a substance causes a decrease in temperature</li><li>Explain the relationship between energy transfer and temperature change</li><li>Describe how scientific theories about thermal energy have changed over time</li><li>Make predictions</li></ul></td><td>Practical notes 8.2b(1), 8.2b(2)
Worksheet 8.2b</td></tr>
</table>

Number of 40-minute periods	Student's Book topic number	Learning episode	Learning and thinking and working scientifically focus	Downloadable lesson resources
			• Make risk assessments to identify and control risks • Carry out practical work safely • Draw conclusions and thermal energy transfer by interpreting the results of an investigation	
3-4	8.3 Conduction, convection, radiation and evaporation	8.3a Starting point		Practical notes 8.3a
		8.3b Conduction	• Use the particle model to describe how energy is transferred by conduction • Carry out practical work safely • Collect and record observations in a suitable form • Draw conclusions by interpreting results • Comment on the strengths and weaknesses of the particle model as a model for energy transfer by conduction	Practical notes 8.3b(1), 8.3b(2) Worksheet 8.3b
		8.3c Convection	• Use the particle model to describe how energy is transferred by convection • Compare the transfer of energy by conduction and convection • Make observations • Draw conclusions by interpreting results • Comment on the strengths and weaknesses of the particle model as a model for energy transfer by convection	Practical notes 8.3c(1), 8.3c(2) Worksheet 8.3c
		8.3d Radiation	• Describe how energy is transferred by radiation • Recognise that thermal energy transfer by radiation does not require a medium. • Compare the transfer of energy by conduction and radiation • Describe the effect of surface colour (black or white) and texture (dull or shiny) on the emission, absorption and reflection of radiation • Explain how transfer of thermal energy by conduction, convection and radiation can be prevented • Carry out practical work safely • Make risk assessments for practical work to identify and control risks	Practical notes 8.3d(1), 8.3d(2) Worksheet 8.3d

Number of 40-minute periods	Student's Book topic number	Learning episode	Learning and thinking and working scientifically focus	Downloadable lesson resources
			• Make observations and measurements • Draw conclusions by interpreting results	
		8.3e Cooling by evaporation	• Describe that evaporation of a liquid causes cooling of the liquid left behind • Describe evaporation in terms of the escape of more energetic molecules from the surface of a liquid • Relate evaporation to the consequent cooling of the liquid, and the cooling of a body in contact with an evaporating liquid • Explain how transfer of thermal energy by conduction, convection and radiation can be prevented • Suggest a testable hypothesis • Plan an investigation to test a hypothesis • Take accurate and precise measurements • Carry out practical work safely • Collect sufficient measurements and record them in an appropriate form • Draw conclusions by interpreting results	Practical notes 8.3e(1), 8.3e(2) Worksheet 8.3e
1	End of chapter review		• To review the learning points of the topic • To test understanding through answering questions	Worksheet 8.4
3–4	9.1 Floating and sinking	9.1a Starting point		
		9.1b Floating and upthrust	• Group materials into those which sink and those which float by testing • Recall that upthrust is a force that pushes up on objects in water or air • Explain in terms of balanced and unbalanced forces why some things float, and others sink • Make careful observations and measurements and present results in a table • Make conclusions from collected data • Suggest a testable hypothesis based on scientific understanding	Practical notes 9.1b(1), 9.1b(2) Worksheets 9.1b(1), 9.1b(2)

Number of 40-minute periods	Student's Book topic number	Learning episode	Learning and thinking and working scientifically focus	Downloadable lesson resources
		9.1c Floating and density	• Sort materials in order of density • State that an object will float in water if it is less dense than water • Describe the forces on an object immersed in water and how this affects its apparent weight • Explain how to work out if something will float by considering its density • Make predictions of likely outcomes for a scientific enquiry based on scientific knowledge and understanding • Describe trends and patterns in experimental results • Draw conclusions from data	Practical notes 9.1c, Worksheet 9.1c(1), 9.1c(2)
		9.1d Floating in air	• Identify the forces on buoyant object • Describe how the density of the gas in a balloon determines whether it will float or sink • Make predictions based on scientific knowledge and understanding	
1	End of chapter questions		• To review the learning points of the chapter • To test understanding through answering questions	Worksheet 9.2
3	10.1 Voltage and resistance	10.1a Starting point		
		10.1b Measuring voltage	• To measure voltage using a voltmeter • To know that a voltmeter must be connected in parallel across a component • To use apparatus safely and appropriately	Practical notes 10.1b
		10.1c Electrical resistance	• To calculate resistance • To investigate the effect of resistors in circuits • Carry out investigations safely • To use apparatus appropriately	Practical notes 10.1c
		10.1d Resistance, voltage and current	• Calculate resistance • Measure resistance or lengths of wire • Make risk assessments for practical work to identify and control risks • Collect and record measurements in an appropriate form	Practical notes 10.1d(1) Worksheet 10.1d(1), 10.1d(2)

Number of 40-minute periods	Student's Book topic number	Learning episode	Learning and thinking and working scientifically focus	Downloadable lesson resources
3	10.2 Measuring current in series and parallel circuits	10.2a Starting point		
		10.2b Series and parallel circuits	• Draw and make series and parallel circuits • Compare the brightness of lamps in series and parallel circuits • Investigate the effects of switches at different points in circuits • Describe and evaluate models and discuss their strengths and weaknesses	Practical notes 10.2b
		10.2c Making measurements in series and parallel circuits	• Use an ammeter to measure current in series and parallel circuits • Use a voltmeter to measure voltage in series and parallel circuits • Record observations and make conclusions	Practical notes 10.2c(1), 10.2c(2) Worksheet 10.2c(1), 10.2c(2)
1	End of chapter questions		• To review the learning points of the chapter • To test understanding through answering questions	Worksheet 10.3
2-3	11.1 Loudness and pitch	11.1a Starting point		Practical notes 11.1a(1), 11.1a(2)
		11.1b Why do sounds have different pitches and loudness?	• State what is meant by the frequency of a sound wave and relate the frequency of vibration to the pitch of sound • State what is meant by the amplitude of a sound wave and relate the amplitude of vibration to the volume of sound • State that frequency is measured in hertz (Hz) or kilohertz (KHz) • Draw and interpret waveforms for different sounds to compare their frequency and amplitude • Describe trends and patterns in results • Make conclusions by interpreting results • Describe how the movement of coils in a slinky spring is analogous to the movement of particles in a sound wave	Practical notes 11.1b

Number of 40-minute periods	Student's Book topic number	Learning episode	Learning and thinking and working scientifically focus	Downloadable lesson resources
		11.1c The frequency range of human hearing	• State that frequency is measured in hertz (Hz) or kilohertz (KHz) • State that humans cannot hear sound with a frequency of more than about 20,000 Hz and less than about 20 Hz • Make conclusions by interpreting results	Practical notes 11.1c
		11.1d Musical instruments	• To apply knowledge and understanding of how sound waves are produced to suggest factors that affect how different pitched sounds can be produced • Make conclusions by interpreting results • Suggest a testable hypothesis based on scientific understanding • Make predictions of likely outcomes based on scientific knowledge and understanding • Plan an investigation to obtain appropriate evidence to test a hypothesis, including control of variables • Evaluate the strength of evidence and how it supports the prediction	Practical notes 11.1d(1), 11.1d(2), 11.1d(3) Worksheets 11.1d(1), 11.1d(2)
2	11.2 Interference	11.2a Starting point		
		11.2b What is interference?	• Describe the amplitude of waveforms using the terms crest and trough • State that interference occurs when two waves overlap • Describe in terms of loudness what happens when two sound waves reinforce each other or cancel each other • Explain how two sound waves can interfere to create a louder sound or silence • Use scientific knowledge to predict what will happen when two sound waves combine • Use the analogy of water ripples combining to explain how two sets of sound waves interact to reinforce or cancel each other • Discuss the strengths and weaknesses of this analogy	Practical notes 11.2b(1), 11.2b(2)
1	End of chapter questions		• To review the learning points of the chapter • To test understanding through answering questions	Worksheet 11.3

Number of 40-minute periods	Student's Book topic number	Learning episode	Learning and thinking and working scientifically focus	Downloadable lesson resources
4	12.1 Evidence for plate tectonics			
		12.1a Starting point		
		12.1b Continental coasts	• To describe how the coastlines of some continents have shapes that appear to fit together • To describe how the layers of rocks exposed on some coastlines appear to match, even though the coastlines are long distances apart • To describe how the movement of tectonic plates suggests that the continents were once joined together • Describe how scientific hypotheses can be supported by evidence • Evaluate the strength of evidence and how it supports or contradicts a prediction	Practical notes 12.1b Worksheet 12.1b
		12.1c Volcanoes and earthquakes	• To explain how and why the locations of most earthquakes and volcanoes are found along plate boundaries • Describe how scientific hypotheses can be supported by evidence • Make predictions for likely outcomes • Describe examples where scientists' unexpected results have led to improved scientific understanding	Practical notes 12.1c
		12.1d Fossils	• To identify and describe some extinct species of animals and plants which can be used as evidence of plate movement • To describe the limitations of using fossils as evidence of plate tectonics • Describe how scientific hypotheses can be supported by evidence • Present and interpret results	Practical notes 12.1d Worksheets 12.1d(1), 12.1d(2)
		12.1e Magnetic materials	• To describe how the direction of the Earth's magnetic field reverses from time to time • To explain how magnetised rocks on the ocean floor provide evidence for plate tectonics • Make predictions of likely outcomes of a scientific enquiry • Plan an investigation to obtain appropriate evidence	Practical notes 12.1e

Number of 40-minute periods	Student's Book topic number	Learning episode	Learning and thinking and working scientifically focus	Downloadable lesson resources
2	12.2 Explaining plate movement	12.2a Starting point		
		12.2b Convection currents	• To investigate a model of the convection currents in the mantle • To describe how this model suggests that tectonic plates move • Make and test predictions based on scientific knowledge and understanding • Carry out practical work safely, including the assessment of risks	Practical notes 12.2b Worksheet 12.2b
		12.2c The effect on tectonic plates	• To apply understanding of convection currents to explain why tectonic plates move and the different types of boundaries between them • Discuss strengths and weaknesses of the plate tectonics model	Worksheet 12.2c
1	End of chapter questions		• To review the learning points of the chapter • To test understanding through answering questions	Worksheet 12.3
5	13.1 The carbon cycle	13.1a Starting point		
		13.1b Photosynthesis and feeding	• To state the word equation and conditions for photosynthesis • To describe how plants make their own food through photosynthesis • To describe how animals directly or indirectly obtain their food from plants • Plan a scientific investigation • Describe the equipment required for an investigation and use it appropriately • Make conclusions by interpreting results	Practical notes 13.1b Worksheet 13.1b
		13.1c Respiration	• To state the word equation and conditions for aerobic respiration • To describe how plants and animals use stored food to provide energy • To describe the role of bacteria and other organisms in decomposition • Plan a scientific investigation • Describe the equipment required for an investigation and use it appropriately • Carry out risk assessments	Practical notes 13.1c Worksheet 13.1c

Number of 40-minute periods	Student's Book topic number	Learning episode	Learning and thinking and working scientifically focus	Downloadable lesson resources
		13.1d Combustion	• To state the word equation and conditions for combustion • To describe how combustion can occur naturally • To describe how humans make use of combustion • Carry out risk assessments • Make predictions of likely outcomes for an investigation into the products of combustion	Practical notes 13.1d
		13.1e The complete carbon cycle	• To describe the complete carbon cycle • To describe how human activities are increasing the amount of carbon dioxide in the atmosphere • Understand that the model of the carbon cycle is simplified and more evidence is being used to improve the model	Worksheets 13.1e(1), 13.1e(2)
2	13.2 Impacts of climate change	13.2a Starting point		
		13.2b Droughts	• To describe drought, its causes and its effects • To model the effects of drought on a plant • Make predictions of likely outcomes for a scientific enquiry • Collect, record and summarise sufficient observations and measurements	Practical notes 13.2b Worksheet 13.2b
		13.2c Floods	• To describe flooding as a likely effect of prolonged heavy rainfall • To understand how increasing global average temperatures can lead to more frequent and severe flooding • Describe trends and patterns in results, identifying and explaining any anomalous results • Make conclusions by interpreting results	Worksheets 13.2c(1), 13.2c(2)
		13.2d Sea level change and sea flooding	• To describe how rising sea levels due to climate change are placing cities on coasts at greater risk of sea flooding • To describe how cities can protect against sea flooding • Present and interpret observations and measurements appropriately	Practical notes 13.2d Worksheets 13.2d(1), 13.2d(2)

Number of 40-minute periods	Student's Book topic number	Learning episode	Learning and thinking and working scientifically focus	Downloadable lesson resources
		13.2e Other extreme weather events	• To describe other types of extreme weather event • To explain why extreme weather events are more likely because of climate change • Sort, group and classify events as changes in climate or as extreme weather events • Evaluate the reliability of secondary information sources	Practical notes 13.2e
1	End of chapter questions		• To review the learning points of the chapter • To test understanding through answering questions	Worksheet 13.3
3	14.1 Collisions	14.1a Starting point		
		14.1b Ideas about how the Moon formed	• To describe three ideas for how the Moon formed • To describe evidence for and against each idea • Evaluate the strength of evidence and how it supports or contradicts a prediction • Make conclusions by interpreting results, explain the limitations of the conclusions and describe how the conclusions can be further investigated	Practical notes 14.1b Worksheet 14.1b
		14.1c The moon and its evidence	• To describe in more detail the evidence in support of the collision theory for the formation of the Moon • To describe how this evidence has been collected • Evaluate the strength of evidence in support of the collision theory for the formation of the Moon • Identify and control risks	Practical notes 14.1c
		14.1d Asteroid collisions	• To describe the evidence found on earth for collisions with asteroids • To describe how asteroids can be caused to collide with Earth • Make predictions of likely outcomes for an experiment to model the impact of an asteroid • Evaluate experiments and suggest improvements • Make risk assessments to identify and control risks	Worksheet 14.1d

Number of 40-minute periods	Student's Book topic number	Learning episode	Learning and thinking and working scientifically focus	Downloadable lesson resources
		14.1e The effects of asteroid collisions	• To describe the consequences of asteroid collision with the Earth • To describe evidence from past asteroid collisions, including climate change and mass extinctions • Make predictions of likely outcomes • Classify asteroid collisions according to their effects using secondary information	Practical notes 14.1e Worksheet 14.1e
2	14.2 Observing the Universe	14.2a Starting point		
		14.2b Nebulae	• To describe nebulae as clouds of interstellar dust and gas • To explain that some nebulae are stellar nurseries • Suggest a testable hypothesis • Suggest how to obtain appropriate evidence to support or disprove a hypothesis	Practical notes 14.2b Worksheet 14.2b
		14.2c Star formation	• To describe how stars form in nebulae • To describe the life cycle of a massive star that may eventually cause a nebula to form • Describe a model of the life cycle of a star • Suggest how to obtain appropriate evidence to support the model	Practical notes 14.2c Worksheet 14.2c
1	End of chapter questions		• To review the learning points of the chapter • To test understanding through answering questions	Worksheet 14.3

Cambridge Global Perspectives™

Below are some examples of lessons in *Collins Cambridge Lower Secondary Science Stage 9* which could be used to develop the Global Perspectives skills. The notes in *italics* suggest how the Science activity can be made more relevant to Global Perspectives.

Please note that the examples below link specifically to the learning objectives in the Global Perspectives curriculum framework for Stage 9. However, skills development in a wider sense is embedded throughout this course and teachers are encouraged to promote research, analysis, evaluation, reflection, collaboration and communication as general best practice. For example, there are numerous investigations and experiments in which students work in groups which give the opportunity for them to collaborate positively and engage in teamwork.

Cambridge Global Perspectives	Learning Objectives for Stage 9	Collins Cambridge Lower Secondary Science Stage 9
RESEARCH	Information skills • Identify and accurately reference a wide range of print and multimedia sources and use them to locate relevant information and answer research questions	*There are a number of activities which give students experience of scientific research. In each case, support and guide students to find and reference a range of suitable sources to answer the questions. Examples are:* • Chapter 3.2b, Activity 3.3: SB p43, TG p50 • Chapter 4.1d, Activity 4.2: SB p73, TG p66 • Chapter 12.1, Activity 12.3: SB p219, Worksheet 12.1d, TG p173 • Chapter 13.2e, Activity 13.3: SB p246, TG p190 *Some worksheets give further opportunities for students' own research enquires.* • Worksheets 5.3, 6.4, 7.6, 8.4, 9.2, 10.3, 11.3, 12.3, 13.3, 14.3
	Conducting research • Select most suitable methods and conduct research to test predictions and answer a research question	• Chapter 1.2b, Activity 1.5: SB p15, TG p33 *Guide students to turn their hypothesis into a research question as part of step A1.* • Chapter 2.1c Activity 2.1: SB p29, Worksheet 2.1c, TG p40 *Guide students to write a research question as part of their plan in step A1.*
ANALYSIS	Interpreting data • Identify patterns and trends in graphical or numerical data in order to support an argument	• Chapter 4.1c, Activity 4.1: SB p70 *After completing step A4, ask students to use their tables and graphs to explain how and why there have been population changes.*
	Making connections • Make some links between causes and consequences of an issue at personal, local and global levels	• Chapter 13.1e: Worksheet 13.1e(1), TG pp184–185 *Ask some students to make their presentations prepared in step 4 of the worksheet. Discuss the impact of increased levels of carbon dioxide globally and nationally.*

Cambridge Global Perspectives	Learning Objectives for Stage 9	Collins Cambridge Lower Secondary Science Stage 9
	Solving problems • Recommend an appropriate course of action and explain possible consequences for a national or global issue	• Chapter 4.1d, Activity 4.3: SB p74, TG p67 *Following the presentations, discuss the actions suggested and those that will have the greatest impact on the problem nationally and globally.* • Chapter 13.1e: Worksheet 13.1e(1), TG pp184–185 *Ask some students to make their presentations prepared in step 5 of the worksheet. Discuss appropriate strategies to reduce carbon dioxide levels in the atmosphere.*
REFLECTION	Teamwork • Relate benefits and challenges of teamwork to personal experience of working together to achieve a shared outcome and identify targets for improvement	• Chapter 4.1d, Activity 4.3: SB p74, TG p67 *Focus on step 4 of the activity. Elicit the points groups have noted and discuss the value of teamwork.*
	Personal learning • Identify skills learned or improved during an activity and consider strategies for further development	• Chapter 2.1c Activity 2.1: SB p29, Worksheet 2.1c, TG p40 *As part of step 3, discuss how the experiment could have been improved before considering how the topic can be further investigated.* • Chapter 4.1b, Teaching and learning ideas (point 5): TG p65 *Develop this activity to include a discussion of how personal learning could be developed. This is an activity that could be used following other learning episodes.* • Chapter 4.1d, Activity 4.3: SB p74, TG p67 *Focus on step A4 of the activity. Elicit the points groups have noted and discuss appropriate learning skills.* *The* Self-assessment *sections in the Workbook at the end of each chapter encourage reflection. Focus on the skills section of the tables* (You should be able to:) *and encourage discussion of what skills students still want to improve.* • WB pp14, 21, 35, 43, 55, 71, 94, 111, 118, 128, 140, 152, 163, 176
COLLABORATION	Cooperation and interdependence • The team assign roles and tasks with an appropriate rationale and respond flexibly when required to help each other achieve a shared outcome	• Chapter 1.2d, Teaching and learning ideas (point 9): TG p34 *At the outset, encourage students to assign roles appropriately and ask them to justify their decisions. At the end, ask them to consider how well the division of labour helped achieve a successful outcome.* • Chapter 4.1d, Activity 4.3: SB p74, TG p67 *At the outset, encourage students to assign roles appropriately and ask them to justify their decisions. At the end, ask them to consider how well the division of labour helped achieve a successful outcome.*

Cambridge Global Perspectives	Learning Objectives for Stage 9	Collins Cambridge Lower Secondary Science Stage 9
COMMUNICATION	Communicating information • Present coherent, well-reasoned and clearly structured arguments with detailed referencing of sources where appropriate	• Chapter 3.4b, Activity 3.6: SB p54, TG p56 *Following step A2, discuss the features of the best leaflets, posters, etc. Consider what arguments and features communicate the information most successfully.* • Chapter 4.1d, Activity 4.3: SB p74, TG p67 *Focus on the group presentations (step A3). Insist on referencing of the research sources in the presentations. Discuss the features of the best presentations.*

MAKING A GARDEN

MATTHEW WILSON

MAKING A GARDEN

MATTHEW WILSON

Quadrille
PUBLISHING

Contents

Introduction

Just why would anyone want to make a garden? Why would they want to take a part of the planet and shape it, plant it and nurture it, enduring the pain of failure, suffering the trials and tribulations of the weather and investing not inconsiderably of themselves emotionally and, in some cases, financially too? After all, it's not as if we actually *need* gardens, is it? At least, not in the same way that we need a home to live in, food to eat or friends to talk to, perhaps.

Over the last two decades in my life as a professional gardener I have met hundreds of people for whom the making of a garden was every bit as important as the security of the four walls of the home. Not as essential, no, but absolutely as important because for them, as well as for millions of others, a garden is a mode of expression, a means of exploring the artistic merits of colour, form, texture and space in a way that is accessible (anyone with a garden can garden) and ever-changing, thanks to the interaction of the weather, the growth and death of plants and the waxing and waning of the seasons. It is this love affair with gardens that is at the heart of this book.

Over the last 10 months I have met six more couples for whom the making of a garden has become an all-encompassing preoccupation: the owners of the six gardens featured in Channel 4's *Landscape Man*. Each of the garden projects was on a grand scale, but otherwise they were as different from one another as could be – from a cliff top in Guernsey to a valley in Wales, a castle and formal grounds in Essex to a paddock carved into dramatic canyons and ponds in Devon. Some of our garden owners wanted to return to nature, grow vegetables and rear livestock or simply indulge themselves in an overwhelming love of flowers;

others wanted, in the very best way, to show off, to make a beautiful, impressive place for friends and family to enjoy. In some cases their financial commitments have been significant, in others almost nothing has been spent at all, but in every case the overwhelming desire to create something living, breathing and tangible has been the same.

What I have learned from my career to date and in particular over the last 10 months is that making a garden successfully, in terms of creating a beautiful outcome that is also fit for purpose, is far from an easy proposition – there are numerous pitfalls at every turn that are enough to sink seasoned professionals, let alone novices.

Making a Garden sets out to help you overcome, and ideally avoid altogether, the many costly and heart-rending mistakes that can befall the process of making a garden. It does so by taking you through the thought processes that are so important when considering what you want from your outdoor space, using the six gardens featured in *Landscape Man* to illustrate the many different and sometimes conflicting demands that are part of the process, allied with examples from my own experience. From there I take you into the visionary aspects of garden making, unpeeling the aesthetic elements that make up the whole and the various merits thereof. Finally the book looks at numerous examples of previously designed gardens, how they were put together, who for, and why.

I believe that the making of a garden, whether large or small, should always be a pleasure, one that has the potential to reward in ways that have a depth that few activities in modern life share. I hope that *Making a Garden* will help you to enjoy those pleasures more readily, and help map the route to a successful garden-making experience more clearly.

THOUGHT PROCESS

1

'All nature is but art unknown to thee'
Alexander Pope

Human beings are, by nature, ingenious yet practical animals. We strive to make our personal environment better by improving the way it works, either by acquiring the items that deliver an improvement or by designing something that fits the bill. Along with our natural inclination towards filling our lives with stuff, these factors form powerful motivators, driving everything from the acquisitive desire for a new car to the realisation that putting up a set of shelves would really help to solve a storage problem.

Collectively we are now more aware and appreciative of good design than ever before, as well as being far less willing to accept the shortcomings of bad design. The realisation that the 'perfect' position for the car steering wheel is also the one that happens to obscure the top half of the speedometer, causing constant craning to read the speed – this and a hundred other instances of design shortcomings contribute to the many irritations of modern living.

The golden rule of design is that form follows function but, when something really works, the partnership between the two is seamless. I have an ancient carpenter's plane bought by my grandfather. It's beautiful because of the warm glow of the rosewood handle, burnished with the patina of use and age, and the quality of the oil-blackened steel, crafted with as much care as a sportsman's rifle. In this very practical object, form and function live together perfectly, simply because of the quality of the design and the standard of the materials used.

And yet, despite our appreciation of really good design, we rarely apply the same principles to our gardens. Instead, gardens tend to evolve from the application of practical considerations rather than any overarching design. For example, a washing line is a prerequisite in most gardens and its dual requirements (a sunny, open spot and a decent length between two points) dictate its positioning, on the longest linear plane between the house and boundary fence. And once the washing line is in place, the realisation dawns that the constant to and fro to hang up the washing is creating a muddy track down the middle of the lawn, so a path is made of paving stones and the garden is entirely bisected. In time, the need for that great overspill of home storage, the shed, becomes more pressing. And what better place to put it than at the end of the path already constructed for access to the washing line?

The resulting garden isn't really a garden at all, just an open space equipped for practicalities – all function and no form. If plants do make it into the equation, they are relegated to the garden's fringes for fear of making it feel small. But it doesn't have to be like this. It is perfectly possible to accommodate practicalities and aesthetics in a garden, no matter the size. In fact, a designed garden will actually work better ergonomically, simply because it has been designed and all the requirements of the owners carefully considered and applied.

The key to achieving this happy balance between beauty and usability is to spend some time thinking through what you want from your garden: what elements simply *have* to be in it, how you intend to use it and how much time and money you want to spend on it – both creating it in the first place and thereafter. But you should also allow yourself to dream, taking inspiration from nature, other gardens or even seemingly unrelated elements: architecture, music or art. I've been inspired by items as disparate as a 1950s architecture poster and the front grille of a 1970s Lada car (the former inspired the layout for a whole garden, the latter the design of a timber fence partition) and am always looking to the non-garden world for inspiration in colour, form and style.

This chapter will act as your guide to the thought processes behind designing your garden in a way that fulfils as many of your needs and desires as possible. It also reveals how the owners of the six gardens in *Landscape Man* have dealt with the combination of practicalities, aesthetics, prevailing conditions and ethical considerations in order to make their own dream gardens come true.

Who and what is the garden for?

OPPOSITE Gardens have the capacity to be complete works of art, their creation and care requiring a blend of art, craft and graft that develop over time. The main outcome of a thoroughly considered thought process is that practicalities and beauty sit side by side and ensure that the garden is everything it can be.

This may seem to be the most obvious question of all, yet it is usually the one that isn't answered early enough in the design process. Quite simply, without a clear understanding of 'who' and 'what' the garden is for, it will be impossible to create an outdoor space that successfully meets your needs. It is the horticultural equivalent of contemplating buying a two-seat sports car when you have a family of six (though for some there might be an ulterior motive for doing just that!).

The answer to the 'who' part of the question may not necessarily be fixed, as circumstances can change. These need to be factored in: there is, after all, little point in planning a garden for two if you hope to become a family of three, or four, or five. The 'what' element will probably throw up a number of different answers as you will need to consider the many different roles you would like your garden to fulfil. Do you want to create a space for outdoor dining, an area for children to play, or simply a beautiful garden to look at, relax in and unwind? Because needs, tastes and interests change and develop over time, the answers to this question may not be fixed either. And in a living environment like a garden there is the not inconsiderable matter of plants growing and changing as well, all of which can affect quite fundamental aspects of the space – light levels, space,, outside views and so on – and how it can be used.

But there is no doubt that time spent thinking about what you really want from your garden is time well spent, considering such questions as how you currently use your garden and how that might change if the garden changed. These are exactly the kinds of questions you should expect a professional designer to ask you as part of the design process. Remember that, in taking all the many practical considerations into account, you should also allow yourself to dream a few dreams and take a few flights of fancy. After all, gardens are for dreaming too...

Accommodating children and pets

ABOVE My old cat Bruce seemed to have been bred for garden destruction. He would wait patiently for the beautiful flower heads of ornamental onions (*Allium*) to begin opening, before attacking each one in turn!

OPPOSITE Children love to get their hands in the soil, digging up worms and creepy crawlies and watching seeds they have sown grow into mature plants. Setting aside an area just for children to garden is a great way to make this happen.

Gardens form a vitally important connection between people and the environment. For children in particular, the garden is the place where they may experience, very likely for the first time in their lives, such exotic creatures as worms and bumblebees, the untrammelled joy of getting really, really muddy, and the little miracle that comes when a seed is sown and grows into a plant.

But designing a garden to accommodate the needs of children, with their love of playing, digging and running around can be challenging. The simplest way to achieve this is to give them an area of the garden that is theirs and theirs alone, to look after and tend. This might be a simple raised bed where a few veggies can be grown, or a big pot for rearing a monster sunflower. And if you want to really engage children with the outdoor environment then there are a couple of 'must haves'. A compost heap or wormery full of slippery, wriggly worms, will be enthralling to youngsters, as well as being a really good lesson in the responsibilities we have towards waste and recycling. A wildlife pond is an endless source of fascination – a miniature universe full of incredible animals – and can be made perfectly safe, even for the tiniest of toddlers. Setting aside even the smallest of spaces in the garden for your children will give them a sense of responsibility and ownership. Not only is this great fun for them, and hugely educational, it will then leave the rest of the garden free to be developed for the grown-ups.

Whilst it might be relatively easy to accommodate the needs of children in a garden, pets are a completely different matter. For one thing there is no reasoning with a bouncy, wilful moggy, and not much point in trying to encourage the use of a specific area of the garden as you might with a child. As a result, there will always be a compromise when one of the clients of a garden designer is four-legged, as it simply isn't possible to design out every challenge in the garden.

There are, however, a few practical things that can be done to help offset the worst that pets can do. Taking your dog for a walk, rather than just letting it loose in the garden, is an obvious one. If you do have to let the dog use the garden for when nature calls, it's perfectly possible to train it to use just one spot. There are also charcoal-based additives that can help to neutralise the acid in dog urine, which is the cause of yellow patches in lawns. Cats are a different proposition as they are almost impossible to train, but leaving aside an area of nice, loosely dug soil – in effect an outdoor litter tray – can help to encourage their use of a specific area for toilet purposes. And avoiding plants like catnip (*Nepeta* species), which is basically like cat LSD, is probably a good thing.

Outdoor dining and entertaining

OPPOSITE This garden has been designed with the focus squarely on outdoor dining, the small space dominated by the large timber dining table. There are some nice touches here: the repetition of colour in the blue tea light holders and cushions, as well as the form of simple, rectangular planters.

There is little point in having a garden if you never use it, and without doubt one of the best ways to use a garden is to spend time living in it, whether that be entertaining, alfresco dining, or simply sitting out and enjoying the elements. Applying thought to how to design the areas where these activities take place will ensure they work in the way you want them to. For example, if you want to sit out in the sun it makes sense to have seating on level ground in a sunny spot in the garden. It sounds obvious, but you'd be surprised how easy it is to get wrong. And if eating outdoors is a priority, tackling a few practicalities by applying thorough design principles will make it even more of a pleasure: outdoor lighting, a weatherproof electricity supply, a surface for food preparation and cooking and a waterproof cover to sit under will all help transform your garden into the most popular restaurant on the street.

An outdoor dining space can become the hub from which the rest of the garden radiates, as well as being a thing of beauty and interest in its own right. A seating area can become a focal point that draws the eye, while a path that is at heart a practical feature to take you from the back door to the shed can become a route through the garden that feels like a journey, rather than just a simple A to B. These elements usually form the bulk of the hard landscaping features in the garden, and are therefore the ones with the biggest costs attached, so getting them right is even more important.

RIGHT A bespoke seating area offers the opportunity to integrate the design of the garden with that of the seating. In this instance the shape of the circular seating and table is echoed in the clipped box balls, whilst the table itself sits in a circular paved area.

LEFT This simple dining area is made a little more special by raising the decked area, creating a step up and giving the space its own distinct 'flavour' from the rest of the garden.

No time to garden

OPPOSITE Designing a garden where the clients have no time at all to garden will inevitably have an impact on the balance between hard and soft landscaping, with the emphasis falling on paving and decks and the type of plants selected for comparative ease of maintenance.

Gardening is an interactive pastime that engages us on many different levels – sensually, intellectually and physically. Those of us who find ourselves trying to juggle busy lives can find this interactivity to be, on occasions, a little too demanding. And yet we still want to have an attractive garden just as much as those who have the luxury of more time to spend.

As a result, the time-poor gardener often ends up with a space that is largely ignored, before being tidied up from time to time with a few hastily bought plants, the results of frantic forays to the garden centre. Another supposed 'solution' (though it is anything but) involves laying the entire garden to lawn. This seemingly easy option will result in you having a garden that is not only horribly dull to look at, but also deceptively high in maintenance – all that grass will still need cutting, unless you want to end up with knee-high hay. Paving the whole space over, by contrast, is guaranteed to bring on advanced depression; surely we spend enough time looking at concrete and tarmac without having it fill our gardens?

The good news is that it is possible to have a beautiful garden that is relatively easy to maintain, though such a space requires more rather than less planning in order to be successful. Lawns are notorious for requiring lots of input to keep them looking good and many are much bigger than they need to be; if you only plan on using your lawn to set up a table and a couple of chairs for summer drinks then make it the appropriate size, or dispose of it altogether. A well-planned balancing of hard and soft landscaping materials can also help to keep maintenance to a bearable level, without creating something that looks like a concrete jungle. Whilst plants often strike fear into the hearts of 'non-gardeners', especially those with limited time on their hands, they only ever cause problems when they have been incorrectly chosen or badly cared for. There are tens of thousands of plants for every kind of growing condition that can happily thrive together, beautifying the garden whilst taking very little trouble to look after. In fact, there are a large number of plants that are so successfully adapted to specific situations that they can get by with no input whatsoever – no watering, feeding, spraying or anything else. Using structural plantings of largely evergreen subjects is another effective way of providing greenery with comparatively little effort, whilst avoiding seasonal plants in favour of those with greater longevity will help save time too.

Site considerations

ABOVE The area where you live will have a microclimate that has a direct impact on the types of plants you can grow. Repeated heavy frosts, extreme wet and cold or drought and heat each require a different palette of plants in order to garden successfully.

Whatever your ambitions for your garden, the climatic and cultural (growing) conditions of the site will have the biggest say in what you will actually be able to achieve. You might be dreaming of a lush, tropical garden but if you live on the side of a windy, exposed hill with almost no sun and regular winter frosts then the chances of your dream becoming a reality are slender to non-existent, as the growing conditions are simply not right for plants with big, soft foliage.

There's a lot to be learned from taking in your surroundings and looking at how other people garden in your area and what plants they use. You'll probably find, for example, that the plants on one side of your street vary from those on the other, as a consequence of the orientation of one side of the road being distinct from the other. There are plenty of clues in nature too, especially when it comes to pointers about soil type and growing conditions, as wild plants tend to be quite particular about where they will grow. A good wildflower book can help you to understand this better, and might even become the starting point for a great hobby; I like nothing better than nosing around in hedgerows looking at wildflowers.

The most important thing to realise is that there is no such thing as the 'average' garden, or 'average' conditions. A side return to a building in a busy city (like the one on the opposite page) will have a set of environmental conditions – wind exposure, desiccation, noise and chemical fume pollution – that are quite different from those someone gardening in a rural situation. And yet you may find they have share some unexpected similarities; perhaps that rural garden backs onto an expanse of field that is regularly sprayed with chemicals by a noisy tractor? What is certain is that it will save a lot of time, money and energy to spend time observing and getting to know the climatic conditions of your garden before you do anything else, rather than going to great efforts only to discover that everything is in the wrong place, or does not work with the conditions that exist in your garden.

Case Study Guernsey

RIGHT The growing conditions on Guernsey are typical of seaside locations. Persistent winds laden with salt spray, high light levels and changeable weather call for a very specific type of planting.

Climate and microclimate

When considering your site, you should make gaining an understanding of the climatic conditions that are prevalent in your garden a priority. These conditions are a combination of the bigger climatic picture – rainfall, summer temperatures, degrees of frost in winter – along with the microclimate that exists in your garden. There's a distinction between the two, because the wider climatic conditions, the macroclimate, don't always translate down to what happens at garden level, which can be incredibly varied and specific, hence it being referred to as microclimate.

To begin understanding your garden's microclimate, start by asking yourself the question, 'What direction does my garden face?' North, south, east or west – this will have the biggest say on what you can grow because it will affect the amount of direct sunlight that falls on your garden. This in turn affects the temperature, which impacts on how quickly the soil warms up, when plants start into active growth and so on.

The physical built environment around a garden also influences the microclimate. A side return to a building might cause a funnelling effect of the wind, which in turn might have an adverse effect on any plants near the

BELOW The dappled shade of a deciduous tree in an otherwise sunny space can create a microclimate quite distinct from the rest of the garden: cooler, drier (thanks to the tree roots) and slightly more sheltered.

side return, or make it an uncomfortable place to sit in. Solid walls can cause the wind to vortex or eddy, which is especially unpleasant if the wind picks up dust or soil.

And what about that most ubiquitous of features, the humble garden fence? A fence is, of course, designed to divide a space, and consequently its two sides face in opposite directions. Although the fence might be just a matter of a few centimetres wide, those two sides will have completely different compass aspects – if one side is south-facing the other is north-facing, if one faces west the other will face east. So within that very short distance two completely different microclimates will exist, which in turn means that what thrives on one side of the fence is unlikely to grow well on the other side, despite the distance between the two being so tiny.

The only way to really understand the conditions in your garden is to spend some time observing it. What you see might throw up some real surprises; an otherwise shaded garden may have an area that is in full sun through the middle of the day, for example. This information doesn't just influence the choice of plants you grow, but also the design of the garden – an area of full sun might be the perfect location for a seating or dining area, whilst a wall bathed in dappled sunlight might benefit from a change of colour to a more light-reflective hue. A heavily shaded spot where the frost sits during winter might not be a great place to sit out in, but could be brought to life with the planting of some late-flowering grasses, which look great all through winter and are at their absolute best when rimmed with frost.

ABOVE Knowing where the sunniest, most sheltered spot in your garden is can help to inform where you position a seating or dining area to make the most of the conditions you want to enjoy most.

The soil in your garden

Although it isn't especially exciting stuff at first glance, without soil there wouldn't be much life on Earth. And although an intricate knowledge of the ecology of soil isn't really essential in order to design a successful garden, a basic understanding is most certainly needed, as is an appreciation of how to manage soil to get the best out of it.

The important stuff to remember is this. Healthy soil is made up of mineral matter (in effect ground-down rock), which comprises roughly 50–60% of the whole, and organic matter at around 5%. Between the two, these form the solid part of the soil. The remainder is a roughly equal mix of air and water, both of which are vital for keeping the soil healthy and ensuring good plant growth. In order to keep soil healthy, and therefore the plants that grow in it happy and flourishing, these component parts need to be maintained at their relative amounts.

Soil types vary from heavy, wet clay to light, fast-draining sand, and the nature of the soil in your garden will affect what you should put in it to improve it for growing, as well as the type of plants you should use. The soil in your garden will also be more or less acidic or alkaline and this is measured using the pH scale. This has as big an effect on what kind of plants you can grow as the climatic conditions and soil type, as some plants are lime-haters, others acid-lovers and so on.

Soil types are incredibly variable, sometimes changing characteristics within a matter of metres. But providing you have a basic understanding of the soil characteristics in your garden you should be able to start making some improvements to the growing conditions. Whilst soil improvement is a massive subject, there are some broad rules for the most common soil types.

Sandy and silt soils

Sandy soils drain well and are quick to warm in spring. They are nice and light to dig and cultivate, but can be harder to 'improve' than heavier soil, as nutrients often wash through the top soil. Consequently they are less suited to plants requiring nutrient-rich soils and better suited to plants from hotter and/or dryer environments. Heavyweight soil improvers such as farmyard manure are ideal for light, sandy soil and can be dug in or, better still, applied to the surface as mulch every year, in spring and autumn.

Clay soils

Clay soils can be hard work; heavy and wet in winter like a wobbly blancmange, hard as concrete in summer. Clay soils tend to be hard to cultivate and poorly drained, which can cause plants to rot off in the ground, or 'drown' in the wet soil. But clay can be harnessed, so long as it's dug at the right time when the soil is neither too wet nor too dry, and the correct soil improvers are used – grit to improve drainage and lightweight organic matter such as an ordinary garden compost or spent mushroom compost to improve structure.

LEFT The soil conditions in the garden at Hedingham Castle in Essex – moist but well-drained acidic loam – are suitable for growing ericaceous (acid-loving) shrubs such as *Pieris*, which would not grow successfully in alkaline soil.

Soil basics

Whatever type of soil you have, there are a few basic rules that should be adhered to in order to ensure that it stays in good condition and is therefore the best it can be for growing plants in. You should only ever cultivate soil when the conditions are suitable, which means avoiding periods when the ground is very wet, frozen or extremely dry. Digging in dry weather will simply remove whatever moisture is left, while any amount of disturbance or weight put onto wet or frozen soil can spoil its underlying structure, as well as the vital balance of air porosity that helps to keep it healthy.

To keep feeding the soil with the organic matter it needs, apply a mulch every spring and/or summer to the surface of the soil. This will help to keep moisture in, suppress weeds and feed the soil as, over time, the organic matter gets drawn down by soil-dwelling animals, in particular earthworms. Applying organic matter to the surface in this way also helps to reduce the amount of digging required, too much of which is actually a bad thing for the soil as it can disrupt and kill soil animals and cause problems to the soil's structure.

CREATING A GARDEN FROM SCRATCH

Clive and Debbie Morris have taken a long time to set down roots in a place they can truly call home. Their business in property development has meant that they have tended to spend a limited time in any of their homes, instead preferring to improve the quality of the property before selling up and moving on. As a consequence, they have never really engaged with the idea of a garden being something to experience and nurture. They bought their current home – a large detached property in Suffolk including a 24-acre garden that was once part of the Ickworth Estate – with the aim of improving it and selling it on. However, within a very short space of time they had fallen in love, both with the house and, more specifically, the garden.

Clive and Debbie's inspiration for their garden has part come from Ickworth Estate itself, a formal Italianate garden with Victorian additions. They see it as providing a place for family and guests to enjoy and be entertained in, as well as establishing a suitable setting for the house. In Clive's words, their aim is to put the life back into what had become a dead heart.

To help achieve their dream, the Morris's have engaged RHS Chelsea Flower Show gold-medal-winning designer Thomas Hoblyn. His plans for the garden are in essence a modern take on the classical landscaped garden estate. These include using structural plantings (such as avenues of trees) to emphasise existing vistas to the surrounding countryside or create new ones; a modern take on the classical rose garden; as well as the creation of a Victorian stumpery – a grotto made from the stumps of felled trees planted with ferns and other shade-loving plants – inspired by one at Ickworth.

But by far the most ambitious of the Morris's plans is the proposed transplantation of a complete RHS Chelsea Flower Show garden, which Thomas has designed for the 2009 Show. Chelsea Show gardens are impressive, expensive installations aimed at showing off a designer's skills, just as the big fashion shows of Paris, Milan and London showcase the talents of fashion designers. But what they aren't designed to do is last more than a week, the length of time over which the show takes place. Dismantling this show garden, recreating it in Clive and Debbie's garden and reinterpreting it to work, long term, in a very different space is a real design challenge and takes the idea of creating a garden from scratch to a new level!

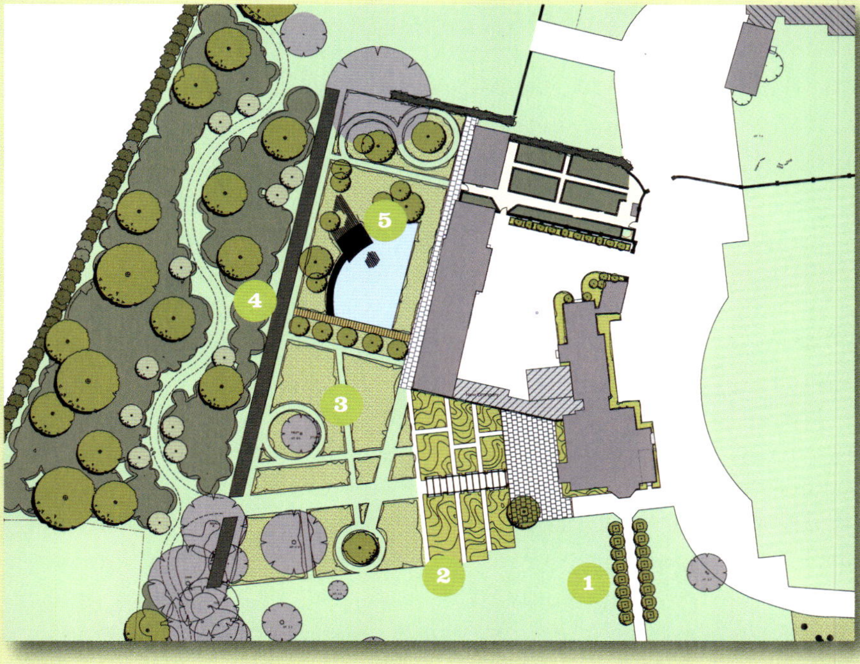

ABOVE Forming an axis from the house out into the countryside beyond, a hornbeam avenue (1) will eventually link with an avenue of poplar and a garden rotunda. Adjacent to the main terrace of the house is a mixed rose and perennial garden (2) that is linked to Thomas Hoblyn's Chelsea Flower Show garden (5), the culmination of the formal garden area, by a simple area of meadow grasses and wildflowers (3). The existing small area of woodland has been improved with thinning out, new plantings and the creation of a stumpery garden (4).

The formal garden

Historically, a formal garden is one based around geometry: straight lines, often interlocking, that frame important vistas or focal points; neatly ordered and edged beds, frequently planted with a restrained palette; as well as the liberal use of structural planting in the form of trees, shrubs and hedges.

As Western garden and landscape design developed over the centuries, it was subject to the influence of art, poetry, science, architecture and social developments too. During the Elizabethan era the comparative stability of the country affected the architecture of grand houses to such an extent that it also changed the way gardens were designed. No longer needing fortifications or castellation as a means of defence, grand houses could instead be built with big windows. which were ideal for looking down onto a garden from and led to a rise in interest in the Knot Garden:a formal, symmetrical arrangement of low, clipped hedges surrounding plantings of fragrant herbs, designed to be viewed from above. There was a practical aspect to these gardens too – the herbs could be cut and strewn on the floor to suppress the widespread unpleasant odours that wafted around.

'In the contemporary sense, the notion of a formal garden is more to do with borrowing stylistic cues ... than an attempt to re-create a classical formal garden experience in its entirety'

From the late 17th century, landscape design theory began to be increasingly informed by the experience of the 'Grand Tour', the tradition by which young men of means would travel Europe as an educational rite of passage, taking in important historical sites and developing their appreciation of classical and renaissance art. As well as accumulating knowledge, these young men would collect works of art and commission new works as copies, including garden statuary and furniture such as balustrades, columns and pillars – grand sculpture that demanded a suitably grand garden setting.

In modern times, as designers have felt less constrained by tradition, the factors that define formal and informal gardens have become increasingly blurred. In the contemporary sense, the notion of a formal garden is more to do with borrowing stylistic cues relating to geometric precision, framed views and vistas and architectural features, rather than an attempt to re-create a classical formal garden experience in its entirety. The design for Clive and Debbie's Suffolk garden includes a number of these classic formal garden features, such as tree avenues and rose gardens, which have either been updated or reinterpreted for use in a contemporary garden.

LEFT This photograph of the pleached hornbeam avenue under construction shows how heavy and impermeable to water the heavy clay soil in Clive and Debbie's garden is. The excavated trench shown is being prepared for a new hard path between the hornbeams.

ABOVE The completed pleached hornbeam avenue, showing the timber frame onto which the hornbeams have been trained. This frame will stay in place for up to five years, whilst the hornbeam establish, before being removed.

Using trees in the garden

OPPOSITE Clive and Debbie's greatest enthusiasm is for trees – the pair have planted numerous long-lived hardwood trees such as oak (top left and middle right) to screen views of a nearby road and form avenues, clumps and woods, and have also invested a lot of effort in preserving the garden's existing old trees such as *Cedrus atlantica* (bottom right), which has been inoculated with beneficial fungus to help prolong its life. Where old trees have died (bottom left), they have been left standing to benefit the many animals that live in or feed on dead timber.

We humans have a bit of an odd relationship with trees. On the one hand we admire them and are somewhat in awe of their presence, longevity and, on occasion, massiveness. On the other we resent them for blocking the light from our windows, fear their questing roots lest they end up causing our homes to subside and curse the seasonal chore of leaf collecting. But to see trees in this way is to miss the point, which is that they are utterly and completely amazing organisms. They manufacture oxygen, filter pollution, provide homes for wildlife and can produce food, building materials and fuel. And the fact that they can do all of these things whilst putting up with the kind of abuse that we mete out on them – digging up their roots, hacking into their branches and mutilating their trunks – makes them all the more astounding.

In Clive and Debbie's garden, trees are used in a number distinct ways and, though the size of their garden is such that the numbers and size of the trees involved is pretty big, the same rules apply in locations where space is at a premium and when using much smaller-growing trees in very limited numbers or singly. Around the fringes of Clive and Debbie's garden, large clumps of big forest trees are used to screen elements of the surrounding landscape – other properties adjoining the estate as well as a road that runs in the view line of one of the boundaries. The tree clumps are grouped in such a way as to break up, but not entirely block, the line of the horizon. This important lesson applies to small gardens too. If you try to conceal something with a block of plants, you'll only emphasise the shape of the object you're trying to hide, whereas if you break up the outline with different shapes it should successfully disappear.

Elsewhere in Clive and Debbie's garden, trees are used not to hide things but to really show stuff off. To one side of the house an axis of trees extends out into the garden and through an adjoining paddock. Initially formed of an aerial hedge of pleached hornbeam (*see page* 31), the line is then extended with an avenue of poplar out to a pair of mature oaks which flank a large rotunda. The aim is the equivalent of shouting 'look at me', as the combination of tree avenues guides the eye towards the rotunda. This principle of using plants for emphasis, either to draw attention to a particular feature inside or outside the garden is particularly effective and can be applied to a garden of any size.

A MOST AMBITIOUS GARDEN TRANSFORMATION

ABOVE Keith and Ros Wiley bought a four-acre paddock in 2004 and have set about transforming it into a very different kind of garden, with a limited budget and largely without additional help.

Keith and Ros Wiley have spent a lifetime making gardens. For 25 years Keith worked as Head Gardener of a public garden in Devon, before leaving to create his own garden from a four-acre grazing paddock nearby. As well as making a garden, Keith and Ros needed to build up their nursery business on the same site in order to make a living. The first three years were pretty tough. Between them the pair took almost no time off; toiling away seven days a week to make the garden and build up the nursery business, whilst living onsite in a fairly basic timber cabin. Ros, a talented landscape artist, hardly picked up a paint brush during this period, spending all her time weeding and potting up plants instead.

Creating a garden from scratch can often be a difficult endeavour, but Keith's approach to making this garden has added to that challenge enormously. The paddock that Keith and Ros purchased was a perfectly decent field with level ground and a gentle slope towards the south – exactly the kind of conditions that most people would choose. However, instead of taking advantage of the prevailing conditions and working with them, Keith decided to completely transform the topography of the site. Using an excavator, he has carved out canyons, created pools of water and hillocks of soil, some over 4 metres tall. The result is quite extraordinary; plants sweep up and down over the hillocks, through which winds a network of paths.

In the centre of all this Keith has created a courtyard garden, flanked by block-built walls that have been rough rendered with a coloured plaster in the style of a Mexican *hacienda*. Inside the courtyard, the space is divided into four roughly equally sized beds, separated by a cruciform path with a huge stone trough acting as a centrepiece. Here Keith plans to create an exuberant planting scheme based around four 'groves' of large pineapple trees (*Cordyline* species).

To top it all off, Keith has excavated two huge ponds that extend along the whole southern boundary of the garden. The trouble is, Keith and Ros have very little help or funding with which to make all of these ambitious plans come to fruition, and as well as trying to complete the garden – which they need to open to the public to help support their business – they also have the added challenge of trying to keep their nursery running at the same time.

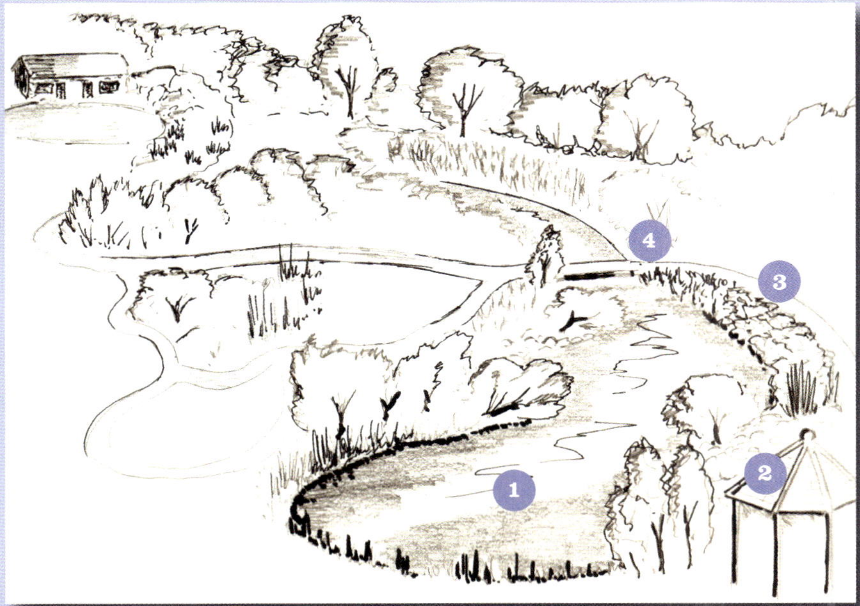

ABOVE The major design feature of Keith and Ros's garden is a series of ponds and bog gardens (1) that are linked by narrow streams and waterfalls and fed entirely by rainfall. A planned summer house (2) will provide a vantage point from which the ponds can be enjoyed, while new paths (3) created around the ponds allow the richly-planted fringes and banks to be glimpsed close up. A series of bridges (4) provide crossing points over the water features as well as additional structural interest.

The naturalistic garden

ABOVE Keith's greatest enthusiasm is for flowers, especially the overwhelming displays of wildflowers seen in nature, the impression of which he has sought to bring back to his own garden in Devon.

The term 'naturalistic garden' is a rather loose one that has come to prominence in recent years, with the upsurge in interest in gardens that are planted in tune with nature and which consequently need less frequent intervention to keep them healthy and looking good. In design terms there are no rules that apply specifically to a naturalistic garden – it is more an ethos than a style – and so a naturalistic garden could in principle be created within the context of formal design principles, or just as easily be a wild, woolly, unplanned space.

Although all gardeners have to exert a certain element of control over nature to keep their gardens in shape, the naturalistic gardener will go to great lengths to ensure they have a thorough understanding of the soil type and microclimate in their garden before choosing the appropriate plants for those conditions, rather than just buying plants they like the look of, planting them and hoping for the best. A naturalistic garden will, by and large, use little or no additional water for irrigation, minimal or zero pesticides and no fertiliser. The plants selected will tend to be very natural-looking too, as they will usually be wild species (not necessarily native to this country but native to somewhere on the planet) or hybrid plants that share most of the characteristics of their wild parents – nectar and pollen richness for example. As a consequence of management and plant selection, a garden made in this way will often be buzzing with wildlife and tend to look more a part of the landscape than something imposed upon it.

'There are no rules that apply specifically to a naturalistic garden – it is more an ethos than a style'

When Keith started making his naturalistic garden, his aim was to create a facsimile of the wild landscapes that have influenced him and his style of gardening for years: the panorama of European Alpine meadows, the bleak beauty of the Mojave Desert and Death Valley in California, and the bright blossoms of the South African Fynbos. What Keith is attracted to in all of these wild places is their vast array of flowers, the blooms of which often last no longer than a few weeks but arrive in a great flush of colour that has the power to overwhelm the senses. By creating hillocks and canyons for these plants to grow on, their colourful flowers are immediately brought up to head height and so can be looked at directly, recreating that early childhood sensation of being lost amongst a multitude of flowers.

BELOW Comprising of canyons and valleys carved from the flat field with an excavator, Keith and Ros's garden is full of flowers and buzzes with wildlife. Despite the poor sub-soil, consisting of shards of stone mixed with clay, there is plenty of nutrition here for growing a wide variety of plants.

Ponds

OPPOSITE Ponds don't just look good, reflecting light, sky and the plants around them, they are also hugely important for wildlife; providing food and habitat and consequently attracting a wide range of species to the garden.

One of the key features of Keith and Ros's garden is a series of ponds that run along the bulk of the garden's southern boundary. Although these aren't technically big enough to be classed as lakes (the definition of which is a body of water with a surface area greater than one acre) they are no shrinking violets, the largest being 25 metres long, 5 metres wide and 2 metres deep. They form the culmination of the garden as an experience, the final big surprise in a garden that has more than most. They also serve a practical purpose too, with Keith planning to use them as informal reservoirs, to provide irrigation water for the nursery and garden, as needed.

While a garden pond may seem a fairly innocuous feature – a simple hole in the ground filled with water and ornamented by a few plants – it really is anything but. A well-designed and executed pond, whatever its size, will begin to attract wildlife within days of being built. First to arrive are usually passing birds, lured by the opportunity for a bath and a drink. The pond's wet margins also provide plenty of mud; the principal building material from which barn swallows and house martins make their nests. Once the water starts to settle and naturalise (which is pretty soon if the pond has been responsibly filled with rainwater rather than from the tap), creatures that live in or on the water will appear – water boatmen with extended legs like oars to propel them across the water surface, whirligig beetles that spin endlessly and seemingly uncontrollably, as well as myriad tiny insects and bacteria that help to keep the water balanced.

'A well-designed and executed pond, whatever its size, will begin to attract wildlife within days of being built'

In time, if the pond has been planted with the correct density of plants to keep it healthy (a mixture of floating, submerged oxygenating plants as well as water lilies and plants that provide shelter around the water's edge) then even more exciting and exotic-looking creatures will show up: insects such as dragonfly larvae, equipped with vicious mouth parts designed for snapping up prey, as well as great diving beetles that appear as if from nowhere to scour the water's murky depths.

Starting from scratch versus taking on an existing garden

OPPOSITE Sometimes inheriting an existing feature might seem like a double-edged sword: the benefit of a mature feature is often welcome, but how do you go about including it in a new design? In this instance an existing mature tree has become a central feature in a new design, from which the rest of the garden now radiates.

For an enthusiastic gardener the idea of taking on an empty blank canvas of a space and, from it, creating a new garden, is an exciting one full of promise. For the uninitiated and inexperienced, however, it has the potential to feel like an overwhelming challenge. Starting a garden from scratch offers as many opportunities as it does potential pitfalls. On the one hand, there are none of the limitations caused by the sorts of features found in a mature garden. Nor is there the temptation to live with something that you just know is in the wrong place, like a patio or a path, simply because it is already there and it would be too expensive, difficult, or time-consuming to move it.

An empty space can throw up its own set of distinctive problems, however. There is nothing to give a garden any sense of being rooted in a new garden, nothing with scale to provide presence or hide unsightly neighbouring structures. The only way to counteract this, if there is a need to do so quickly, is with hard cash – buying in maturity in the form of big trees and shrubs or garden structures with scale. There can also be problems that arise from the conditions on the site: a newly built property (one of the most common instances where a new garden is taken on) can often have garden 'soil' that is little more than builders' rubble. Often there is little to do but remove the top 30cm or so of soil and replace it with bought-in topsoil, an expensive and messy operation that doesn't always solve the problem, as bought-in soil often doesn't match the prevailing conditions in the area, and can be of poor quality too.

A pre-existing garden is likely to have at least a couple of mature plants, which can give the garden a sense of presence, and may well have other useful features such as seating areas, a garden shed and so on. But what if none of them are to your taste? What if the one tree in the garden is patently in the wrong place? How do you go about choosing which plants to save and which ones to remove? The liberating freedom of the blank canvas can seem a blessing by comparison, but when tackling a mature garden, it is the pre-existing features themselves that should be savoured and built upon. With a bit of thought (and a little work) it might be possible to revamp existing hard landscaping features like paths and patios and give them a new lease of life, whilst that one mature tree in the centre of the garden could, with careful pruning, become a lighter, more open feature around which the new garden pivots.

PLANNING THE PERFECT SPACE

ABOVE Filming on location at Anne and Geoff's garden in Sussex by the tennis court: a part of the garden that was often a source of contention (*see page 44*). After much discussion between Ann, Geoff, myself and their designer, the court was eventually turned into a wildlife pond.

Anyone searching for a garden with real potential could do worse than heed the estate agent's mantra of 'location, location, location'. Anne and Geoff Shaw had been looking for a property in the area of the South Downs in Sussex for almost five years before purchasing a run-down and unloved old house and dilapidated garden in a wonderful setting with spectacular views. Having spent so long searching for the right property, they are understandably anxious to create a great garden – but to do so they must first overcome some pretty big challenges.

Great views often come at the price of exposure to the elements. Anne and Geoff's garden is no different in this respect, being located at the top of a ridge overlooking a valley. Careful thought will therefore be needed to ensure that the garden is designed in such a way as to allow them to enjoy the beautiful scenery without having to retreat into the house whenever the wind blows. The soil in the garden is another problem: heavy clay that will need to be skilfully managed if anything is to be made to grow in it.

But the biggest hurdle the pair must overcome are the local planning rules. Anne and Geoff's home is very old, with parts of it dating back to 1650. As a consequence, the local planning officers are keen to ensure it is developed sensitively, and their scrutiny doesn't stop with the house. The curtilage of the garden – the clearly defined area associated with the house that has always been gardened as opposed to farmed – is very obviously marked by old flint walls that appear in ancient tithe maps of the site. Anne and Geoff have engaged garden designer Ian Kitson to create a plan which includes sweeping, curving walkways and dramatic sunken gardens, along with wildlife pools and wildflower meadows. Crucially, however, this plan proposes alterations to the curtilage.

Altering such a historically important feature as curtilage isn't something to be entered into lightly, especially in somewhere like the South Downs: a pocket of natural beauty in the busy southeast of England so precious that they were awarded National Park status in 2009, which brings with it much stricter planning regulations to protect the landscape from inappropriate or damaging development. Anne and Geoff will therefore have to rely on Ian to help lead them through the tricky business of making a beautiful garden on a site that is both physically challenging and is subject to the approval of the local authorities' ever-watchful eye.

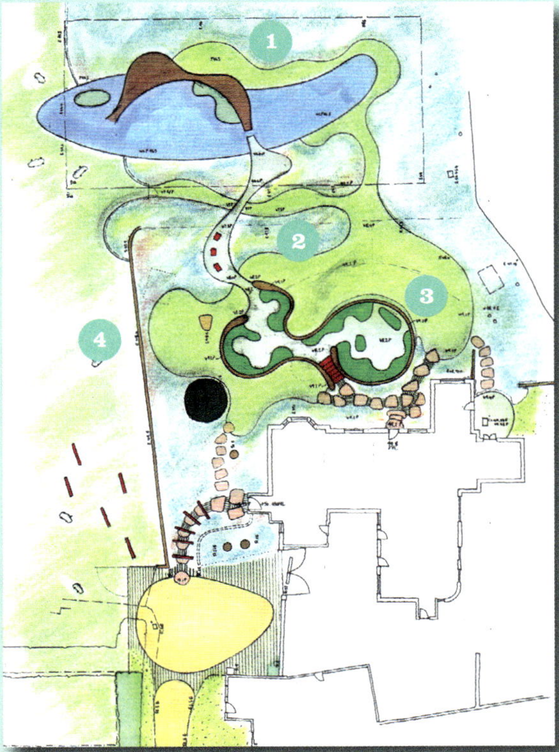

ABOVE The old tennis court (1) has now been redesigned into a wildlife pond that is the highlight of the new design. It features a large timber deck which spans the water, widening enough to include a seating area before 'diving' into the water close to a small wildlife island. The sunken garden (3), in front of the house is another significant new feature, the walls dressed with napped flint in the local vernacular style. A previously steep level change (2) has been graded out using spoil from the excavation of the sunken garden, whilst beyond the old boundary wall a wildflower meadow (4) helps to blend the garden with the wider countryside.

Hide-and-seek

OPPOSITE The garden at Anne and Geoff's home was largely overgrown and near derelict when they bought the property, with just the odd stone sculpture (top right) and a few plants (middle right and bottom left) as a reminder of its gardened past. Designer Ian Kitson's plans for the garden have needed alterations along the way to take into account planning, budgets and permissions. For example, the old tennis court (top left), which was originally destined to be resurfaced and refurbished, has instead been turned into a wildlife pond.

Every garden has a feature or features that we wish we could hide away. Often it is the practical elements of a garden – the wheelie bin, the rotary washing line or the garden shed – the necessary parts of the 'working' aspect of a garden, which are amongst the most aesthetically challenged. And yet, in just the same way as a kitchen or a bathroom, this is a space that has to do things. Anne and Geoff's garden is blessed with two great natural characteristics: it is in an elevated position, and it has fantastic views to the surrounding landscape of the South Downs. But right in the way of the most important view from the house – the one that is seen from the entire rear aspect of the property – a previous owner has built a tennis court.

When compared with more prosaic items such as rotary washing lines, a tennis court is without doubt a luxury item. But it still fulfils a practical purpose, and if you like tennis, and have the space and the money to install a court of your own, then why not? To put one in such a location, however, is a pretty odd choice, similar to putting a garden shed right in front of the rear window of the lounge.

We go to all sorts of measures to try to hide things in gardens. One of the most common reactions is to use colour, either a dark, matt green or plain old black. Yet neither colour works well; both absorb a lot of light, which actually make the object you are trying to hide more rather than less visible. It's far better to either use no colour at all – allowing timber to gradually fade to silver grey for example – or use recessive paint finishes, especially light greys, which 'disappear' against the sky.

The other thing we do is to cover the unsightly object with something of exactly the same size and dimensions, so we cover an ugly building with a solid block of plants, which only serves to emphasise its shape. A more effective way to conceal an unpleasant object is to hide its outline, so rather than using solid blocks of planting to obscure an object, instead try using a mixture of vertically and horizontally shaped plants or structures to break up the outline. Alternatively, make a virtue out of an unattractive necessity. Bins can be hidden away in a neat little cabin with a green roof, whilst sheds can be clad with attractive materials and turned into a feature.

As for Anne and Geoff's tennis court, after much talk of building up levels to hide it, and sinking areas of the garden too, they eventually decided they could live without tennis and ripped it up to make a wildlife pond instead.

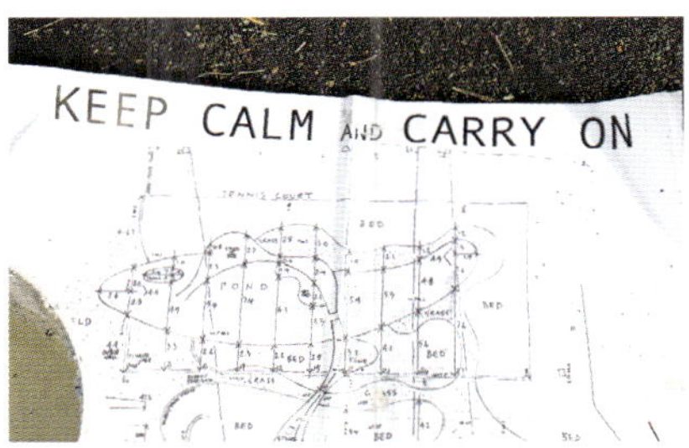
KEEP CALM AND CARRY ON

Wildflower meadows

OPPOSITE Wildflower meadows are vitally important habitats for a huge range of native species such as the common blue butterfly.

Amongst the plans for Ann and Geoff's garden is a wildflower meadow area, extending from near the house out into an existing paddock and then along the entire length and beyond the old flint curtilage wall. Really good wildflower meadows don't just happen by accident; they are actually a consequence of a number of factors coming together, usually over a long period of time – centuries sometimes. Agriculture is the reason for the existence of meadows, and the need to produce hay to provide winter feed for livestock. This ancient practice involves allowing areas of wild grass and flowers to grow uninterrupted until midsummer, when they were then cut and the grass left to dry for hay, which was then collected and stacked, in more recent years being baled using a machine. Once the meadow had greened up and recovered sufficiently from the cut, normally by autumn, sheep would be put on to graze until late winter. This combination of factors, and an absence of any chemical controls or fertilisers, ensured sweet hay for livestock but also a wildflower-rich meadow that wild animals such as butterflies, moths and small mammals could thrive in.

But old hay meadows are not especially productive, and the modern practice of silage production, which involves using large amounts of fertiliser and chemicals to create copious growth of emerald green grass that can be harvested at least twice a year, has had a devastating effect on both the old meadows and the wildlife that once used to thrive alongside them.

OPPOSITE The plans for Anne and Geoff's garden seek to blur the edges between the meadow and gardened areas by mixing formality and informality in the planting.

It's Ann's childhood memories of walking through wildflower meadows in Northumberland that is in part responsible for her desire to have her own meadow in their new garden, along with her love of wildlife. It would seem like the easiest thing to do, but trying to re-create something with so many complex interactions isn't without

problems. Firstly, the soil needs to be 'unimproved'; the fertility of the soil reduced by stripping the topsoil away, exposing the less fertile subsoil. In a smaller garden this might be as simple as stripping off the turf and the top couple of inches of soil. Once this soil has been cultivated it can be sown, ideally using seed collected from a local source as this will contain plants that are genetically specific to the area, the optimum time for sowing being late autumn or, if that opportunity is missed, late winter. Once established, the meadow will need maintaining in a way that echoes the traditional cycles of hay harvesting and sheep grazing, with mechanical mowers taking the place of four-legged ovines!

'The beauty of native flowers and grasses, with butterflies flitting from bloom to bloom and small mammals and amphibians scurrying about among the foliage, is hard to beat'

Whilst trying to recreate a meadow isn't exactly straightforward, it is well worth the effort, even on a small scale. The beauty of native flowers and grasses, with butterflies flitting from bloom to bloom and small mammals and amphibians scurrying about among the foliage, is hard to beat; and, once established in the right setting, like Ann and Geoff's garden, it can look as if it has been there for centuries.

AN ENGLISHMAN'S HOME IS HIS CASTLE

They say that an Englishman's home is his castle, but there can't be many of us who get to say that for real. Jason and Demetra Lindsay can, having inherited a large but rather crumbling Queen Anne house with various barns and outbuildings as well as a fairly unusual garden building – a huge Norman keep over 800 years old!

Jason Lindsay is descended from the founders of Hedingham Castle, continuing an occupation of this site by the family that has lasted for over 800 years. In recent years Hedingham, like many other stately homes in Britain, has suffered from the combined challenges of death duty taxes, gradual deterioration and a lack of people willing to work the land, and sadly now only a fragment of what was once a great estate now remains. The Lindsays moved in during 2005 with only portions of the house habitable, and have spent the intervening time trying to make it a home for their three young children, as well as working hard to make the estate pay for itself by turning it into a venue for weddings and events.

ABOVE Jason and Demetra Lindsay have taken on the daunting task of restoring the gardens at Hedingham Castle, which has been in Jason's family for over 800 years.

With so many drains on the estate's finances, and faced with pressing upkeep problems such as leaking roofs and damp in the walls, the garden has understandably taken a bit of a back seat. But the garden has its own fascinating history and is full of interesting features such as a dovecot, dating from 1720, as well as a brick Tudor bridge that at one point led to a Real Tennis court where, allegedly, Henry VIII once played. The most dominant feature in the garden is a large body of water known as the Long Shank, which frames magnificent views to and from house, the views embellished by venerable lime trees planted to the sides of the water.

But just as the house has suffered the ravages of time and circumstance, so has the garden. The drive to the house, which should show off the grandeur of the house to the visitor, is lined with hugely overgrown laurels and conifers, ruining the effect. And yet the bones of a wonderful garden are still there to find. The overgrown nature of the drive means many beautiful trees, such as a wonderful paper-bark maple (*Acer griseum*), are lost behind a lumpy vista of mundane shrubbery. Needing the garden to work just as hard as the house in order to pay its keep, Jason and Demetra have some tough decisions to make in tackling this overgrown, neglected garden.

ABOVE A new crinkle crankle beech hedge (1) ensures some privacy is restored to the driveway after the completed clearance of overgrown shrubs along it – a move triggered in part to help uncover a particularly beautiful paper bark maple (2). The newly opened vista (3) shows off the house to much better effect, whilst a mixture of wildflower and bulbs (4) lines the drive and contrasts with the tightly mown grass.

Hard landscaping in a neglected garden

OPPOSITE Such was the scale of work at Hedingham that heavy machinery (top left) was required to carry out a lot of it. Pre-existing features, such as the beautiful snowdrop walks (top right) were improved through the clearance of self-sown saplings, additional water features, including a restored informal pond next to the old dovecote (middle left), have been added, whilst the Norman keep (centre) is a dominant garden feature.

Taking on a neglected, overgrown garden can be a daunting challenge, whether a large historic garden like the one at Hedingham or a domestic back garden in any city, town or village. Just working out where on earth to start work can be a headache-inducing, overwhelming experience. As a consequence, overgrown gardens quite often stay overgrown for some time!

To tackle an overgrown garden most effectively, you first need to have a really good understanding of what is already there. Try to work out the original layout of the garden; check to see whether you can find any hard landscaping features like paths and seating areas lurking under the weeds and mess. Depending on how long the garden has been neglected these can be quite well hidden, so a little gentle excavation might be required. Sometimes the route of an overgrown path will be obvious due to yellowing of the grass or weeds that have grown over it, or unevenness compared with the surrounding ground. If you do uncover an old path, you might consider reinstating it, but the chances are that it will either be beyond repair or in the wrong place. In this case the materials could potentially be reused as the base for a new path or patio.

A good understanding of the layout and structure of a neglected garden can also help to highlight why a particular feature or plant is where it is. The presence of an unearthed patio in a seemingly random location might make complete sense once you observe the passage of the sun and realise that it is the sunniest spot in the garden, or the position of a bog planting in one corner might seem surprising until you realise it is where all the water in the garden runs to in times of heavy rain.

'Try to work out the original layout of the garden; check to see whether you can find any hard landscaping features like paths and seating areas lurking under the weeds and mess'

Before you do any kind of serious digging – i.e. deeper than 1m – check with local utility companies and get a plan of the underground utilities. The chances are they are far below the depth that you will ever dig to, but better safe than sorry. And be careful before you take a saw to a rickety fence or a sledge hammer to an old wall, because it may not actually be yours to demolish. Check the deeds of your house, or talk to your neighbours, before taking on a potentially costly and un-neighbourly project.

NEW HOLLAND

Careful
Deep
Water

Working with existing planting

A little judicious 'editing' can work wonders in the garden, often just through the timely removal of a branch, or an individual plant. But working out where on earth to start – what to remove and which plant to take out first – can be extremely difficult, especially if the plants concerned are mature and of sizeable stature. As with hard landscaping features it pays to understand the layout and structure of the garden. Quite often, boundary trees will have been planted for a very particular reason, to help screen an unsightly building or structure. Removing, say, a mature shrub from a the garden boundary to free up space or improve light levels may seem like a good idea, until its gone and you are left with a view of the local meat-rendering plant...

ABOVE The 'Long Shank' at Hedingham is an historically and visually important feature, with sweeping views from the house that have been adversely affected by later additions of planting such as the yew hedge across the main lawn, which interrupts the view.

Add to this the problem that, once you start taking out plants, it can be hard to stop. At Hedingham, Jason and Demetra decided to tackle the overgrown planting on the driveway up to the house to reveal the presence of the building, restore a sense of grandeur to the experience of arrival and to rescue some rather lovely plants that had become lost. They assessed the drive and selected a number of plants that they hoped would be worth keeping once the clearing was underway. However, what they quickly discovered was that once shrubs become overgrown they tend to become misshapen and can often be completely dead on one side. As a result, only a couple of the plants they had selected for keeping were worth saving, and what had been planned as a fairly modest restoration instead became a massive job. Though this situation might sound terrifying, it's worth remembering that sometimes it's better to start afresh with a new planting than carry on with an old, overgrown and potentially unsafe mess. With the right plants, decent soil preparation and aftercare, it will be only a matter of a couple of years before a replacement planting looks established.

Before going too far with removing established plants in your garden, there are also some practical and legal considerations that you must be aware of. Old or significant gardens often have listed status or might be included in a conservation area, AONB (area of outstanding natural beauty)

or national park, in which case you will almost certainly need to apply to your local authority before any significant work takes place to get the advice of their tree and planning officers. Individual trees can sometimes be protected by tree protection orders and there can be quite hefty fines for interfering with such plants. You should also be aware of nesting animals that might be disturbed by felling and clearing – in the UK it is illegal to disturb a nest site. To avoid the latter, try to carry out any clearing work during autumn and winter, outside the nesting season.

ABOVE With an established garden it is important to carefully consider the existing features before taking action. Trees or shrubs might frame a particular view (left), or obscure an eyesore, whilst other plants may hide their charms for much of the year, but be incredibly beautiful when in flower such as this *Pieris* (right).

Thinking green

ABOVE You don't need a big garden or allotment to grow vegetables – the smallest of spaces, can be suitable for growing, as these 'reinvented' containers prove.

We live in a time when concern for the environment is a topic of conversation, driver for legislation and a source of often contradictory and conflicting advice from those who are supposed to be 'in the know'. Most of us would probably like to live in a manner that is less damaging to the planet, but with so many opposing views and contradictory so-called evidence it is hard to see how to do so. In recent years we have all become accustomed to the idea of recycling, and huge investment has gone in to improving recycling facilities and collection of waste, but instead of engendering a more responsible attitude to waste it has merely changed the way we dispose of rubbish – where once we would throw everything into a single bin and then absolve ourselves of any responsibility, we are now encouraged to put our rubbish into five bins instead, yet are still largely blissfully ignorant of what happens to it.

In the face of all of this our gardens, and the way we garden, seem rather inconsequential compared with the seemingly unstoppable challenges of climate change. But our gardens have a very positive role to play in connecting us with nature, wildlife and the wider environment, and in a modest but important way can help us to make a contribution toward reducing our impact on the planet.

Just as gardens have the capacity to be complete works of art, so gardening can be an expression of nature. To grow a plant from seed requires a basic understanding of the natural environment, an appreciation of simple science and a quality that is in increasingly short supply in our modern world – patience. And the resulting plants that grow from a seed sown from our own hand, whether to admire for their beauty or savour

RIGHT If you have the space, giving over a couple of raised beds to growing food can help to reduce your carbon footprint, by reducing the need to buy imported vegetables. It's also great fun, highly educational and enables you to know exactly what has gone into making the food that you eat.

OPPOSITE **Planting trees is a popular way of offsetting your carbon footprint, but the other benefits that trees bring to the landscape – filtering out pollution, providing homes for animals and birds, as well as making our cities, towns and the countryside more pleasant places to be – are all just as important.**

for their taste, gives a satisfaction quite unlike anything else – not the instance gratification of acquisition but an almost childlike excitement, a sense of, 'I grew that!'

Gardens can also be the place where we can make a small contribution toward the health of the planet, or at the very least enjoy ourselves in a way that isn't detrimental to the environment. Plants help to reduce pollution and produce oxygen, and can provide food for humans, which in turn helps to reduce our carbon footprint because the less we need to buy from shops, which in some cases will have been imported from abroad, the less we add to transport pollution. And the ultimate destination for plants – the compost heap or wormery – can be a veritable hub of virtuous activity, helping to dramatically cut the amount of recyclable green waste that we throw away be processing it and reusing it in the garden.

Encouraging wildlife

ABOVE Gardens provide a range of habitats for wildlife through the plants and features that comprise them. But installing bird feeders and habitat boxes, such as this ladybird box, can enhance the wildlife value further.

OPPOSITE Garden plants provide a feast for wildlife, be it in the form of fruit, nuts, seeds or pollen and nectar, essential for animals like the red admiral butterfly.

Aside from the benefits to the wider environment, our gardens can help to support the local wildlife that relies on domestic gardens for food and habitats. Gardens are like corridors for wildlife, linking our towns and cities with the wider countryside, allowing more mobile animals to pass from one to the other. They also form the vital green space in what might otherwise be an entirely built environment – the lungs of the city.

Garden plants can provide food for animals in the form of pollen and nectar which can feed a wide variety of insects including bees, butterflies and moths, and although non-native garden plants can't always provide the very specific food source required by the larval stages of insects (the orange tip butterfly caterpillar will only dine out on two or three native plants, including *Cardamine pratensis*), they can provide food for the adult insects over a longer period, as there are non-native garden plants that flower when native flowers are in short supply.

Trees are a wildlife habitat par excellence, with a mature oak tree providing a home for an estimated 20,000 individual creatures – from microscopic bugs to owls. Very few of us have room for an oak in our back garden, but even a small tree can be of significance to wildlife, especially one that produces fruit and nuts for food. If there is no room for a tree, then shrubs and hedgerows are ideal nesting places for birds, insects and small mammals – and if there isn't much room for them there are nest boxes that can be bought or made to do the same job.

A garden that is friendly for wildlife also tends to be a healthy garden, as without excessive intervention in the form of chemical controls a natural balance can build up, so that the 'pest' insects that nibble plants will be nibbled themselves by predatory animals. There will undoubtedly come a time when some type of intervention is required, but if this is kept to a minimum it gives the garden and nature the chance to look after things first, rather than trying to heal the damage inflicted by impatience. There is so much more joy to be had from a garden that is full of colourful flowers and colourful wildlife, rather than a sterile environment, and if children are using the garden they will adore the adventure they embark on every time they venture out.

LIVING THE GOOD LIFE DOWN IN THE VALLEY

ABOVE Trevor and Sarah Woods have given up an urban life to try to lead more sustainable lifestyles with their three children in the heart of the Welsh countryside. The plot of land and house they have bought is largely derelict, and their budget is very tight – so they'll need to be creative and recycle or reinvent materials to get what they want.

Sarah and Trevor Woods gave up their urban lives in a spa town to pursue the goal of a near-self-sufficient, environmentally aware lifestyle in the countryside. Having sold up their property, they were able to buy a dilapidated, near-derelict farmhouse and five acres of pasture in rural Wales. Their dream is restore the house as a home for themselves and their three young children, and turn the garden into an outdoor space for the whole family to use for fun, entertaining and relaxing, whilst turning the bulk of the land over to the production of food.

In order to realise their dream for the garden, however, Sarah and Trevor have a huge hurdle to overcome – they have almost no budget with which to achieve their aims. Instead they are relying on the generous help of a garden designer friend to give them the guidance they need, allied with a mixture of bartering for products, online trawling for bargains, skip diving (utilising the materials found in skips generated by building projects and domestic waste disposal) and recycling or reinventing the objects they find on site. And whilst they do all this they are enduring their second cold winter in a cramped caravan, which the whole family share with five rescue dogs of varying 'vintages' and sizes.

Sarah and Trevor's garden is located in one of the most beautiful parts of mid-Wales, about 30 miles from the Irish Sea and in a valley surrounded by hills that range from gently rolling to steep-sided, eventually giving way to mountain ranges. It's a beautiful location to live and garden in, complete with a small river that runs through the site and forms one of the boundaries of the property. But gardening here isn't without challenges. The climate can be tough, freezing in winter and exposed to winds. And whilst the rain in Wales may be a cliché, there is no doubt it is true here – the rainfall average is over 1200mm per annum, three times as much as parts of the south-east of England.

So, with climatic challenges, a minute budget and high aspirations, how can Sarah and Trevor make their ambitious plans work?

ABOVE The near-derelict farmhouse forms the fulcrum of the site, and when rebuilding is completed the new garden will radiate from it. The new design integrates an existing green roofed barn that Sarah and Trevor built when they first bought the plot (1). Currently used as storage for most of their possessions, this is set to become the chicken shed. The new fruit and vegetable garden will be created on the site of the existing kitchen garden (2) but will utilise raised beds to improve growing conditions.

To help create their new garden, Sarah and Trevor have been able to draw upon the expertise of designer Jill Fenwick. Jill has created a plan for a new family garden in the area adjacent to the house, which includes an outdoor living space complete with a fire pit to help keep cold evenings at bay (3) as well as an outdoor dining area to enable the family to enjoy al fresco dining (4). An unusual water feature and a variety of structural planting to soften the hard landscaping and architecture of the house complete the plan.

Growing your own

OPPOSITE Sarah and Trevor's sustainable dreams began with the purchase of a near-derelict farmhouse. Growing at least a proportion of their own fruit and vegetables is a key part of their aims, which they hope to achieve by creating a kitchen garden and orchard, whilst livestock also form an important part of the Wood's sufficiency plans.

Trevor and Sarah are determined to make their land work for them by producing as much food as possible for the family, and also providing grazing and living space for the rescue battery hens and Cancun pigs that patrol the pasture. 'Growing your own' has undergone a renaissance in recent years and it's easy to see why – there is something incredibly satisfying about growing your own food.

Taking a tiny seed, sowing it and watching it grow into something that can feed you, your friends and family, and knowing everything that has happened to that crop during its growth, is both rewarding and reassuring. It's also something that everyone can engage in, and is a great way of introducing children to the idea of plants, as well as being an easy way to keep fit. Best of all, you don't need a big plot of the type that Trevor and Sarah have available to do it. A container or two, or a window box, can be enough to grow a range of vegetables, herbs and salad crops, even potatoes.

Trevor and Sarah may have a lot of space in which to grow but that doesn't mean their ambitions are problem free. For one thing, the cold and often wet weather in their area can mean that the growing season is much shorter than in warmer areas. Early in the season the soil can be too wet and heavy to cultivate, making it impossible to sow seeds into. And it can take much longer for the soil to warm up sufficiently, further delaying sowing. Damp and cool conditions also suit particular pests, especially slugs and snails, which can devastate vegetable crops, especially brassicas such as cabbages. There are also a host of fungal diseases that thrive in warm, damp and close conditions once summer arrives and the temperatures rise.

The best way to keep soil in top condition in wet areas is by using raised beds. Often made from timber, either purpose made or hand built, but also available made from a range of materials including recycled plastic, raised beds ensure the growing medium is raised out of the worst of the wet. In effect, they are like oversized boxes full of soil. Because the sides of the raised beds are exposed to the sun, the soil within the bed warms up earlier in the season, drains more readily and can therefore be sown into much earlier. Metal or plastic hoops can be fitted to the beds and covered with clear plastic to create a warmer growing environment, rather like a greenhouse, and once crops are established the plastic can be replaced with a protective mesh that can help to keep pests at bay.

Reinvention in the garden

As Plato once observed, and Trevor and Sarah have discovered over the course of their garden project, necessity is the mother of invention. When you have almost no budget with which to make a garden, reusing, reinventing and recycling objects and materials to give them a new life is a priority. Fortunately there are plenty of opportunities to reuse items in the garden that might otherwise just be thrown away, in doing so giving them a new lease of life and reducing waste to landfill. Gardeners have been good at this for centuries; reusing scraps of baler twine or old bits of wood to tie up runner beans or fill the gaps in a fence, for example.

Scrap materials can be used in a number of ways in the garden, especially as containers for plants. A galvanised water tank or old section of air conditioning duct can make a stylish planting container that will have the look of an expensive purchase but cost almost nothing. A length of iron or plastic guttering, planted with low-growing house leeks (*Sempervivum* species and cultivars) can make an architecturally interesting living sculpture, as can an old copper or steel cooking pan. Even old sanitary ware – toilets, sinks and the like – can be turned into planters, although you will need to decide for yourself at what point curio becomes bad taste!

ABOVE Giving new life to objects and materials that have already served a purpose, rather than throwing them away, is increasingly important as the waste we produce mounts up. A salvage sculpture can make an interesting curio in the garden – even if it's just for the sake of irony!

'There are plenty of opportunities to reuse items in the garden that might otherwise just be thrown away'

To find all these interesting materials, you need to be willing to look in the right places. This can mean an online trawl, a visit to a local scrap metal merchant or architectural salvage, or a bit of peering into builders skips. There are also bargains to be had in the form of purpose-made equipment that other people no longer need. Trevor and Sarah managed to get hold of a solar dome – a type of large glasshouse made as a geodesic dome – for next to nothing as it was no longer wanted and would have cost the owner money to dispose of. They've also used the slate innards from an old billiard table to make a beautiful outdoor dining table, something that (even in a country as rich in slate as Wales) would cost a fortune. Regular checks on recycling websites will often throw up similar bargains.

LEFT Collecting builders' waste for reuse can offer up all sorts of interesting opportunities for reinvention. Waste pipes can become planting containers, offcuts of timber or stone reused, while paving and walling can be lifted, brushed up and re-laid.

BELOW A little time spent on recycling websites can throw up all sorts of great bargains. Sarah and Trevor found this geodesic dome online, inside which they now grow frost-tender veggies such as cucumbers and peppers for the garden.

CONSERVATION ON THE CLIFF TOP

ABOVE The island landscape of Guernsey is dramatic and beautiful, but like many small islands there are pressures on the environment that pose significant challenges.

Of all the places to make a garden, on top of a cliff in the middle of the English Channel must be one of the most interesting and challenging. Vanessa and Ed Adams bought their modest bungalow not for the property so much as for the spectacular vistas. Four acres of cliff top on the south-western tip of the Channel Island of Guernsey overlook a beautifully craggy bay with just a few other properties to spoil the illusion that it might just be an uninhabited island.

Vanessa and Ed see the development of their garden as a chance to create something that is in tune with the local environment, whilst at the same time being a viable family space for themselves and their two young sons. It's a pretty big challenge: the site is overgrown with bracken and brambles, with a few remaining shrubs and trees planted years before – the ghostly remnants of a previous garden. The site also slopes rapidly away from the house down to the cliff path, below which is an almost sheer drop of thirty metres or so to the sea.

The rock that forms the island of Guernsey is granite, one of the toughest of all rocks. Above this the soil on the island 'floats', held down largely by the plants that grow on it. This makes removing weeds something of a double-edged sword – it is important that they do so to enable areas to be cleared and replanted, but at the same time there is a very real risk that, if left for too long, the topsoil will simply blow away in the prevailing sea winds.

Vanessa and Ed must also face the tensions that will arise as a result of trying to create something that is ecologically sensitive but at the same time a viable garden. They want to include practical features in their garden such as play areas for their boys, an outdoor dining area, as well as somewhere to grow fruit and vegetables. But flat space is in very short supply in the steeply sloping site, and the obvious solution to this – putting in a series of terraces using materials like stone or timber – is time-consuming and expensive.

There's no doubting Vanessa and Ed's commitment to preserving and improving their local environment, but how much can their garden be expected to achieve? Intelligent design will be needed to ensure they don't end up with a space that is neither garden nor wild habitat, that meets neither their needs nor their dreams, and that, far from being a shining example of a new way of gardening in tune with nature, might just be a bit of a mess.

ABOVE The area (1) behind the house has been levelled to make a flat play area for Vanessa and Ed's boys to enjoy. On this steeply sloping site flatter areas have been created by terracing, using dry stone walling and railway sleeper timbers (2). A winding path (3) provides a journey through the garden whilst following the contours of the slope, making for a safer, less precipitous route. Timber jetties (4) form the turns in the path, and provide areas to sit and enjoy the spectacular views.

Nurture or nature

OPPOSITE One of the big difficulties, when gardening with nature, is that the aesthetic tipping point between attractively wild and unkempt and mess can be extremely fine. Railway sleeper sets smothered in wildflowers bring together the deliberately planned features of a garden with the randomness of wild beauty.

Gardening is, by definition, the manipulation and subjugation of nature. Or perhaps more accurately the attempted manipulation and control of nature, because nature will always, without question, win out. Even if a garden is only left untended for a matter of days or weeks, the first signs of re-colonisation by nature will appear in the form of a covering of weeds. Left for longer, grasses, brambles and eventually trees will take over from manicured borders – it's the way of nature, and always has been.

The tension between where a garden ends and nature takes over is especially difficult in places where the landscape is under threat, or is of particular ecological sensitivity. Vanessa and Ed Adams' garden in Guernsey is representative of the challenges and sometimes seemingly imponderable questions that face anyone wanting to garden in an environmentally sensitive way. Guernsey wrestles with all the same environmental issues as the rest of the Britain: overcrowding, pollution, waste disposal and so on. But on a small island of just 24 square miles, and with a population of 60,000 (that's 2,500 people per square mile) the problems are amplified. Vanessa and Ed want to respond to these challenges by sensitively managing their bit of cliff top, partly simply for their own enjoyment but also for loftier ambitions, to offer a possible template by which other cliff-top gardens in Guernsey might grow.

The difficulty for anyone wanting to garden with nature is that it's impossible to decide how much of it to allow in, and how much to control and keep out. Whilst we might like to think we are in control of nature, we clearly aren't. Therefore, how can a garden be created without accepting that intervention will have to be made, and that in some instances it might not always be beneficial to nature in every sense?

The answer, to an extent at least, is that it can't. The manipulation required to make a garden will inevitably cause disruption to the natural state of things – from disturbing the soil flora through digging, to removing a tree or a shrub that may have been a nesting site for birds. But the reality is that nature itself can be far more destructive: creating forest fires, blowing up hurricanes and whipping the sea into tsunami. And at the end of the disruption caused by making a garden, the result is something at worst benign, perhaps beneficial, that will provide habitats and food for all sorts of animals, as well as pleasure for humans too.

'Even if a garden is only left untended for a matter of days or weeks, the first signs of re-colonisation by nature will appear'

By the seaside

Gardening by the sea throws up a range of specific challenges that aren't faced by gardeners anywhere else. Sea winds are laden with salt, which can scorch and burn foliage, cause plants to wilt and in the most extreme cases completely defoliate. The strength and persistence of the winds can cause plants to become stunted, one-sided versions of what they should really be, turning big forest trees into bonsai. As they can bent over, broken or even pulled from the ground, and the soil around them blown away into the sky, it can be a challenge to get new plants to establish in such persistently windy conditions. The bright sunlight that comes with a seaside location can cause wilting and limit the range of plants that can be successfully used, as well as washing out soft, pastel colours that would work well elsewhere.

But gardening by the sea isn't all bad news. The high light levels and wide horizons create a purity of light usually found in places like the Mediterranean fringe. Strong, bright colours look beautiful by the seaside and colours that would usually look too bold or overbearing can be used with enthusiasm. Likewise, found materials such as driftwood, beach pebbles and the detritus of the seaside can look terribly kitsch in an inland garden, but look perfectly lovely in their natural environment.

ABOVE A seaside location need not preclude having a garden, but it does require a lot more thought and planning to ensure the correct plants and practices are being followed.

And then there are the plants. Nature has provided a great bounty of those that can tolerate seaside living, as well as a number specialist plants that absolutely have to be by the seaside to thrive. Some, such as sea holly (*Eryngium maritimum*) send down deep, thick, questing tap roots that carry out several functions; acting as a storage vessel for starch and sap, collecting whatever water lurks far down in the ground, and anchoring the plant firmly to terra firma – no mean feat when you might be submerged by the tides or whipped by onshore winds. On the California coast there is a little plant called *sand verbena*, which is quite uncommon, and which has sticky leaves and flowers that cause sand to be 'glued' to the plant, which helps to act as a protective shield from the harsh environment.

Understanding the challenges and responding to them sympathetically, rather than trying to fight against them, is the key to creating a successful seaside garden.

BELOW In Vanessa and Ed's garden, only those plants that are adaptated to the particular seaside conditions, such as *Elymus magellanicus* (centre), native ferns (centre right) and the shoreline-growing sea holly *Eryngium maritimum* (bottom right) will thrive. Seating areas (bottom left) are fitted with timber screens to provide shelter from the wind for those enjoying the superb sea views.

VISION

2

'God is in the details'
Ludwig Mies van der Rohe

You don't have to be a genius to make a garden, but great gardens often feel like the work of genius, effortlessly combining a mixture of art, craft and graft to create a living and changing three-dimensional space. Almost 300 years ago, Alexander Pope adopted *genius loci* (genius of the place) as a central principle of landscape design. His mantra of respecting the individual qualities of a place, developing an understanding of nature and applying that knowledge foremost when designing a garden is still as valid as ever:

Consult the genius of the place in all;
That tells the waters or to rise, or fall;
Or helps th' ambitious hill the heav'ns to scale,
Or scoops in circling theatres the vale;
Calls in the country, catches opening glades,
Joins willing woods, and varies shades from shades,
Now breaks, or now directs, th' intending lines;
Paints as you plant, and, as you work, designs.
Epistle IV, to Richard Boyle, Earl of Burlington

Today we are exposed to design in a variety of guises – from industrial to architectural, automotive to technological. The aftermath of the Second World War created both the opportunity and the market for well-designed, often minimalist products – with everything from pottery, glassware and furniture to electronics, automobiles and architecture becoming the subject of fresh thinking and uncluttered design. Just as with Pope's *genius loci*, the philosophies of the designers behind these products have also influenced us. Amongst these are the ten design commandments of Dieter Rams, the German industrial designer most associated with the groundbreaking Braun products of the 1950s and 60s:

1. *Good design is innovative.*
2. *Good design makes a product useful.*
3. *Good design is aesthetic.*
4. *Good design helps us to understand a product.*
5. *Good design is unobtrusive.*
6. *Good design is honest.*
7. *Good design is durable.*
8. *Good design is consequent to the last detail.*
9. *Good design is concerned with the environment.*
10. *Good design is as little design as possible.*

Rams' principles can be applied to any discipline, even garden design. We want our gardens to be useful, but also aesthetically pleasing. Durability is certainly important, as is a concern for the environment. But what really makes a well-designed garden is a considered combination of the various components (and not just the practical ones such as the shed or the washing line) that go together to form the whole.

If consideration is given to the details such as how space is organised and divided, as well as to the effects that certain important elements such as light, colour and texture have on that space, then your garden can be transformed. This vision process, informed by the practicalities covered within 'Thought Process' will help you to create a template for the design of your garden. It is also the best opportunity throughout the course of the whole process to have some fun and really allow your imagination to run free. Professional designers love to push against the boundaries of a brief, dreaming up all sorts of unusual solutions to the issues posed by its most practical aspects, and this is a great chance for you to do the same.

Space

OPPOSITE Tiny spaces can be incredibly challenging to design– they can easily look overly contrived or be so crammed with 'stuff' that they become unusable. Here a small space is maximised by using vertical and horizontal planting to blur the area's narrow boundaries, whilst a table and chairs act as a focal point to draw the eye from the walls to the centre of the space.

Whether commuting to work on overly busy, traffic-filled roads or shopping in the local supermarket with its jammed isles and interminable snaking queues of shoppers, all of us wish we had a little more space from time to time. We need to feel like free people, rather than caged animals. And yet, as a consequence of modern living, we are constantly battling against an overall lack of space – whether that be a dearth of practical storage space in our homes or just a little elbow room on the train. The space 'poverty' that we suffer from affects every aspect of our lives and results in us valuing what little space that we do have ever more highly.

Our gardens offer us a chance to experience a comparative rarity, the simple, unalloyed joy of free outdoor space. But even here this precious taste of space is squeezed, our homes are now often built closer and closer together and our gardens are crammed into ever-shrinking plots. The proximity of other houses often means our gardens are overlooked, with the fences and hedges of neighbouring gardens becoming intrusive, overbearing structures as we struggle to control what little privacy we have left. And though our gardens may be smaller, our desire for space is not reduced. In these circumstances the need to apply considered and thoughtful design to the garden becomes even more important if the little space that is available is to be used to best effect, with all our practical considerations taken care of in a way that doesn't just meet our needs, but allows, in this ever-more pressurised world, for a few dreams to be dreamed and escapes to be made.

RIGHT Spatial division in the garden isn't just about making room for the things you need; it's a fundamental principle of the aesthetics of design. Even a simple area of lawn can become its own space when viewed through a screen of surrounding plants.

Divisions of space

OPPOSITE Spatial division can be as simple as the allocation of a place for a particular feature, such as an outdoor dining table and chairs or a garden pond; or can be far more considered and precise in execution, using a series of dividing features to create garden rooms that have their own distinct character.

Most gardens are, in effect, boxes. They may vary in shape and size but the majority will typically have four boundaries – these could be fences, walls, hedges, other properties, or a combination thereof.

The natural reaction to the urge to maximise the space within these boundaries is to make the middle of the box as empty and clutter-free as possible: emptiness equals space. So flower borders and trees are pushed out to the edges of the garden and it ends up looking rather like the drum of a washing machine after a wash, with everything flung out to the sides and a gaping hole in the middle. And yet having nothing but space can actually make the garden appear *smaller*; with the boundaries clearly visible, our minds are able to read the space for what it is.

But, if you apply a little trickery and instead try to divide up the box within the boundaries you can deceive your eyes into believing the space is bigger than it actually is. Creating particular areas or 'rooms' within the garden that have a different atmosphere – either through planting, hard landscaping or something as simple as the quality of light – can create a sense of moving through an experience, rather than viewing something from a distance as a detached observer; because what our eyes cannot see is always more intriguing and alluring than what they can.

LEFT This precipitous garden has been deftly divided into a series of mini garden rooms, created by forming the garden into three level areas by means of terraces, each with a distinct aesthetic flavour and strong architectural planting in the form of clipped box.

RIGHT Applying flowing, curving lines within the boundaries of a rectangular outline can help to break down the 'washing machine' effect, where planting is pushed to the outer boundaries of the garden, leaving a dull void in the middle.

'If you apply a little trickery and divide up the box within the boundaries, you can deceive your eyes into believing the space is bigger than it actually is'

ABOVE Using a variety of hard materials including timber decking and stone paving, divided by generous planting areas, helps to create the sensation of a variety of different spaces without the need for vertical barriers.

Vistas and focal points

Case Study Suffolk

OPPOSITE One of the ways that Clive and Debbie Morris intend to give their garden dramatic visual impact is through the intelligent use of focal points. This wire urn from Saling Hall in Essex sits amongst the frothy flowers of native cow parsley, which effectively complement the urn's colour and lightness of frame.

Our eyes are wonderful things, miracles of design with extraordinary power. But they do tend to wander. Even when seemingly fixed on one spot, they are constantly moving, re-focussing and taking in new information. And yet, when there is something interesting or different to see, something that grabs our attention, our roving eyes pause, even if just for a moment.

The power of the well-placed 'eye catcher', as focal points are often called, has long been understood by landscape and garden designers. Many of the great gardens that were created between the late 17th and 19th centuries were designed with follies, temples, bridges and grottoes as focal points, often reinforced by strategically placed avenues of trees that would create narrow views or 'vistas' to lead the eye in their direction.

In the domestic garden, vistas and focal points can be effectively used to do one of two things: to either show something off, or to draw attention away from something that you would rather not make a point of. Though the space available in most gardens is likely to prevent the construction of a hermit's grotto or an elaborate mausoleum at the end of the garden path, a focal point can be almost anything that catches the eye – a piece of garden furniture, a brightly painted plant pot, even the shed. In fact, the focal point doesn't even need to be in the garden, as borrowing interesting features from the landscape beyond will instantly help to make your space seem bigger and more interesting.

'A focal point can be almost anything that catches the eye – a piece of garden furniture, a brightly painted plant pot, even the shed'

But whatever the nature of the focal point, the vital rule to follow is that it is deliberately and consciously framed. There's no point in having half a dozen brightly painted pots dotted around in open space hoping they will act as focal points; that wandering eye will simply read them as a gaggle of lurid pots. And this is where the vista comes in. A grand garden might have sweeping alleys of majestic trees to frame a view, but in a smaller garden a simple vista can be created by making an aperture in a hedge, or arranging a group of pots in such a way that it frames an interesting feature.

By establishing multiple focal points framed from different view points throughout the garden you can really begin to exploit the space, creating numerous points of interest. Even the smallest garden will benefit from a carefully placed focal point. Surprisingly perhaps, even though they will be intimately known to you, these focal points will continue to catch your eye.

LEFT The borrowed view is one of the great classical landscape gardening motifs, but usually in reference to borrowing the view of distant fields, lakes or valleys. Here, on a more modest scale, a pierced brick wall offers views beyond one part of the garden to another with a different planting style. The principle of the borrowed view still applies today, even in a small garden, and can apply to a carefully framed external view or to something within the garden that is deliberately picked out, as here.

Case Study *Sussex*

LEFT In Anne and Geoff's Sussex garden, one of the few remaining features – a faux classical sculpture of a woman – is being recycled into the new garden to continue its role as a focal point, watching over the view to the South Downs.

'By establishing multiple focal points framed from different view points throughout the garden you can really begin to exploit the space, creating numerous points of interest'

ABOVE Vistas are typically associated with gardens of scale, as here at Saling Hall in Essex where a classical, colonnaded building acts as the focal point to a long vista comprised of parallel plantings of shrubs, hedging and trees. But vistas need not be so grand; the combination of deliberately aligned features with a terminal point can be achieved on a much smaller scale. It's the composition that counts, not the size.

Taking a journey

One of the biggest weaknesses of the 'washing machine' garden, where there is nothing but an open space in the centre and a few borders flung out to the boundaries at the sides, is that there is no sense of being able to take a journey through the garden. When everything is on display, there is little or no imperative to step out into the garden and wander around the space, because you can see all too clearly that there is nothing to explore.

And yet, what makes a garden really interesting is combining the deliberately shown off – the focal points designed to draw your eye and the vistas used to frame them – with the concealed, the controlled, the just out of reach or round the corner. In this respect, the alignment and orientation of the humble garden path can play an important role in setting up the sense of a journey waiting to be taken, a little tease to make you want to venture off in search of where it might take you. A curving path automatically encourages the viewer to want to see around the corner, whilst a straight path with right-angle turns can lead the eye in stages through the garden, setting up multiple opportunities for 'events' to take place.

These 'events' can take many forms but each is in essence designed to act as natural punctuation in your journey around the garden. An event might be a focal point such as a bench which, when you reach it, beckons you to sit. If you're really clever, these events will open up the unexpected: sitting upon that bench, for example, could reveal another focal point in the opposite direction, whilst on walking through an archway you might be greeted with an unexpected change in planting style or a previously hidden open space or seating area.

RIGHT In a small garden the space for a journey will always be limited. But a journey can be taken with the eyes, from a static position, as much as on two feet. Here a number of factors combine to create the sense of a visual journey: the terminal focal point of a water feature flanked by *Cordyline*, the inset lawn surrounded by sleek stone paving, and two benches that invite use.

Case Study *Essex*

LEFT A mown grass path through a mini meadow of long grass and wildflowers creates a romantic, bucolic journey, the equal of anything that can be created with big budgets and hard landscaping. At Hedingham Castle the cessation of mowing in large parts of the garden has allowed wildflowers to become re-established and meadow walks to be created.

'What makes a garden really interesting is combining the deliberately shown off ...with the concealed, the controlled, the just out of reach or round the corner'

Vertical space

OPPOSITE Planted alongside a boundary barrier, pleached hornbeam avenues (top) help soften a garden's edges. Features such as pergolas (bottom right) or archways add vertical interest to a garden and serve a practical purpose as plant supports without occupying too much horizontal space, while a stepped series of planters against a wall (bottom left) can be a very effective way of getting both height into a garden and utilising an otherwise 'dead' space.

Given that we humans move through three dimensions, it's interesting that, when it comes to conceptualising our garden space, we tend to take a very two-dimensional approach. Whilst our gardens are measured in linear fashion (80ft x 20ft) or by square footage or metres, the vertical plane remains the great, unexplored frontier – a by-and-large limitless space that we tend to ignore. In part this is a result of our widely held (incorrect) belief that an empty void in the middle of the garden is the best way to maximise our garden space and that any tall features will compromise this. Perhaps there is also a fear of using features or plants with height in case they look out of scale, or cast unwanted shade where we want the maximum possible sun to enjoy.

However, far from shrinking the garden, exploring the vertical plane can help to make it seem much bigger. Vertical features draw the eye upwards and outward, and help to give your garden an interesting extra dimension. These features don't need to be massively tall – plants such as the common foxglove or hollyhock can draw the eye upwards even though neither is taller than 1.8m. Vertical structures can also play multiple roles in the garden; a pergola or archway can provide a vertical accent whilst also acting as a support for growing plants, a means of casting dappled shade and a mechanism for framing a vista or as a focal point; whilst obelisks – if carefully placed and used a recurring feature – can create a sense of rhythmic repetition. Whatever feature you chose, remember that the trick is not to confine it to the peripheries of your space. Instead try using it in the heart of the garden, so that rather than emphasising the smallness of a space it can balance it out, whilst adding drama and scale at the same time.

Five plants to add height

Echium pininana
A huge narrow spire smothered in blue flowers that are loved by bees.
2–4m x 90cm

Molinia 'Skyracer'
Fine, airy stems with a translucent quality and lovely golden autumn hues.
2m x 40cm

Eremurus robustus
Produces an impressive tall spike of pale pink flowers in June.
2m x 1.2m

Digitalis parviflora
Has very narrow spires of rusty brown flowers that are handsome en masse.
1.2m x 30cm

***Juniperus scopulorum* 'Skyrocket'**
An evergreen conifer with a pencil-thin habit. A great vertical structure plant.
3m x 50cm

Case Study Suffolk

LEFT Bringing vertical features into the centre of the garden can seem counter-intuitive; our instinct is to push height to the boundaries and preserve the centre of the garden as open space. But as this mini avenue of pleached hornbeam demonstrates, adding vertical interest to the heart of the garden can add to the space in the garden rather than compromise it.

Scale, proportion and balance

Getting scale, proportion and therefore balance right in a garden can be a tricky exercise, especially when space is at a premium. In a small garden the need to fit in all the practical features required, along with a diversity of desirable elements, can lead to the overall effect being bitty or twee. When the features that form the boundaries or outlook of the garden are substantial, it can be even harder to keep things in scale – if your tiny plot is surrounded by towering buildings or looks out onto vast open fields, what then?

One way to help achieve a sense of balanced scale and proportion is to use the proportions of the house to set out the garden by making a scale drawing of the facade of the house and laying it over a drawing of the garden – a technique popularised by the groundbreaking British designer John Brookes. By doing so, the major features of the house begin to help determine where features in the garden begin and end in a very natural way. This doesn't work in every instance of course, but it's a useful tool in situations where the proportions of house and garden allow.

There are other simple guidelines that can help you to get the balance right. When setting out the dimensions of a planted border, ensuring that for every linear metre the border is half as wide again will help to prevent the creation of thin, 'mean' borders. Though there is a limit to how far this rule works, depending on the overall space of the garden, it is a helpful guide.

RIGHT Achieving the right scale and proportion in a really tiny garden can be a challenge. This very small courtyard garden features a dominant lead tank water feature and ornamental trellis surround painted the same colour. Although seemingly outsized at first glance, the scale of the water feature works – just because it is so bold.

LEFT **Perspective affects scale and proportion as surely as actual physical size, and this can be manipulated to create different visual effects in the garden; creating a falsely elongated vista or foreshortening a view. Here the size of the summerhouse, the length of the path and the height of the hedges in the foreground are all in balance.**

In a really tiny garden it can sometimes help to be deliberately bold with scale, upsizing pots, furniture and plants to create the illusion that the garden is larger, which works well providing some restraint is applied – too many massive pots and huge plants can leave you feeling Lilliputian. But it is certainly better than going the other way and filling a tiny space with tiny things, as this is guaranteed to make the space look very small indeed, and probably rather silly. It can also lead to a loss of cohesion as a clutter of tiny features, all jostling with one another for visual attention, is unlikely to make for a happy blend of elements.

There are instances of course when it helps to play with scale and proportion. Using a slightly undersized focal point, say a bench or garden building at the end of a vista, will make the object appear further away and the vista longer, in turn making the garden feel bigger. Vista lengths can also be cheated by accentuating perspective – gradually narrowing a path, or bringing the two sides of an avenue gradually closer together, will lengthen a view. Bear in mind though, that this technique does need to be carefully executed, as too sudden or dramatic a narrowing will appear too obvious a 'cheat'. Ultimately, the only foolproof way to ensure that the balance and proportion of features in the garden is right is to plan thoroughly, try them out on the ground, and be prepared to change them, if need be, as you put your plan into action.

ABOVE To avoid the visual confusion and general sense of incoherence that can be created in a small space when too great a variation of plants and hard landscaping materials are used, this small courtyard garden has been very simply planted using a restrained palette. The seating area uses paving stones that match the colour and materials of the house walls, with the result that the space appears larger than it actually is. The simplicity of the predominantly green planting, and the textural and repetitive qualities of the clipped box orbs all contribute to creating a calming, cohesive space; whilst the row of mature grape vines gives a sense of enclosure, of looking through plants, without impeding views or shrinking the available space.

ABOVE This small, shaded, courtyard city garden has been designed to make the most of the available space. A pair of large clipped bay trees, planted in containers, flank the entrance to a timber decked seating area, framing the view and delineating the end of one space – an area of stone paving and plants in terracotta pots – and marking out the beginning of another. Creating a focal point between the bay trees, a large clump of African lilies (*Zantedeschia aethiopica*) and slightly offset table and chairs invite the viewer into the seating area. The vertical space in this garden is well used too, with height from the boundary wall topped with trellis and the generous tree and shrub planting at the end of the courtyard.

Light

OPPOSITE The effect of light and shade can be as visually arresting as any flower or foliage combination in the garden. Here an ancient lantern in an alcove and the hatched shadow cast by a pierced timber screen, playing out against a plain, whitewashed wall, are all that is needed to create a beautiful composition.

Light is as fundamental to our existence and that of the majority of animal and plant species on Earth as oxygen. In more prosaic terms, light is vital for our wellbeing, and is a central component of the garden experience.

The amount of natural light that we have in our gardens depends on four factors: orientation, season, surroundings and content. Orientation is the direction a garden faces according to the compass and has a fundamental effect on natural light levels and light quality. As the principal source of natural light is the sun, orientation also influences how warm the garden gets, how quickly it cools off, whether the soil warms early in the growing season and so on. In simplistic terms, in the northern hemisphere a south-facing garden is sunny and warm, a north-facing garden shaded and cool, an east-facing garden sunny in the morning and warm for a few hours before cooling off, whilst a west-facing garden is sunny and warm during the afternoon and evening. There is no right or wrong to the orientation of a garden, just different conditions to understand and work with.

Light levels also vary according to the season, due to the relative position of the sun in our sky. During high summer the trajectory of the sun is at its most elevated, ensuring the maximum amount of light in open, south-facing gardens and the greatest amount of reflected light experienced in otherwise shaded north-facing gardens. By midwinter the sun is at its lowest, and even on a bright day the shadows cast by fences, trees and walls will last all day.

Surroundings can have a dramatic impact on light levels too. An open boundary, or one marked by a permeable barrier such as a chain link fence, will allow light through without casting a shadow, whilst a solid barrier like a timber fence or brick wall will either act as a sun and light trap, with the light and heat being reflected back from it, or will cast shade.

The content of the garden has rather more subtle implications for the intensity, quality and duration of light, as features of any description with verticality – be they shrubs, pots or posts – will cast shade. How deep that shade is will depend on the nature of the object casting it: a tree with a light canopy like silver birch (*Betula* species) casts a pleasant, dappled shade, whilst an evergreen conifer with dense foliage such as the popular hedge plant, yew (*Taxus baccata*), can cast shade as dense as that of a brick wall. The materials employed can also create interesting effects; light-reflective paint surfaces, water, light-coloured stone, and the nature of the plants – from the feathery plumes of an ornamental grass to the glossy leaf of a *Camellia* – will all change the way light is reflected, amplified or subdued.

Playing with natural light

OPPOSITE Very few shaded gardens are completely in shade all day. Here an understanding of where light and shade are cast has influenced the placement of a table and chairs, which sit in a pool of diffused light. The white-painted wall helps to reflect as much light as possible back into the garden.

The interplay of light and shade in a garden is what helps to bring it to life. It also offers us a range of opportunities to exploit; a naturally occurring pool of light can be the perfect spot for an eye-catching feature, lighting it up like a leading actor in a theatre spotlight, whilst complex shadows cast against reflective white paint can be as beautiful to look at as a planted border, the simplicity of the constantly changing shadows creating movement and shifts in mood throughout the course of the day.

Certain plants, such as ornamental grasses, have the ability to seemingly 'catch' light in their flowers or seed heads, whilst the capacity to create intricate patterns of light and shade is one of the key features of certain large-leaved, architectural plants. Through its ability to enhance light and reflect the sky and its surroundings, a body of water can also transform a garden space.

'A naturally occurring pool of light can be the perfect spot for an eye-catching feature, lighting it up like a leading actor in a theatre spotlight'

A gloomy side return, one of the most common of all urban garden experiences, can be turned from a space to be passed through in a hurry to something you will want to show off through the application of bright, light-reflective paint, light-coloured paving materials (which will reflect far more light than a matt, dark-coloured slate), containers and the planting of big, bold, glossy-leaved plants – many of which are big and bold precisely because of their adaptation to shade.

Five light-catching plants

Miscanthus sinensis
An oriental grass with feathery flowers that light up in the low autumn light.
1–1.5m x 1.2m

Chamaerops humilis
This hardy fan palm has great structural presence and will cast intriguing shadows.
1.5–2m x 1m

Lunaria annua
The translucent seed heads of this biennial look like silvery moons.
1m x 30cm

***Clematis tangutica* 'Bill Mackenzie'**
Bears fluffy, hair-like seed heads after orange peel coloured flowers.
5m x 2m

Fatsia japonica
The large, very glossy leaves of this shade-loving plant will reflect a lot of light.
2–3m x 2–3m

TOP LEFT In this shaded spot a number of elements are working hard to maximise the available light. The vertical water feature has been made from highly reflective metal, whilst the running water and the smooth, polished paving both work to bounce back as much light as possible.

BOTTOM LEFT Here a curtain of highly reflective plastic sheets, coated with a silvery finish, acts like a wall of mirrors to improve light levels. The timber deck has been cleverly designed to match the dimensions of the plastic curtain sheets, creating the illusion that the two features are one.

ABOVE A body of water is the ultimate in light-reflective surfaces. Here a superbly executed formal pool shows not only the visual impact of water, but also – with the stepping stones that seem to float just above the surface – the striking way it can interact with its surrounding materials.

'Through its ability to enhance light and reflect the sky and its surroundings, a body of water can transform a garden space'

Shade

OPPOSITE **Gardening in deep shade can seem like a daunting prospect, but there are a host of plants that will thrive in shade, many of which have beautiful architectural qualities and superb foliage texture and colour.**

If there is one climatic condition guaranteed to strike fear into the heart of the gardener, it's shade. Thoughts of half-dead plants in dusty, dry soil, loitering in gloomy half-light are not usually the stuff from which gardening dreams are made. But there really is no need to be scared of shade; instead, all that is required is an understanding of how to manage the conditions it brings, because there are actually thousands of plants that will thrive in it.

Shade affects the growing conditions in gardens in a number of ways. Because the soil gets less direct sunlight, it is usually slower to warm up, which affects how quickly plants start their season's growth. The lower light levels also mean that plants have to be adapted to shade in order to be able to make food by photosynthesising; plants that aren't will turn up their toes and die when thrust into the gloom and left to get on with it.

To succeed in a shady garden, use those plants that are adapted to growing in shade (there are thousands), and look at implementing strategies that may help improve light levels, such as the use of reflective materials and paint finishes. Light levels can also be boosted through the manipulation of plants – woody plants, for example, can have the bulk of their foliage thinned out to create an open branch structure that allows more light through, or hard pruned back to the ground to boost light levels at the start of the season, when many lower-growing plants need it most to start into growth.

Light-coloured or variegated foliage plants can also provide a light-boosting opportunity. A group of the silver-, white- or golden-leaved plants, such as the silvery false forget-me-not *Brunnera macrophylla* 'Jack Frost' in a dark corner can be as good as a pool of sunlight. Careful placement of variegated plants is required, however, as too many of them, especially with a mixture of golden and silver variegation, can look disastrously incoherent.

Five plants for shady gardens

Dryopteris erythrosora
This handsome border fern has bronze-pink young fronds.
60cm x 40cm

Anemone hupehensis
A lovely autumn-flowering perennial that quickly colonises in shade.
60cm x 40cm

Hakonechloa macra 'Aureola'
An attractive yet understated grass with deep green foliage.
35cm x 40cm

Galanthus nivalis
Snowdrops are one of the year's first flowers and are always a joy.
10cm x 10cm

Geranium macrorrhizum
Has fragrant foliage when crushed and attractive flowers.
50cm x 60cm

Full sun

OPPOSITE Amongst the many thousands of plants that thrive in full sun are some real specialists. *Allium* (top left) cope with drought – as do the majority of bulbous plants – by vanishing below ground before the weather gets too hot. Many of the palms, including *Chamaerops humilis* (bottom right) have a hard outer coating to their leaves, like a plastic shell, that helps them to hold on to moisture, while *Aeonium* (bottom left) have thick fleshy leaves to store water and starch.

Whilst gardens that benefit from full sun may seem easier to deal with in design terms than shaded gardens, there are actually just as many challenges to face. Materials have to be selected that aren't blindingly harsh in bright sunlight, whilst a comfortable retreat from the heat and sun for the most delicate plants has to be provided. Managing the soil can throw up challenges too, as unrelenting sunshine can cause rapid drying out, which then leads to plants wilting.

However, just as with a shady space, appropriate plant selection is key to creating a successful garden in full sun. Unsuitable plants (often those with very big, soft leaves or lots of soft, sappy growth) can wilt in minutes in a really hot garden, whilst containers will need a continuous round of watering unless they are planted with suitably drought-resistant plants, in which case they might go for weeks without a drop of water, even in really hot weather.

Strongly architectural plants such as *Yucca*, *Agave*, *Aloe* and *Chamaerops* excel as they have both the presence and the appropriate aesthetic to succeed in full sun – the glare of which brings plants and features into sharper relief and bleaches out subtle colour. These plants are at their best in the garden when used sparingly; a planting featuring a few key structural plants along with a ground storey of lower-growing plants is the best way to showcase these extreme drought specialists. It is also worth remembering that such powerful architectural plants also cast strong shadows that can be as attractive as the plants themselves when thrown against a white-painted wall or sheet of reflective water.

Five plants for full sun

Lavandula lanata
Like all lavenders, this species is a must-have for its fragrant foliage and flowers.
75cm x 90cm

Yucca gloriosa
The 'Spanish Dagger' is an imposing plant with large, sword-like leaves and huge flower spikes.
2m x 2m

***Perovskia* 'Blue Spire'**
This upright shrub has silvery-white foliage and powder-blue flowers.
1.2m x 1m

***Cistus* 'Blanche'**
This perrenial shrub is easy to look after, fast-growing and very colourful.
1.5m x 1.5m

Santolina chamaecyparissus
This cotton lavender makes neat, rounded bushes with finely divided foliage of silvery-white or grey-green.
2m x 2m

Using artificial light

BELOW Artificial lighting can be as subtle and complex as the effects created by natural light and shade. In this garden a range of lighting sources and styles have been combined to create general background light as well as targeted light to pick out specific features.

When the sun goes down, the lights need no more go out in the garden than they do indoors. Artificial lighting allows the use of the outdoor space to be maximised, extending the garden experience beyond the daylight hours, but at its best it can bring a completely new and distinct quality to the garden. Whilst natural light can be manipulated to some degree, the deployment of artificial lighting is completely in our control. At its most basic, this can be the illumination of the space to provide safe access after dark. But with considered application, artificial lighting becomes something far more significant, capable of highlighting specific features or providing a continuum from the indoor space, or creating an entirely distinct outdoor experience that emphasises the special nature of gardens after dark.

Outdoor lighting, or 'lightscaping' as it is sometimes known, can transform the experience of the garden at night. A pool of water, for example, primarily acts as a reflective body during the daytime, bouncing and amplifying light from its surface. But at night, artificial lighting placed

within the pool completely changes its nature, causing it to be a source of light, rather than a mirror to it. Artificial lighting can also be used to highlight plants or features that are largely hidden during the day – a tree, shrub or container can become the centrepiece for the entire garden with appropriate lighting. And the mode of lighting can become a feature in its own right; tea lights or candles arranged in an alcove or lined up along an outdoor dining table or floating on the surface of a pool can bring a little nocturnal magic to the garden.

ABOVE The repetition of this row of clipped evergreen shrubs becomes even more effective after dark, thanks to the addition of simple uplighters that accentuate their form and shape.

LEFT Even the most simple, low-level lighting can have an impact if it is positioned well. A double alcove with tea light holders is effective because of the uncluttered simplicity, whilst a down lighter illuminates the light-catching flowers of *Pennisetum villosum*.

Colour

OPPOSITE **This extraordinary mixture of *Bougainvillea* overhanging a garden wall in Marrakech, demonstrates how sometimes a riotous combination of colours can work against all the odds.**

Of the many reasons why humans are attracted to making and caring for gardens – the opportunity to create, the connection with the environment, healthy living and so on – a love of colour must come close to the top of the list of motivating factors. Flowers offer near-limitless colour variation, foliage less in scope but often different in tone, and the materials, paint finishes and features we use in the garden offer yet more variety. Even without flowers or other adornments, a garden will still brim with the most ubiquitous of outdoor colours – green.

A love of colour may well be one of the principal reasons for wanting to garden, but colour can also be a point on which tastes diverge wildly. Many of us will have a favourite colour, to wear or simply to admire, and a hue that we avoid at all costs. Years ago an acquaintance of mine used to restore and sell old sports cars which he would often paint in lurid shades of lime green, turquoise and vivid tangerine. His theory was simple, and often worked; whilst nine out of ten of his potential buyers would be abhorred by the garish colours, one person would fall head over heels in love and simply have to have the car.

'A love of colour may well be one of the principal reasons for wanting to garden'

ABOVE **Whether complementary or contrasting, some colours seem especially well suited to one another. Here steel blue becomes even steelier when set against a foreground of rusty orange-brown.**

In a garden, the use of colour is arguably even more complicated, not least because colours are rarely static or stable in this living, changing environment, affected as they are by variations in light levels and, if the colours in question are those of plants, also by their conditions. Indeed, light has the critical part to play in how we 'read' colour, as it is wavelengths of light that we see as colour. Early and late in the day, when light is often diffused due to the relative position of the sun, many colours appear more vibrant, with blues and purples often taking on almost fluorescent qualities. It is at these times, at dusk and dawn, that garden and landscape photographers are often busiest, taking advantage of the clarity and softness of the light and the vibrancy and intensity of colour.

Hot and cold colour

OPPOSITE **Blue is definitely a cold colour, as the wonderfully intense *Meconopsis grandis* (top right) demonstrates. Combined with a mixture of silvers and purples, it can form effective cool complementary plantings (bottom right). Hot colours such as vibrant red, vivid yellow and burnt orange will show well even in strong light (bottom left); whilst placing a group of hot colours together (top left) can help foreshorten vistas.**

Although the colours experienced in the garden can often be complex and highly variable, for the sake of brevity and simplicity in design terms they are usually arranged under two main headings: hot or cold. Hot colours include red and its various hues, along with shades of orange and magenta. Cold colours include blue, violet, cyan and turquoise, as well as shades of green. The 'fringe' colours, yellow and purple, are variable in that they can be either hot or cold depending on the colours they are associated with. White and silver, though usually thought of as cold colours, can in fact be placed under either heading, again depending on how they are used.

Hot colours are progressive in the landscape – that is to say they usually appear closer than they actually are, thanks to the vibrancy of their hue – whilst cold colours are recessive, appearing further away. These qualities can be played with to deploy some special effects; a short garden can be made to appear longer by massing cool colours at the far boundary, whilst a long garden can be foreshortened with the use of plenty of hot colours at the end of the longest view. Various permutations of this can be applied to make wide gardens seem narrower, narrow gardens wider and so on. Whilst this technique isn't completely foolproof, by and large it works.

The way we interpret colour is influenced by light levels and background colour, which also impacts on the relative heat or coolness of colours. A mass of hot reds and purples can look dark and turgid in shaded conditions, whilst cooler colours will sing out of the gloom, providing a light contrast against their dark surroundings. For this reason, plants that rely on pollinating insects such as moths that fly at dusk and dark often have light-coloured flowers – as well as being fragrant in the evening – to act as a beacon to draw the insects in. Conversely, in really bright, sunny conditions, especially where there is a lot of reflected light, pastel shades can appear washed out and faded into blandness, whilst bold, hot colours really stand out.

'A mass of hot reds and purples can look dark and turgid in shaded conditions, whilst cooler colours will sing out of the gloom'

Complementary colour

OPPOSITE Complementary colour schemes have an overwhelming sensation of being easy on the eye. Their soothing, harmonious effect is the same whether the scheme is hot, like this one, or cold (below).

How we use colour in the garden will, inevitably, be heavily influenced by personal taste, but there are some opportunities and pitfalls that it helps to be aware of. Complementary colour schemes can often help to create a sense of unity and cohesion in a garden, which can be especially effective in a small space where the aim is to make the garden feel less cramped. This doesn't mean using a single colour throughout, but a range of colours that sit beside one another well; blue with purple, violet, lavender and silver, for example. The overall effect is often cool, restrained and measured, but there is no reason why a complementary colour scheme can't be hot and bold – it is entirely down to the composition and content.

A complementary colour scheme should major on subtle shifts in tone and hue rather than dramatic changes. Fortunately the flower colours of plants tend to have seemingly infinite variability; within what could be described as the colour blue, for example, are indigo, purple, lavender, sky blue, royal blue, powder blue, turquoise and so on. The same is true of pretty much every colour in the garden, even plain old green (which of course is anything but). But this variation alone is not enough to ensure that a complementary colour scheme is successful, as there needs to be some contrast otherwise the effect will be at best flat, at worst bland and forgettable

'To ensure that a complementary colour scheme is successful there needs to be some contrast or the effect will be at best flat, at worst bland and forgettable'

If you're a plant enthusiast and your garden is small, then the desire to have a variety of plants will often outweigh the golden rule that, by and large, less is more. If plants with similar colour qualities are grouped together, then the risk of creating a colour bombshell is avoided, with the added bonus that the space will probably seem larger, thanks to the unifying colour scheme. Repeating plants or features with the same colour can also enhance the sense of travelling on a visual journey through the garden, by drawing the eye through a composition or along a feature such as a path.

Contrasting colour

BELOW In this composition, from Jardin Majorelle in Marrakech, Morocco, the contrast between the deep-blue wall colour – a signature of the garden – and the eye-wateringly vibrant flowers of *Pelargonium* couldn't be more stark. But the pot colour has also been carefully selected; fresh, spearmint green being as much a contrast to the blue background as the pinkish-red flowers.

Contrasting one or a number of colours against others creates a very different feel to a complementary scheme. Done well it can bring a sense of excitement and vibrancy to proceedings, the effect of differing colours intermingling throwing up some startlingly delicious results, but when badly executed the results can end up looking like an explosion in a paint factory. To avoid this potential disaster, the most effective contrasting schemes often utilise a limited palette of different hues, with just a scattering of loud colours. But then sometimes the whole point of a contrasting scheme is to take the book of good taste and burn it in a great conflagration of clashing colours.

In some instances the plant will provide its own contrast within the confines of its flower or foliage, be it a subtle, muted combination such as the magenta petals and dark-centred flower of *Geranium psilostemon*, or more flamboyant blends like the foliage of *Actinidia kolomikta*, with its mid-green leaves splashed with white and dusky pink. But plants don't work in isolation, of course, and the integration of hard landscaping materials with the colours of flower and foliage offers plenty of opportunities to exploit. Cool, gray slate or weathered cedar, teamed with the silvery, sword-like leaves of *Astelia chathamica* can create a crisp, clean look, particularly effective in semi-shaded conditions. In brighter light the use of warmer-coloured materials will counteract the bleaching effect of the sun, a pinkish-terracotta wall can be the ideal counterpoint for blowsy, impossibly bright-coloured *Bougainvillea* flowers.

RIGHT When it comes to colour, taste can be a very subjective thing. This stained-glass window (in the cathedral of St Stephen, Budapest) is a cacophony of colour, yet somehow it still manages to be beautiful. Assembling such an array of colours together in a garden isn't always successful as so much depends on the depth of colour – the subtleties of shade and hue, the shape and size of the plants and their flowers, the light levels and so on.

ABOVE Orange and purple make for one of the best contrasting colour combinations in the garden, as the intensity of one highlights and makes more vivid the vibrancy of the other. Here the orange, papery bloom of *Papaver ruprifragum* hovers above the starry purple flowers of *Allium christophii*.

'Sometimes the whole point of a contrasting scheme is to take the book of good taste and burn it in a great conflagration of clashing colours'

Seasonal variation and intensity

OPPOSITE Autumn is the season when many plants provide a spectacular last hurrah, their leaves displaying a range of vivid colours as they undergo a process of chemical change.

Seasonal variation of colour is one of the natural highlights of both the garden, and the wider natural landscape, with certain colours and hues dominant and others recessive at various times throughout the course of the year. We're all attuned to nature's brilliance during autumn, when in a good year, and in the right place, the vibrant shades of turning foliage rivals any other natural phenomenon, but each season has its own colour 'flavour', the hues that help define it from the rest.

In the garden this seasonal variation is evident in the dominance of particular colours at different time of the year. During early spring, yellow is in the ascendance, with daffodils, pussy willow, *Mahonia* and *Forsythia* amongst the many yellow-flowered plants at their peak, before giving way to the blue of bluebells, *Salvia*, *Camassia* and alpine *Campanula*, accompanied by the almost translucent freshness of newly unfurled foliage. By summer the hotter hues take centre stage – reds, purples and violets amongst them – before autumn kicks in with golden yellow, orange and deep, burgundy red. These colours aren't dominant to the exclusion of all others, of course, but nevertheless they help give each season a distinct colour range and feeling.

'Each season has its own colour "flavour", the hues that help define it from the rest'

ABOVE Fruit trees are full of blossom in spring, creating a very distinctive seasonal display that will change as the season progresses and fruits form and begin to ripen.

For some plants the ageing process brings out a completely new dimension to the colour of the plant, or changes its colour altogether. Autumn leaf colour is a consequence of the decline of the structure of the leaf in readiness for winter, causing the green chlorophyll to dissipate and the other colours within the leaf – gold, yellow, red – to be seen. This process affects all plants, not just big forest trees, and there are plenty of grasses and perennials that are just as colourful as their bigger brethren. Autumn is also the season for colourful fruits, berries and nuts, which come into their own as leaves fall and flowers fade, many of which will persist for some months until the combination of frosts and hungry animals or foraging human hands finally sees them off. Beyond autumn and into winter the top growth of most non-woody plants dies, and so changes colour again. Ornamental grasses like *Miscanthus* and *Molinia* take on a straw-coloured hue, eventually bleaching almost to white, whilst flowering perennials turn dark brown or black, often topped with attractive seed heads that are as pretty to look at as they are important as a food source for animals.

Texture

OPPOSITE When the brightest of colours are stripped away, leaving subtler hues of green and straw-yellow, the less-celebrated qualities of texture can come into their own. Here *Allium* 'Firmament', *Stipa tenuissima* and *Geranium* 'Irish Blue' demonstrate the impact of textural contrast.

I suspect that texture is probably the last thing we think of when it comes to gardens. Colour and fragrance are probably top of the list, followed perhaps by light, and maybe space – or at least a preoccupation with the lack of it. And yet the textural qualities of plants and materials are just as important, and enjoyable, as these more obvious elements.

The term 'texture' is most often used in relation to the physical feel of an object, whether it is smooth or rough, hard or soft, sticky or dry. But texture can also refer to a compositional outcome, principally in musical terms where the effect of layers of sound or tonal variation is often described as influencing the 'texture' of a piece. The term, with this meaning, is equally applicable in the garden, where the interaction between the various components that make up the whole (say the combination of soft, silky oriental grasses with the coarse surfaces of igneous basalt rock) can have a sense of cohesion that is the visual equivalent of an orchestral symphony.

'The various components that make up the whole can have a sense of cohesion that is the visual equivalent of an orchestral symphony'

Plants offer plenty of textural interaction in their foliage, flowers and stems, with every experience on offer from rabbit soft to razor sharp, rough as sandwood to slippery as patent leather. Hard landscaping materials have just as much to offer texturally. A smooth, diamond-cut slab of sandstone or a piece of dressed slate will be just as irresistible to touch as a soft, velvety leaf. And the texture of hard materials can also change according to the weather conditions; little pools of water will collect in roughly dressed stone, but roll off smoother finishes, for example.

To me the tactile qualities of objects and plants are just as pleasurable as the visual and olfactory, but it isn't just how things feel texturally that matters in a garden; it's how texture changes the look of things too, and in turn how this influences the scale of a space. The influence of textural changes in a space, though often subtle, can be highly effective. A few years ago a study was made into the effect of paving surfaces on burglary rates in residential estates. It found that replacing tarmac with coloured block paving discouraged people from going beyond a certain point. The coloured paving created an effective visual, though not physical, barrier. In a garden, similar textural tricks can be played out too: changing the scale of paving materials from large, smoothly finished paving stones to smaller, more detailed paving such as a stable yard paver (about the size of a brick but divided further into six raised squares) can encourage a slowing down, or dictate a change of direction in a path.

Texture in plants

OPPOSITE Texture in plants comes in many guises. The peeling bark of the London plane tree (top right) displays a mixture of textures and colours. Sea hollies (bottom right) have dry, spiny bracts, *Euphorbia* (bottom left) thick leaves covered in tiny hairs. The sword-like leaves of *Astelia* (middle left) are coated with a dusty bloom, while chard (top right) has large, leathery leaves.

Some plants just cry out to be touched. The flowers of the grass *Pennisetum villosum* are as soft as cotton wool, the leaves of lamb's ear (*Stachys byzantina*) are like crushed velvet, whilst another grass, *Stipa tenuissima*, is quite rightly known as the hair grass, as its leaves are as fine and pliable as a fulsome head of hair.

These tactile qualities are not the result of the plant's desire to be fondled of course, rather a consequence of the need to adapt to particular climatic or regional conditions. The nap of thick hairs on the leaves of lamb's ear form a protective barrier that traps moisture from the air and reduces water loss, whilst the needle-thin leaves of hair grass are as narrow as they are to prevent excess water loss. Extreme drought specialists need to carry as much moisture as possible in their leaves, stems or roots, which often become swollen with sap and starch as a result. *Aloe vera* and *Agave americana*, for example, have thick, fleshy leaves that feel more like polished wood, plastic, or old, burnished leather to the touch than living plant material. Other plants have adopted the trick of keeping their leaves and even their flowers covered with a thin layer of wax, sticky enough to feel noticeably tacky to the touch, in order to prevent water loss from wind.

Some plants have developed textural qualities further – into defensive armour, forming spines, barbs and thorns. The formidable *Colletia paradoxa* has entirely disposed of conventional foliage and instead has flattened stems equipped with fearsome spines, whilst a huge array of cacti have tooled up with monstrous barbs and thorns designed to challenge all but the most tenacious of herbivores. While these plants may not be top of the touchy-feely list, there is no doubting they bring a strikingly different textural dimension to the garden.

'*Aloe vera* and *Agave americana* have thick, fleshy leaves that feel more like polished wood, plastic, or old, burnished leather to the touch than living plant material'

Flowers are as texturally variable as foliage; however, their tactile qualities are often overlooked, perhaps as a result of their striking visual qualities. Rose petals have the most exquisitely silky feel, whilst the central boss of *Echinacea* has a stiff, bristly quality. The bottlebrush plant *Callistemon rigidus* has soft, brush-like blooms that are soft to the touch and are wonderfully exotic to look at, whilst other flowers, such as *Bracteantha bracteata* (known as the golden everlasting daisy) have a papery texture that feels more like parchment than plant.

LEFT Some *Allium* flower heads are very large, about the size of a grapefruit, but each is comprised of masses of smaller individual flowers that are loved by bees and that give the plant its distinctive textural quality.

LEFT *Echinops* flowers have a wonderfully tactile quality. Dry and rather stiff to the touch, the individual petals spring back into place when rubbed with a finger.

ABOVE The most obviously visible part of the *Bougainvillea* flower is in fact a bract, which has a very distinctive papery texture that contributes to its long-lasting colour in the garden – in effect it is a living dried flower.

RIGHT ***Nigella damascena*** **has a papery quality to its flowers, but also incredibly fine, filigree foliage that gives the whole plant a floating, gossamer-like texture that contributes to its aesthetic suitability for cottage garden plantings.**

LEFT The flowers of the aptly named bottlebrush plant ***Callistemon rigidus*****, an Australian native, are perfectly described by the common name, both in shape but also in texture, which is like a soft brush.**

LEFT The flowers of *Astrantia major* are comprised of white- or pink-tinged bracts that have a stiff, papery texture and surround the flower. These bracts are designed to make the plant more attractive to pollinating insects by making the flower seem bigger and therefore more worth a visit – it's the plant world's equivalent of an advertising hoarding.

Texture and materials

If you've ever been on a country walk and have crossed an old-fashioned timber stile, your hand will probably have rested on a wooden upright so well worn by thousands of preceding hands that it was as polished and smooth as a chestnut. A few inches further along the stile, the near-virgin wood, untroubled by all that handling, will likely be almost as rough to the touch as it was when it left the mill. It's just one small example of how the texture of materials can change; in this case through repeated handling, but also from weathering and from different treatments during manufacture that lead to a variety of textural finishes.

RIGHT Using hard landscaping materials with different textural qualities and sizes is an effective way of altering scale in the garden. Here a smooth-finished paving stone gives way to a much rougher-hewn granite sett, in a variety of sizes, which in turn yields to a timber deck.

LEFT Using the same material in different forms can help to unify the garden by providing continuity of materials whilst also giving different textural qualities. Here slate is used in two forms: smooth finished paving stones for walking on, and small crushed shards as a mulch for plants.

BELOW The contrast between the solidity of the timber comprising this raised bed, and the soft, billowing grass overhanging it, makes for a very effective composition. The rusted iron capping is a nice design touch that adds another textural dimension.

Stone is an extremely versatile hard material which can be worked in a number of ways to create a variety of textural effects – a good understanding of which opens up the possibilities for their repeated use as part of the composition of the design throughout the garden. The finish of different types of stone also affects its light-reflecting qualities and the rate at which it sheds or holds water, which in turn changes its appearance and colour – wet slate, for example, has a glossy, almost oily appearance that changes significantly when it is dry and takes on a slightly chalky hue.

The texture of timber is similarly affected by the manner in which it is finished. A milled, planed oak post has a smooth, precise quality suitable for an application where clean lines are important, whereas the same piece of timber, split and roughly finished, will have a far more rustic feel, ideal for a country garden or the boundary between a cultivated space and the wild ground beyond. Similarly, timber also changes its textural quality through weathering, with raw softwood losing some of its roughness thanks to the gradual softening by wind, sun and rain; whilst perhaps one of the most dramatic examples of the manner in which materials can change is the way that green oak (oak timber that is sawn and dressed whilst still 'wet' with sap) develops 'shakes', cracks and fissures that appear along and across the grain of the wood over time.

Combining textures

BELOW **Combining the relative textural qualities of plants and hard landscaping materials opens up all sorts of design possibilities to exploit. Deeply cut fern fronds can complement the small-scale complexity of a pebble mosaic path (left), whilst large slate paving stones can provide an effective contrast with interlacing panels of tiny-leaved creeping plants such as *Soleirolia soleirolii* (right).**

The textural interaction between different hard landscaping materials, their various finishes and the plants around them is also significant. Smooth, highly finished stone, laid in geometrically precise lines, takes on a strong, contemporary air when paired with strongly architectural plants like grasses, exotic palms, glossy-leaved evergreen shrubs or succulents. Conversely, a meandering gravel path with uneven width along its length will, when combined with the softer, smaller or deeply-cut foliage of cottage garden plants such as *Nigella damascena*, *Delphinium* and *Lupinus* result in a far more informal, relaxed look.

Mixing different textures can also create an interesting aesthetic tension, or help to soften or sharpen a feature. A large 'statement' rock, placed in the bend of a path or at the end of a vista, will have a distinctive character if left uncluttered as a monolith to be enjoyed in its own right, but when planted with an airy, soft-textured grass such as *Nassella trichotoma*, *Hordeum jubatum* or *Eragrostis airoides*, it takes on the role of a dynamic textural counterpoint.

Crevice planting offers an opportunity to closely combine the different qualities of stone and plants, using small-leaved succulent plants such as *Sempervirens*, planted in between the crevices formed by standing slate or sandstone on edge, which are then packed out with growing media. The succulents quickly colonise around the crevice stone so that the two appear to be a single entity. Likewise, containers can be selected for their ability to contrast or complement the texture of the plants within them; a copper pan, smooth-textured but with the textural and colour patina of verdigris, can be combined with the bronze-leaved, low-growing *Sedum*, which complements the patina of the pot whilst providing a counterpoint to it, thanks to the size and glossy finish of the foliage.

The passage of time, and the effect of foot traffic, handling and weathering all alter the textures of stone and timber too, softening out hard edges and encouraging the colonisation of the ultimate textural softener – and indeed the most natural and effortless combination of plants and materials – mosses, algae and lichen.

LEFT In Tom Hoblyn's Chelsea Flower Show garden, timber wave forms contrast strongly with large 'pebbles' (in fact cushions designed to look like stone). The garden was later rebuilt as part of Clive and Debbie's Suffolk garden.

BELOW The coarse vertical foliage of *Equisetum hyemale* creates a textural curtain behind this beautifully executed rill, and contrasts with the hard materials of cut stone and crushed slate that comprise it, as well as the smooth pebble stepping stones beyond.

Shape and form

OPPOSITE Sometimes the simplest features make the boldest statements in the garden. Here the combination of strongly contrasting colour and bold shape and form are all that is needed to make an arresting composition.

A shape is an enclosed area, distinct and distinguished from its surroundings and usually considered to be two-dimensional; it has height and width. Form, by contrast, is the organisation of visual elements into three dimensions; in addition to height and width, form has depth – or at least the illusion of depth. A knowledge of the many different shapes and forms that can be found within the garden is as crucial to the success of your design as an understanding of colour or texture, and will help to bring a sense of dynamism and life into a space.

Amongst the many factors that affect our perception of shape and form is the viewpoint or position from where we see an object, which can highlight certain features and obscure others. This can change as we move around an object, getting closer to it or further away, or viewing it obliquely or head on. It is also the visual phenomenon that, in part, is exploited by *trompe l'oeil* (trick of the eye), where a flat surface, often a wall, is painted to appear as a three-dimensional object, often as an outdoor view to create the illusion that a space is much larger than it actually is.

The character and content of the space around an object can also affect how we perceive shape and form. A solitary oak tree, standing tall in an open landscape, has far more visual impact than an oak tree in a wood, where it becomes absorbed as part of the fabric of the whole. Likewise, a garden seat, as part of a group, is understood to be merely one item in a set of furniture, whereas separated out and positioned at the end of a vista it instead becomes a focal point, or 'eye-catcher'.

Light too has a crucial role to play in how we perceive shape and form. If an object has a smooth side and a rough side and is lit head on, both textural qualities will be apparent. But if lit from the smooth side, the rough side will be in deep shadow and to all intents invisible. Light can also change the character of an object by making it appear more or less worn and aged, with softer, head-on lighting far more flattering than hard, shadow-casting side lighting.

Core plant shapes

ABOVE The daisy flowers of plants from the Asteraceae family, such as *Echinacea* are amongst the most recognisable of all flower shapes.

There are a number of core shapes that are recognised in plants, and a good understanding of these will enable you to combine them in ways that will lend maximum visual impact to your garden.

Fastigiated, columnar or pencil-shaped plants, such as *Taxus baccata* 'Fastigiata', with their upright, relatively tight forms, create strong dominant outlines within the garden. These are the shapes that make a statement and can be used to create 'punctuation' (marking the beginning or end of a path or a gateway) to frame a vista, or to create a sense of rhythm in a space through their repetition. At the opposite end of the shape scale are the creeping, horizontal plants that grow outward along the ground rather than upright. These plants have many useful practical applications; they are good for covering ground and suppressing weeds and also help form the important lower level for the overall composition of a planting scheme. In between these two groups are a variety of spreading, arching and cascading shapes, which can be useful for everything from filling space (forming natural alternatives to topiary) to making a graceful focal 'pinch point' or turn in a planting that encourages you to look around the corner.

At a much more detailed level are the shapes and forms that comprise a plant's individual flowers and foliage. From the classic daisy to the complex, bilateral symmetry of orchid blooms, flowers have distinct, often beautiful shapes. Mixing these differing shapes can be an effective way of adding life and interest into your garden, with clever combinations such as the marrying of *Echinacea*'s fat, daisy flowers with the slender flower spikes of the powder-blue *Perovskia* 'Blue Spire', or the mixing of cascading blooms of *Laburnum* with plump, round *Allium* flower heads providing the little touch that can lift your garden from the mundane to the sublime.

5 plants with distinct shapes

Malus sargentii
A lovely, slow-growing, small-spreading tree with white blossom and red fruit.
4m x 5m

Armeria maritima
Forms a tight bun of foliage with pretty pink flowers.
20cm x 30cm

Cotoneaster horizontalis
This low-growing shrub spreads outward parallel to the ground.
1m x 1.5m

Calamagrostis emodensis
A beautiful grass with pretty, arching flowers.
60cm x 60cm

Cupressus sempervirens
Pencil-thin and tall, this is the quintessential tree of the Tuscan countryside.
15–20m x 1–6m

LEFT The vertical flower spikes of plants like the foxtail lily, *Eremurus*, can be used to create the vertical element in a planting, or celebrated for their shape, form and presence by allowing them the space to stand out alone.

'Mixing differing flower shapes can be an effective way of adding life and interest into your garden'

Manipulating plant shapes

OPPOSITE Fruit trees are pruned and trained for aesthetic, practical and cultural reasons; elaborate pruning styles such as this freestanding pyramid restrict the overall size of the plant, improve fruiting by encouraging lots of lateral flowering wood, and also make a very effective feature plant.

RIGHT Some plants have the capacity to take pruning and training to such an extent that they can become quite unlike the plant in its native state, as these pad-pruned hornbeam *Carpinus betulus* demonstrate.

Along with the exploitation of naturally occurring plant shapes, the manipulation of plants by pruning and training offers another level of opportunity for creativity in the garden. Perhaps the most common form of garden plant manipulation is the planting and shaping of hedgerows. Whilst these are regularly pruned into roughly 'A'-sectioned shapes and used as boundary-marking rectilinear forms, by planting and cutting them to create interlocking shapes of varying heights they can instead be made to frame a particular vista or object, or sculpted into visually intriguing, undulating shapes or serpentine forms.

Individual plants can also come under the influence of the clippers to interesting effect. Topiary has something of a mixed reputation – historically evocative and impressive in execution when seen in an appropriately grand location, it can seem out of place and twee in suburbia unless executed with absolute conviction and skill. But the less extreme version of topiary, where plants are clipped into strongly architectural shapes as opposed to stylised objects, can be highly effective at creating the same kind of effects of punctuation, framing and repetitive rhythm that can be found when using plants which naturally possess distinctively strong form.

There are almost no limits to the ways in which plants can be manipulated into seemingly impossible shapes. Street trees are often dramatically pruned not just to restrict their growth but also to provide a suitable amount of shade. Climbing plants such as roses, vines and climbing hydrangea are usually trained against wires on a timber or steel framework in order to enhance their flowering, fruiting or simply to support their growth; whilst fruit trees are commonly trained so that they can be accommodated into smaller spaces – often (particularly with the more elaborate forms of espalier training) with the most strikingly effective aesthetic results.

'There are almost no limits to the ways in which plants can be trained and manipulated into seemingly impossible shapes'

Case Study *Essex*

ABOVE The garden at East Ruston Old Vicarage in Norfolk makes great use of formal topiary shapes contrasted with loose, billowing grasses. Here, clipped box pyramids and box balls frame a central topiary holly, all softened by ornamental grasses and bold orange *Dahlia*.

RIGHT Two huge blocks of yew (*Taxus baccata)* stand like sentinels either side of the top of this flight of steps, the ascent of which is echoed in the clipped box to either side that steps up in concert with the risers of the stone steps. The combined effect magnifies the impact of what would otherwise be a fairly modest feature, turning it into something with significant presence.

LEFT Repeated clipped box balls, so closely planted as to appear connected to one another, create an undulating sea of topiary that is further amplified by dappled sunlight.

Fragrance and sound

OPPOSITE
The fragrance of flowering plants, such as sweet peas, provides one of the greatest pleasures in the garden and can be added to through appropriate plant selection. The sound, colour and movement of visitors, such as birds and insects, are what makes a garden truly come alive.

Whilst the most obvious pleasures of a garden – colour, form and texture – may well be the easiest to express visually and are therefore the ones we often consider first and foremost, fragrance and sound are just as vital in contributing to our overall sensual enjoyment of the garden. Whether experienced up close with a nose pressed against the flower of a pungent beauty, or as a scent caught on the air, the source an unseen mystery that invites greater exploration, fragrant plants are one of the garden's true joys. And though we seek refuge in their supposed 'quiet', gardens are far from silent. Instead they are alive with the sounds of buzzing insects, birdsong, the trickle of a water feature or the rustle of the wind blowing through the foliage of an ornamental grass or the leaves of a tree.

The enhancement of these two sensual pleasures within your outdoor space can often be the result of other changes you will have instigated, for a variety of reasons, in the process of making your garden. Encouraging wildlife into the garden by planting pollen- and nectar-rich plants in the garden will attract more bees, butterflies and moths to arrive; whilst putting out bird seed, erecting nest boxes and resisting the urge to use chemicals to control pests will see more birds descending upon the garden in pursuit of prey, seed or habitat, as well as ensuring that there is a greater chance of them hanging around. Including more fragrant plants in the garden makes perfect sense, not least because in many cases the most fragrant plants are often also those that are rich in pollen and nectar, covering two desirable outcomes at once: a beautifully fragrant garden and a host of handsome insects that will visit and enjoy them. And creating features that are designed to make a noise, either to help block out a louder, unwanted sound or simply to be enjoyed as a pleasure in their own right, should always be a consideration when designing a garden.

Fragrance

OPPOSITE When placing fragrant plants in the garden, try to keep the nose-to-bloom distance at a minimum by choosing a location – a house wall near the kitchen door, by a seating area or, as here, on an arch – where you will enjoy them most.

Smell is the human sense most associated with memory, so it is unsurprising that we value fragrance in plants highly and often seek out fragrant plants in favour of those without scent, even if they may be slightly poorer in terms of flower colour or size. Plants produce scent as a means of attracting the insects that many of them rely on to unknowingly pollinate their blooms. Consequently there are some plants that have developed especially powerful fragrance at particular times of the day, or specific times of year, to ensure that the right type of pollinator is wooed. Night-scented stocks (*Matthiola bicornis*) and evening primrose (*Oenothera* species) both belt out their fragrance as night falls to attract the last of the day-flying insects to stop for one last sup before they head home, and also to ensure the first moths on the wing are given a good reason to visit them first. Winter-flowering plants, like winter sweet (*Chimonanthus praecox*) and *Viburnum farreri* have to cast their scent over a long distance to ensure that, on the odd dry, sunny winter day whatever pollinators emerge – principally bumblebees and species of moth – are made aware that a feast is available.

Although insects are the intended beneficiaries of flower fragrance, we too can profit from their generosity. The key to placing fragrant plants in the garden is to put them where they are most likely to be enjoyed. A sunny seating area is the perfect location for surrounding with swathes of lavender, for example, as both the flower and foliage can be enjoyed, with a leisurely pass of the hand. Old-fashioned climbing roses, like *Rosa* 'Paul's Himalayan Musk' may only have a brief flowering period, but it's a pungent one, and this rose is perfect for training over an archway or beneath the first-floor windows of a house. As lovely as scent is in the garden, it's even better wafting into the house, so clustering fragrant plants near the house doors will ensure at least some of their smell permeates the home. This is particularly important with winter- and spring-flowering subjects that, during prolonged periods of poor weather, might be completely overlooked unless they are close enough to be enjoyed through an open window or door.

'The key to placing fragrant plants in the garden is to put them where they are most likely to be enjoyed'

Sound

OPPOSITE Ornamental grasses are amongst the finest garden plants for bringing sound to the garden, either when in active green growth or during autumn and winter when they begin to dry and become more straw-like.

ABOVE Sound in the garden can come from a variety of sources, some of which, like water features, are controllable, whilst others like wind chimes or the chatter of birdsong are less easy to manage, but equally valued.

Though few of us ever get the opportunity to appreciate the luxury that is a complete absence of sound, we all strive to escape from the hubbub of modern living from time to time. Whilst indoors we can shield ourselves from unwanted noise with double glazing, thick curtains, insulated walls and snug-fitting doors, in our gardens there is no such barrage of barriers to block out the aural intrusion.

Planting rows of trees and hedges in attempt to screen out excessive noise is, thanks to the way that sound waves travel, of limited or no effect. Though the sound might be slightly muted or disrupted, screening out loud noise adequately would require very deep planting indeed. Instead, often the best way to block out unpleasant background noise is to create an alternative, diversionary sound – a 'foreground' noise that you actively want to listen to. Within the garden the principal means for doing this are with water and plants.

Running-water features need great care to get the sound levels right. Too much water flow and the volume can be too loud, leading to the sound becoming overwhelming. Too little flow and not only will the purpose be lost, the noise can become another source of irritation. A well set-up waterfall or fountain will ideally create an even sound that is loud enough to be enjoyed but not so loud as to be intrusive. There are other ways to create diversionary sounds with water too, an exotic example being the Japanese *shishi odoshi* – traditional bamboo rod bird scarers that seesaw with water, tipping into a stone bowl with their distinctive 'tuk' sound.

Plants can create sound in the garden too, but only in concert with the wind. Ornamental grasses and bamboos are well regarded for their aural qualities, creating a noise not unlike the rushing of water as the wind passes through their foliage. The wind can also be harnessed to create sound with calming wind chimes and Aeolian harps, though these need very careful handling if you want to avoid creating a continuous cacophony of noise, or something overly kitsch.

However, perhaps the most interesting sounds in the garden are those that aren't intentionally noisy, as such, but help form a part of the overall composition. The crunch of gravel under foot, the explosive pop of the drying seed pod of a Californian poppy, or the languid buzzing of a fat bumblebee – these sounds alone give reason for making a garden.

REALISATION

3

'One who plants a garden, plants happiness'
Anon, Chinese proverb.

The physical making of a garden can be a lifetime process, the gradual accretion of knowledge invested back into the garden in the form of a series of improvements and developments that can be never-ending. Essentially this continuity is what, by nature, gives the garden its beauty, pleasure and spirit. The novelist and gardener H.E. Bates wrote that '*the garden that is finished is dead*' and he was absolutely right. A garden can never be finished; all the time there are plants growing and dying, seasons changing, natural weathering taking place, personal taste and circumstances developing and changing, not to mention all the many changes that are occurring to the external environment, whether as a result of property development within the area, or as a result of inevitable climate change.

Having read the 'Thought Process' section of this book you will already be aware of all the practical requirements you have for your garden, and will have also formed a good idea of which desirable elements you would like to include. The realisation process brings together these considerations, informed by an understanding of the components that comprise the garden experience described in the 'Vision' section, to make a whole. Though this might sound relatively easy and straightforward, in practice it is a venture full of potential pitfalls and tricky questions, because making a garden is, for most of us, as out of our day-to-day experience as rewiring a house or plumbing in a new bathroom. However, so long as you are armed with a simple understanding of how to go about turning your dreams into reality, there's no doubting that it can not only be great fun, but also hugely rewarding.

The key to minimising potential problems and maximising the enjoyment of the process is in preparation and planning, whether you undertake all or part of the work on your own, or with the support and expertise of a professional designer. Putting the practical and desirable elements of the garden into some kind of deliverable order to ensure that making one part of the plan happen doesn't have a negative impact on, or even preclude, another part of the plan taking shape is vital. Without a plan it is all too easy to make mistakes, as some of the garden owners in *Landscape Man* found out. Even with a plan, and the support of a designer and the range of professionals that are employed in making a garden, things can, and often do, go wrong. Sometimes this can be the result of a simple mistake – Clive and Debbie's hornbeam avenue being incorrectly plotted out led to the whole tree avenue needing to be dug up and moved – other times, as in Anne and Geoff's case, a change of plan was necessitated as a result of the local planning authority refusing permission for certain proposed design elements. Whatever the reason for change, however, by having a plan in the first place you will able to clearly see the impact that one event has upon another, giving you the best opportunity to take the appropriate steps to put things right.

With a deliverable plan to work to, as well as a clear idea of what it is you want from your garden (and what you don't) there is every chance that you will be embarking on the beginning of what could be a long and extremely rewarding journey. And as your garden develops and grows, and elements wither and die as must always be with a garden, you will be given endless opportunities to redress those elements that don't work, to enhance those that do, and to accommodate changing needs. It's rare for mistakes to be made in a garden that cannot be resolved, and on occasion it is these mistakes which make the garden more interesting and are what help us to learn – although I always immodestly claim that such happy accidents are by design! This is what makes gardens, and gardening, such a rewarding exercise, providing you with as many challenges and opportunities as you want to take on and allowing as much rein to your imagination as you are willing to give.

Going it alone or engaging a designer?

Throughout the history of gardening there has been a generous smattering of knowledgeable and enthusiastic amateurs who, by establishing an overarching philosophy for their gardens and driving their development intelligently, have created great gardens on their own. Indeed, if you have the confidence and knowledge to design and realise your own garden plans then this can be an incredibly rewarding experience.

But what exactly do you need to know in order to be able to make your own garden a success? To an extent, the level of knowledge required to develop and execute your own plans depends on the type of garden you desire. A garden dominated by hard landscaping with minimalist planting, for example, will need a thorough understanding of the qualities of the materials you intend to use and the ways they can be composed on the ground to best complement each other, as well as an ability to demonstrate restraint in execution. Conversely, if you want a plant-filled garden, then you'll need an understanding of what will and won't grow where, what to do to provide optimum growing conditions, as well as having the confidence to compose with them to make the best use of colour, texture, form and shape.

The accrual of the knowledge required to plan a garden of any type can take time, whilst the ability to visualise it and then make it happen can be something that comes naturally to some people and not to others. But for every great garden made solely by its owner is a garden made by a designer working under the benefaction of an enlightened patron. Of the six gardens featured in *Landscape Man*, Keith and Ros were the only owners not to actively seek the help of a designer, and even though Keith is a professional gardener with more than 30 years' experience, he still sought assistance and employed craftsmen for those elements of the garden development that he felt were outside his skills. Clive and Debbie, Anne and Geoff and Sarah and Trevor all sought help from professional garden designers to create plans for their gardens and help project-manage the realisation stage, whilst Jason and Demetra and Vanessa and Ed had design advice from me, as well as additional assistance from a number of specialists and craftsmen.

So what are the pros and cons of going it alone versus employing a designer? A designer will of course cost money, their fees usually based on a percentage of the total contract value, this after having developed draft plans, often just sketches, which are usually done 'at risk' – i.e. the designer takes a punt that you will like what they are proposing and decide to engage them to take the work forward. Some gardeners do not charge for an initial consultation, some do. Personally, I would think twice about parting with any money 'up front' before you see what the designer is proposing.

In return for the designer's fees, you will gain a huge amount of expertise and support through the design and realisation process, which can be worth several times more than the fees charged in peace of mind and ease of process, and might actually end up as a quantifiable cash saving too, as a designer should be able to foresee many of the potential pitfalls and risks that can come with any project. A designer will also be able to take whatever ideas you have gathered, be they a simple list of likes and dislikes or more developed thoughts, and interpret and embellish them in their brief.

Working with a designer will also bring with it the wealth of contacts, craftsmen and contractors that he or she has assembled over the course of their career, taking away another time-consuming aspect of garden realisation, the trawl in search of reliable contractors that can often feel like a lottery. A really good designer will be able to hold your hand throughout the process of making your garden from brief to completion, and as such I would always recommend seeking out their help even if for only part of the process.

Budgeting

Through a lack of understanding of the total price of the elements that go into making a garden, many projects go way over budget or fall short of expectations. Usually this is due to unrealistic costing of the various components that combine to make the whole, rather than one specific area. Breaking your budget down into the following sub-sections (where relevant) should ensure that you should have every angle covered:

The site survey
Consultation, design, specification and possible project management/overseeing
Structural engineer for any complex or major construction work
Site clearance including debris removal (skip hire etc.)
Site infrastructure works (drainage, services etc.)
Hard landscaping (materials and labour)
Soft landscaping (including turf, plants and labour)
Soil preparation
Accessories such as lighting, irrigation, sheds, furniture, water features etc.

Putting it all together

OPPOSITE This simple sketch plan effectively illustrates the key measurements that you will have covered in a basic survey of your site.

If you decide not to use the services of a designer, or have a design drawn up and then choose to progress the realisation of the design yourself, the time will come when you will need to take the ideas, needs and desires for your garden from whatever paper form they take and make them happen on the ground. Ideally you'll have two paper documents from which your garden will spring – a drawing, and an action plan. The drawing can be as simple or as complex as you want it to be, from a 'back of a fag packet' sketch to a fully detailed, modelled, AutoCad (a computer-aided design) drawing that can be manipulated on-screen and stripped back or built up in layers to reveal different elements of the plan.

In some cases an AutoCad plan can be complete overkill, where all that is required is a simple planting plan, or a sketch that shows the relative positions of features that are 'self-designing', such as an off-the-shelf patio paving kit or a garden shed. But in other cases a plan of great detail can be essential, especially if dealing with complex level changes and where they relate to things like the damp-proof course of a house or existing drainage systems. If your garden requires the sort of detailing required by an AutoCad plan, then it makes complete sense to call in a professional designer, who will organise a survey of the garden before developing a plan drawing.

Creating a sketch plan

If all you need is a simple sketch plan, then it's possible to do it yourself, and if so then you'll need to start by making a basic survey of the site. To carry out a basic survey, you need a sheet of graph paper, two long tape measures, two wooden marker pegs and a hank of ranging line (which you can buy from a builders merchant). Find the longest distance between two points in your garden that can be measured in a straight line – often the boundary of the garden formed by a fence, wall or hedge – measure it and draw it on the graph paper, and ideally mark it on the ground too, with the two marker pegs and ranging line. This is what is known as the A–B line, and from it you can measure off, with a reasonable degree of accuracy, all the main fixed objects in the garden – existing paths, plants, walls, sheds and so on. This basic survey will give you such information as the overall size of your garden in square metres and the relative positions of major features, onto which you can draw whatever new features you have decided on – a new seating area, plantings, and so on. Over the next few pages you'll find examples of sketches that I have made for various gardens, some of which have then needed further, more detailed plans; others have been realised just with a pencil sketch. Take time with the drawing stage, experimenting

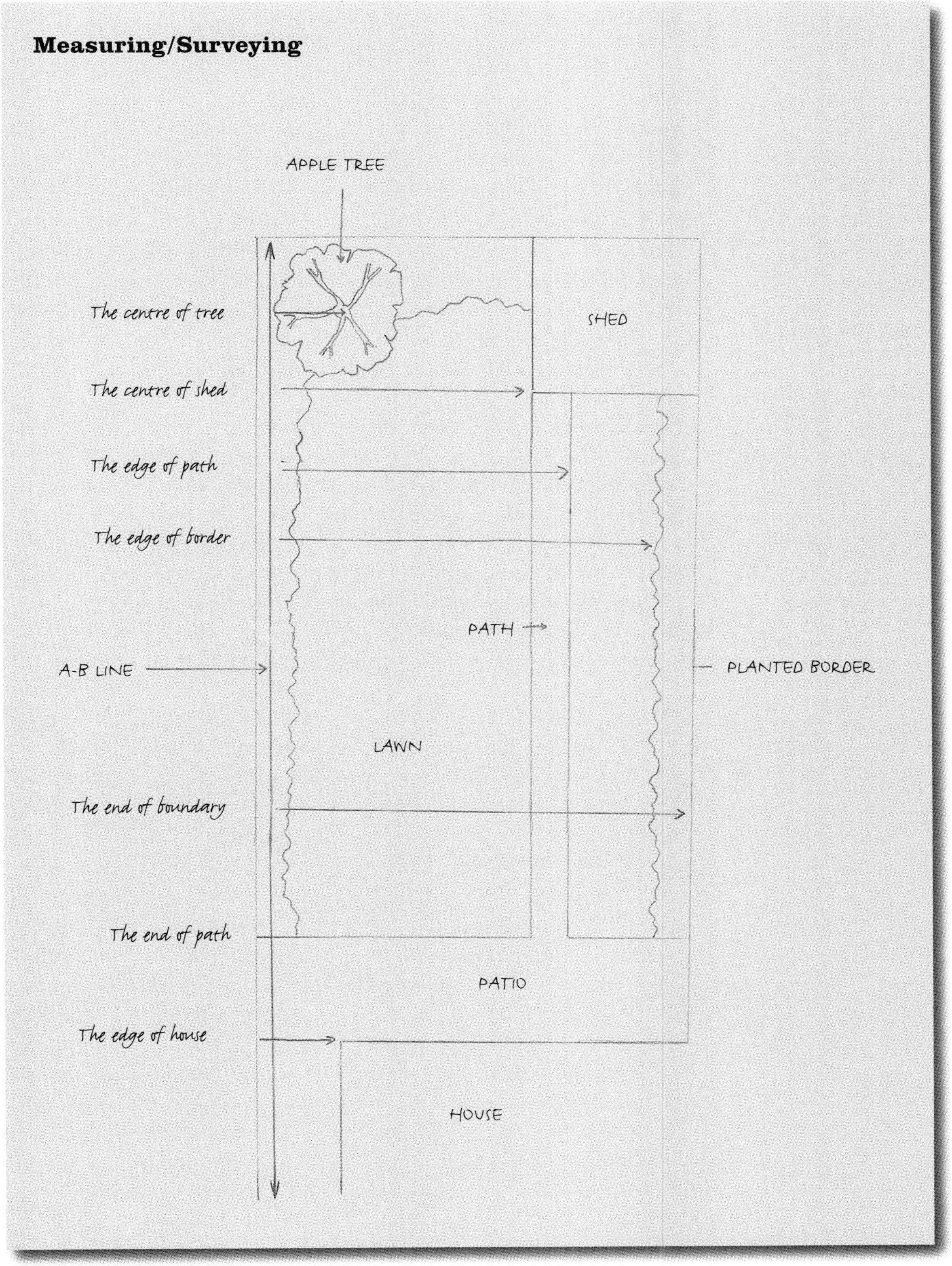
Measuring/Surveying
APPLE TREE
The centre of tree
SHED
The centre of shed
The edge of path
The edge of border
PATH
A-B LINE
PLANTED BORDER
LAWN
The end of boundary
The end of path
PATIO
The edge of house
HOUSE

with different approaches to each feature that you want to include, before settling on a final version. Most professional designers will spend a great deal of time working up ideas in pencil and ink before taking them further to a drawing board or AutoCad.

When you come to plotting out where the new features will go on the ground, you can use the A–B line to accurately position them by measuring

RIGHT This sample action plan for laying a patio covers all the necessary steps from researching the designs to laying the surface itself. When laying a patio surface, remember that if the maximum depth of excavation is more than 80cm you will need to check for underground services by getting a copy of the building services plan for your house from one of the utilities providers, or hiring a portable CAT scan device. If you are in any doubt at all about what lies beneath the surface, then don't dig!

Sample Action Plan – Laying a Patio

1. Research patio designs/materials online and through suppliers' catalogues.
2. Finalise design and sketch onto plan.
3. Measure out your new patio on the ground from sketch plan.
4. Visually check for any underground services (e.g. manhole covers!).
5. Estimate sub-base materials by multiplying overall size of patio (e.g. 20 sq m) with depth of sub-base (e.g. 30cm). Allow for an extra 5%.
6. Estimate finished surface materials.
7. Place orders for both sub-base and final surface, geotextile underlay, and for sand and cement or environmentally sustainable alternative (e.g. ScotAsh).
8. Book hire of cement mixer and wacker plate compactor.
9. Work out whether you will be using excavated spoil. If not cost in disposal.
10. Mark out patio, double-check and then excavate.
11. Install plastic duct pipes (x 2) under the patio for any future cabling.
12. Install geotextile fabric underlay to stop water/spoil transfer from sub base to soil.
13. Lay sub-base, consolidating in layers with wacker plate compactor.
14. Lay the final patio surface.

Case Study Suffolk

LEFT This computer generated sketch plan for Thomas Hoblyn's Chelsea garden demonstrates the benefits of 3D programmes in creating a detailed representation of a finished garden.

off from it, and then using something like a line marker paint (from a builders merchant) or dry sand to plot how each feature looks. Don't be afraid to make adjustments to the plan at the marking-out stage, as the visual impact of a feature can only ever be fully assessed on the ground, not on a paper plan.

Drawing up an action plan

An action plan is in essence a check list of what needs doing, and when. This might sound blindingly obvious but unless you have really thought through how to do things, and in what order, all sorts of nasty errors can crop up. A typical mistake, and one that I have made myself more than once, is to lay the base – or worse still the finished surface – of a patio only to realise that you should have put in a water pipe or electricity cable, to run a water feature or lighting. The resulting digging up of a section of perfectly laid patio, which almost never seems to go back down again properly, will remind you forever more of the need for an action plan. Likewise an action plan will prevent you from unnecessarily redoing work over and again. There's not much point in spending hours digging over borders if you then need to walk all over them to install the timbers for a pergola or put up a shed.

An action plan doesn't need to be a great big, in-depth document – it can be like a shopping list, or perhaps more accurately a holiday luggage list – just a simple A to B plan that helps to ensure that you get to where you want to go with your plans with the minimum of fuss and costly mistakes.

Realised gardens

Case Study Suffolk

OPPOSITE Chelsea Flower Show gardens such as Thomas Hoblyn's 2009 garden are designed to delight and inspire for a very brief period of time – the week long duration of the show. But although ephemeral, the realisation of such gardens is the result of meticulous planning and painstakingly detailed construction and planting.

The following 32 pages illustrate a number of different 'realised' gardens in both sketch and completed, photographed form. The sketches have all been made as part of the design process, as a simple means of expressing the design concepts that were then considered for the finished garden. In some cases the sketches were worked up further and replaced by more detailed, modelled plans; in others, the ideas were taken and mapped out on the ground without additional development.

Regardless of how good an artist you may or may not be, sketching out your ideas over the graph paper survey you have made of your garden (*see Creating a sketch plan, page 148*) really is the best way to help you to focus your thoughts and consider what works and what doesn't, as well as allowing you to address such issues as scale, proportion and balance between various components. This is much easier to do than marking your plans out straight into the garden would be, and alterations are certainly much simpler to make with a pencil eraser than when something has already been built on the ground!

Each of the garden sketches is also accompanied by an additional drawing that picks out a particular planting scheme within that garden. The suggested plant combinations here have been chosen both to look good together and to fulfil a specific aesthetic quality in keeping with the design of the overall garden space, as well as being suitable for each garden's particular growing conditions.

Urban modern The design for this small city garden needed to work hard and be well organised in order to deliver the desired effect of a stylish contemporary space that can accommodate outdoor dining, structural planting, a number of different seating opportunities and a degree of privacy from the surrounding properties. The result is understated but effective, and a space that should be easy to live with for years to come whilst not so minimalist as to be unchanging or lacking energy.

1. SEATING This bespoke table and benches comprise the most significant feature in the garden, as is often the case in smaller spaces. Consequently it is far more than just a place to sit, as it is also a focal point and structural device – so the need for it to be as attractive as possible is obvious. Here the elliptical table and benches nestle together in a cohesive, and visually neat way.

2. SCREENING
A row of three olive trees, *Olea europaea*, create a strong structural presence along one boundary of the garden. The repetition helps to create a sense of harmonious unity, whilst the verticality of the trees helps to screen out the adjoining garden.

3. PLANTING
A simple, low planting along the path to the seating area helps to offset the large expanses of paving in the garden and also brings greenery close to the most used part of the garden.

4. GRAVEL
The central area of the garden, where traditionally a lawn would be, is occupied by a low-maintenance granite-chip gravel surface which can be used as an additional seating/dining area. By retaining the shape and size of where a lawn would once have been, the balance of the garden is maintained.

5. PAVING The variety of paving surfaces in this garden helps to give textural interest, changes in scale and also avoids the sensation of a 'concrete garden', which can be an unfortunate side effect of making a hard landscaping, heavy, low-maintenance space. Here a blend of granite chips, stone paving and timber decking ensure variety.

Elegant classic

Elegant classic Gardens can often be as much about illusion as reality, using the horticultural equivalent of smoke and mirrors to cheat scale, give the illusion of space and detach the garden from whatever its setting is, however mundane, and place it in a more engaging context. This small town garden has aspirations of classical grandeur that could easily prove to be 'ideas above its station', but for the confidence with which those aspirations have been realised. The result is actually rather restrained and pretty, a calm and restful space.

1. RETREAT
This small garden building is punching well above its weight in achieving a number of aims in its slender form: a focal point/eye-catcher and practical space in which to repose and enjoy the view across the garden back to the house. Its design is interesting, with a steeply swept roof line that is bordering on being too steep but that somehow manages to work, sitting as it does on slender pillars.

2. HEDGING
The position and scale of the hedges in this garden creates a very definite garden room at the centre of the site, comprised almost entirely of lawn, whilst not compromising the views to and from the garden building or closing down the overall space.

3. LAWN
Though a lawn can be a liability in small gardens, damaging easily under footfall and proving tricky to maintain, here it is hard to envisage what else would work in a garden where the references to classical influences are subtle, but still important.

4. CLIPPED BOX
The simple device of repeated pots of clipped box creates a unifying presence through the garden, as well as introducing the suggestion of that quintessential classical garden feature, topiary.

5. STEPS
In a rectangular garden the obvious orientation for a flight of steps is at a right angle with the boundary. Here the steps are to one side and canted at an angle, which adds to the sense of a journey through the garden.

Mediterranean style

1. SEATING Seating areas abound in this garden. The orientation of the different areas allows for the sun to be enjoyed, or avoided, whatever the time of day.

3. BOLD PLANTING Architectural plants make a strong visual statement in this garden, including ferns, palms, olives and fastigiate junipers. The combination of plants adds to the Mediterranean flavour and provides year-round interest.

2. CLIPPED BOX A pair of clipped box balls in oversized terracotta long tom containers act as sentinels or gate posts for the larger seating area, helping to divide the garden into particular areas or 'zones'.

Mediterranean style Bright, clean contemporary lines are mixed in this garden with a blend of strongly architectural plants and a high proportion of containerised plants, to create a garden that has the simplicity of modernist design in its hard landscaping but the eclecticism of less-formalised gardens in its planting, and a strong reference to the sunny courtyard gardens of the Mediterranean. The repeated use of seating areas in different locations – there are three in what is quite a small space – instantly create a journey from one space to another, the spaces as much decreed and delineated by the seating and tables as by any physical feature.

4. SOMEWHERE TO EAT The smaller of the two seating/dining areas has been located adjacent to the distinctive water feature to make best use of its cooling presence. There is a neat colour reference to the larger seating area: the colour of the table top matches that of the seats around the rectangular table.

Rooftop container garden

Rooftop container garden The hidden green lungs of our cities, perched on building tops, rooftop gardens provide an important opportunity for many urban dwellers to connect with the green world. Because the soil on a rooftop is usually restricted to containers, and the exposure to sun and wind even more unrelenting than at ground level, maintaining moisture levels and preventing desiccation is a constant battle. This rooftop garden demonstrates just how much can be achieved in a space that would normally never see the greenery of plants.

1. SCREENING Vertical planting, in the form of containerised trees, helps screen not only surrounding lateral views but also overlooking views from several storeys up. These trees also help to provide much-needed shelter from buffeting winds.

2. SHELTERED SEATING The seating area in this garden has been very deliberately located in the lee of the adjoining building to take full advantage of any shelter it provides from the wind.

3. CONTAINER PLANTING Every plant in this garden has been grown in a pot, as there is no soil to plant into here. Clustering the containers together creates the sense of a planted border and allows you to try out a number of planting combinations – if you don't like one then simply move the pots around!

4. TIMBER DECKING The deck in this garden functions much as a typical grass lawn would; it is an open space which breaks up the overall composition and can be flexibly used. It's practical too, as timber decking is much lighter than a stone or concrete surface and so ideal for a rooftop.

Low-maintenance indulgence

This roughly south-facing garden in a small town in Wiltshire is typical of many rectangular gardens with timber fences forming the boundaries to each side. The owners, a couple who have owned the property for many years, have cultivated a serious interest in plants and gardens and, with the advent of retirement, now want to indulge themselves.

One serious challenge with the garden is the presence of two semi-mature trees, a lime (*Tilia cordata*) (N) and fastigiate hornbeam (*Carpinus betulus* 'Fastigiata') (I) both of which are subject to Tree Protection Orders (TPOs) from the local authority, prohibiting felling or heavy pruning without written permission. The consequent design incorporates both trees and celebrates their presence by removing garden elements that are compromised by large trees – stone paving that can become slippery during leaf fall, lawns that are impossible to keep looking good without huge investment of time and money, large ponds that can become clogged with leaves – and instead emphasises the benefits of having plants with such presence in the garden.

A: Steps
These steps lead from the first-floor living room into the garden. There is access from the ground floor too.

B: Outdoor dining/entertaining
The presence of the trees within this garden means that this area of hard paving near the house is the only open spot with decent sunlight through the day, making it suitable for turning into a garden room for entertaining and relaxing in the sun. At its centre is the only bit of natural stone in the garden, a bespoke feature made from ground and polished granite edged with rough-cut granite setts.

C: Bench seats
These built-in timber benches form permanent seating in the garden and are backed with fragrant plants, a simple planting comprising *Trachelospermum jasminoides, Lavandula* 'Twickel Purple', and *Rosa* 'Rosarie de l'Hay'.

D: Pots
Framing the long view up the garden are a pair of extra-tall 'long tom' pots, planted with box *(Buxus sempervirens)* clipped into balls. To either side are dividing fences with scalloped ends that drop to meet the rim of the pots. Rather than solid fences these have large apertures between each horizontal slat, to allow as much light through as possible.

E: Deck path
By using cedar decking, a path has been made from one end of the garden to the other without the need for excavation, which could have seriously damaged the roots of the two feature trees. Cedar is an attractive, low-maintenance choice of timber for such a path, and laying sections of deck in blocks that alternate from horizontal to vertical alignment creates additional interest.

F: Water features
Three zinc trough water features line the path, each equipped with a pump and filter to keep the water clear of weed. The reflective quality of water helps to enhance the light levels beneath the shade of the trees, and the relative shallowness of the troughs makes it easy to keep them clear of leaves.

G, H & K: Planted areas
These areas have been planted with shade- and semi-shade-tolerant plants including; *Anemone hupehensis, Geranium endressii, Convallaria majalis, Asplenium scolopendrium, Rodgersia podophylla, Pachyphragma macrophylla, Epimedium youngianum* and *Kirengeshoma palmata*.

I: Hornbeam tree

J: Storage
The very necessary, but not especially aesthetically pleasing, garden shed is here made into part of a dividing barrier that, along with a stub fence opposite, creates a third 'room' in the garden.

L: Planted area **(*see page 166*)**
Underneath the hornbeam tree is a planting scheme adapted to the conditions its canopy and roots create – shade and dry soil respectively.

M: Garden pavilion
This simple, timber pavilion is a tongue-in-cheek reference to a nearby stately home that the owners often visit. It forms the culmination of the garden and provides a seating spot to enjoy the views back into the garden.

N: Lime Tree (*Tilia cordata*)

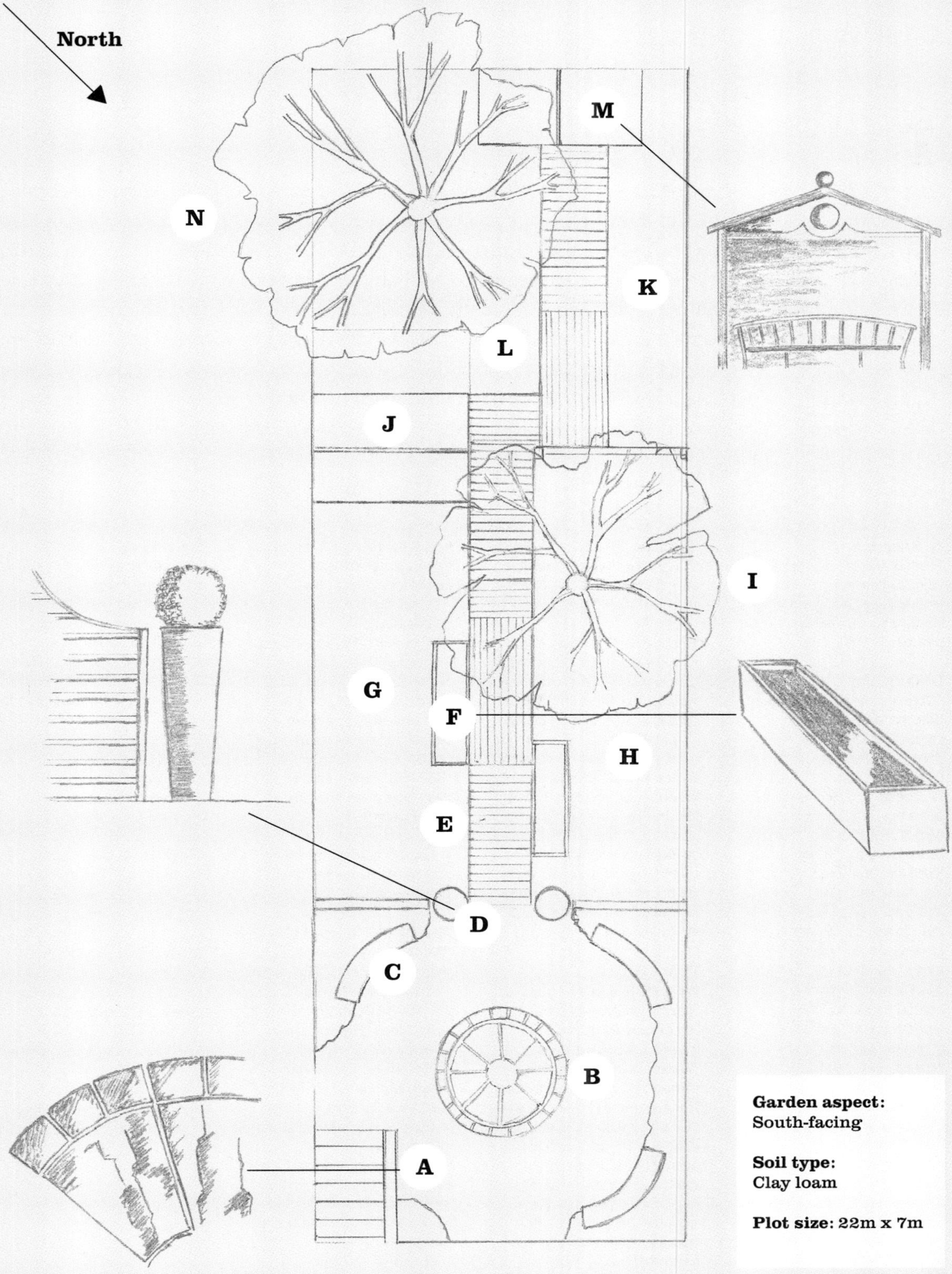

Garden aspect:
South-facing

Soil type:
Clay loam

Plot size: 22m x 7m

Making a difficult space work

This very small garden in west London provides several big challenges. Being north-facing, it is in shade until mid-afternoon, when the absence of any large structures on the western boundary enables the sun to cast its light across the garden until sunset. The presence of two large trees to the northern boundary means the soil at this end of the garden is quite dry and nutrient deficient, whilst the small size of the garden makes delivering the owners' wish list – which includes outdoor dining, covered seating, a barbecue and cooking area and somewhere for composting – very difficult indeed.

The resulting design is based on three interlocking circles of diminishing sizes, which enables the comfortable integration of features such as the barbecue and dining area to be orientated toward the late-afternoon sun. A shallow rill, a narrow water course that runs like a gully, creates movement and sound through the hard landscape. The planting features a lot of lush foliage plants, mostly permanent perennial plants but with some half-hardy plants such as banana and castor oil plant.

A: Paving
Double doors open onto this paved area of the garden which features container planting.

B & D: Planted areas
These areas have been planted with a minimalist semi-shade-tolerant planting of *Hakonechloa macra*, *Hosta sieboldii*, *Deschampsia cespitosa* 'Goldhange' and *Geranium phaeum album*.

C: Outdoor dining/ entertaining
In a small garden it can be tricky to accommodate any kind of storage for tables and chairs. Here a built-in bench and table, fashioned from heavy-gauge oak, have been constructed as permanent features, which can be softened and enlivened with colourful scatter cushions when in use. The table and chairs are made from timber for continuity of materials.

E: Water rill
Essentially a shallow gully through which water flows, which can be easily stepped over, this rill has an additional reflective quality as the gully has been formed from stainless steel. It also adds a gentle background sound of running water to the garden, helping to mask the noise of nearby traffic.

F: Outdoor dining/ entertaining
This barbecue and food preparation area has been made using the same size oak as the built-in bench and dining table.

G: Snowy Mespilus
Paired Snowy Mespilus (*Amelanchier lamarckii*) help to 'pinch' the point at which the two main circles join, helping to give the sense that the garden is divided into two distinct zones, without reducing the space in this cramped garden.

H: Covered curved seating
A covered, curved seating area, open at the sides, from where the rill emanates, creates a pleasant sensation as the water is right by the feet when sitting under the cover. As it is west-facing, the open front is exposed to the evening sun. A green sedum roof helps to minimise the visual impact of the structure, as well as maximising the greenery and potential wildlife habitats within the garden.

I: Sculpture
This feature sculpture acts as an 'eye-catcher' as it is located on the main vista through the garden.

J: Planted area **(*see page 167*)**
A planting of full sun loving species emphasises the effectiveness of composing plants with varying forms to create a sense of movement.

K: Concealed composting
This composting and utility area has been tucked away behind the covered seating to minimise its visual impact.

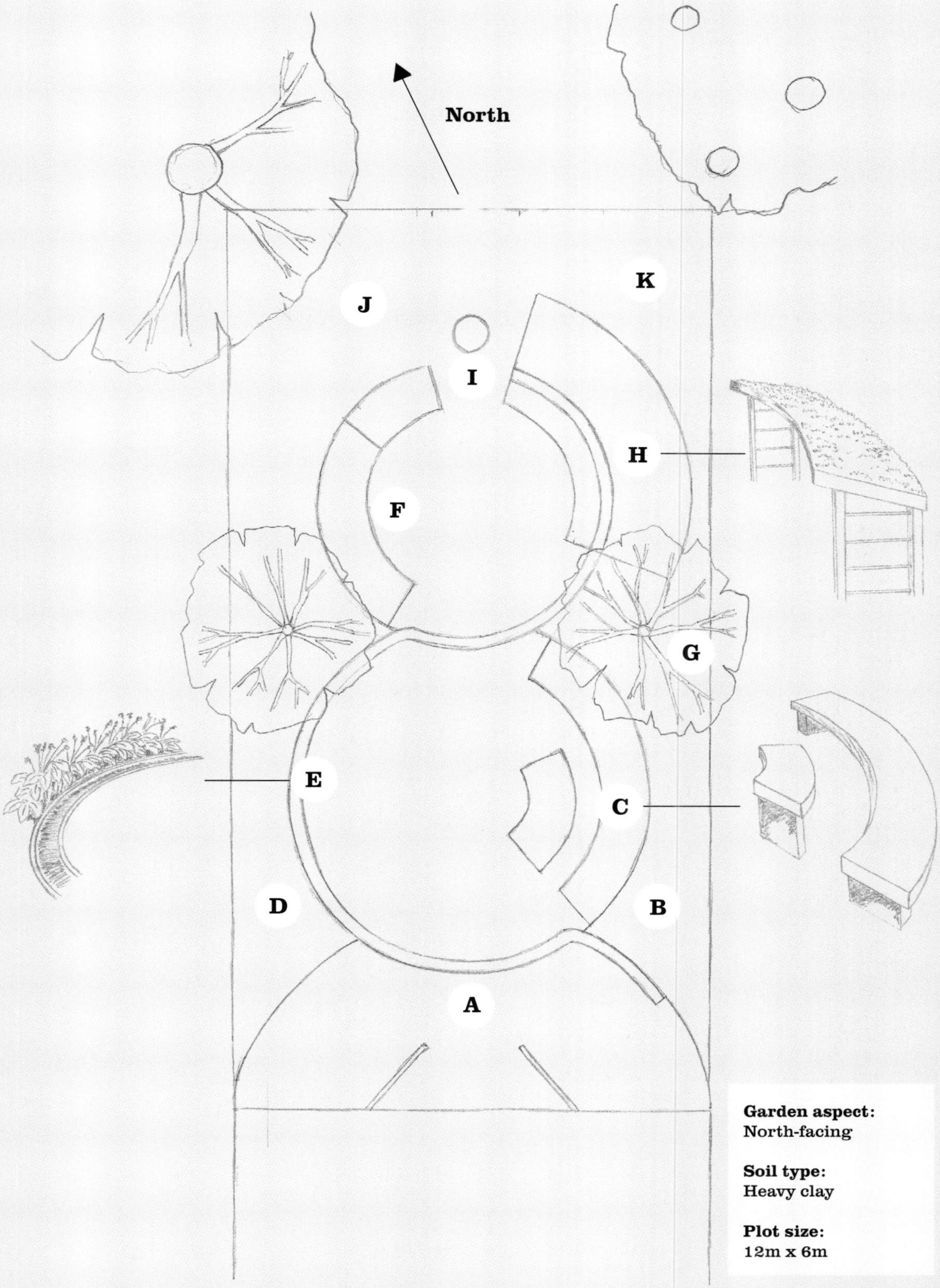

Garden aspect:
North-facing

Soil type:
Heavy clay

Plot size:
12m x 6m

Low-maintenance indulgence (p162)

Large, mature trees are wonderful to have in the garden, but they pose challenges if you want to try to grow anything underneath them as the combination of dry soil and shade creates challenging conditions, while the roots of the tree can be as impenetrable to dig around as concrete.

To improve the soil before planting, generous layers of grit – to improve drainage whilst enhancing moisture retention – as well as spent mushroom compost were applied to the surface over a period of six months, allowing weathering and soil-dwelling animals such as earthworms to help break down and absorb the compost into the soil. Plants were purchased small (nothing larger than a 1 litre pot) so that they didn't require vast amounts of watering. All the plants are adapted to shade, with the aim of the planting being to create an idealised woodland glade with the emphasis on the texture and form of foliage, sprinkled through with flowers. Bulbs are drifted throughout the planting, crossing over and through groups of plants to provide a more natural effect.

Planted area L

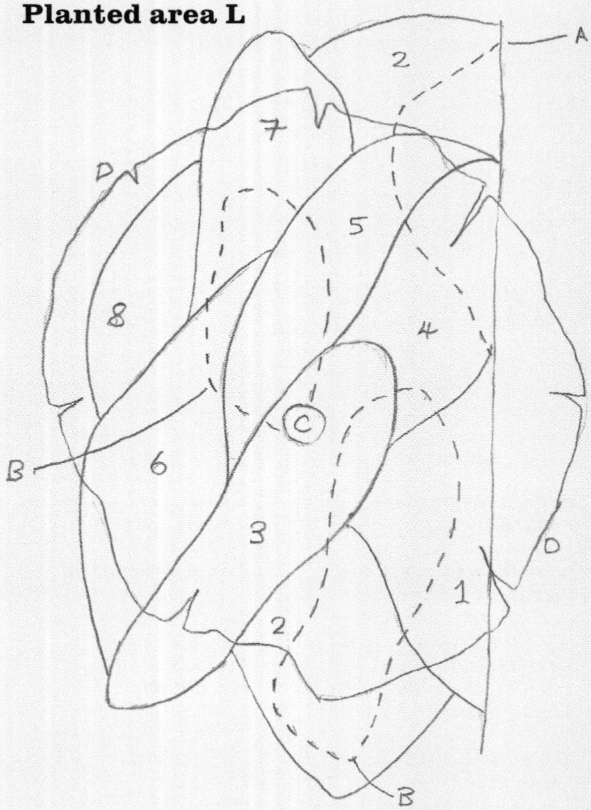

1. *Tellima grandiflora* Strong growing with coarse, mid-green leaves and plenty of foamy yellowy green flowers. Seeds itself around if the conditions are good for it to do so. 40cm x 30cm

2. *Brunnera macrophylla* 'Jack Frost' The leaves are splashed with pewter silver, lighting up dark shaded spots. Pale forget-me-not blue flowers. 60cm x 35cm

3. *Pulmonaria* 'Blue Ensign' Lovely, intense blue flowers early on in the season before the lime tree is in leaf. Looks lovely with spring bulbs. 25cm x 25cm

4. *Polystichum setiferum* 'Herrenhausen' One of the best ferns for dry shade, and adds great foliage interest and architectural strength. 1.2m x 50cm

5. *Vinca minor* 'Gertrude Jekyll' A tough, trouble-free ground-covering evergreen, very compact with pretty white flowers that is ideal for filling a space where little else grows. 10cm x indefinite

6. *Hosta* 'Big Daddy' Big, bold, blue-green leaves (28cm long). Suitable for sun or shade. 60cm x 1m

7. *Chaerophyllum hirsutum* 'Roseum' Adds an uncultivated, wild look to a planting. with finely cut, fern-like foliage topped with dusky-pink flowers. 60cm x 30cm

8. *Onoclea sensibilis* This lovely fern has very light-green fronds that emerge with a pinkish bronze tinge in spring and turn butter-yellow in autumn. 60cm x indefinite

A. *Narcissus* 'Hawera'
B. *Narcissus* 'Jetfire'
C. Existing Hornbeam (*Carpinus betulus* 'Fastigiata')
D. Tree canopy

Making a difficult space work (p164)

Though composing plants effectively can seem daunting, there are some compositional 'rules' that it pays to follow. This planting, viewed head on, demonstrates how to use a mixture of low-, mid- and tall-growing plants with varying forms to create a sense of movement that wouldn't exist if all the plants were the same or similar size, shape and form. The dotted line shows how the plants have been ordered according to their height hierarchy. As this part of the garden was the only one getting full sun, and the owners wanted minimal-maintenance input, the plants used are drought-tolerant and require only limited intervention to keep looking good. There is a subtle purple/bronze colour theme running throughout the planting, sometimes just a hint of colour, other times very boldly represented. There are also plenty of bulbs that love hot, dry conditions strewn throughout the planting, including *Sternbergia lutea*, *Colchicum autumnalis* and a variety of *Crocus*. To get the soil in good shape a 25cm layer of sharp grit was dug into the top 40cm of soil, along with a 30cm layer of spent mushroom compost.

Planted area J

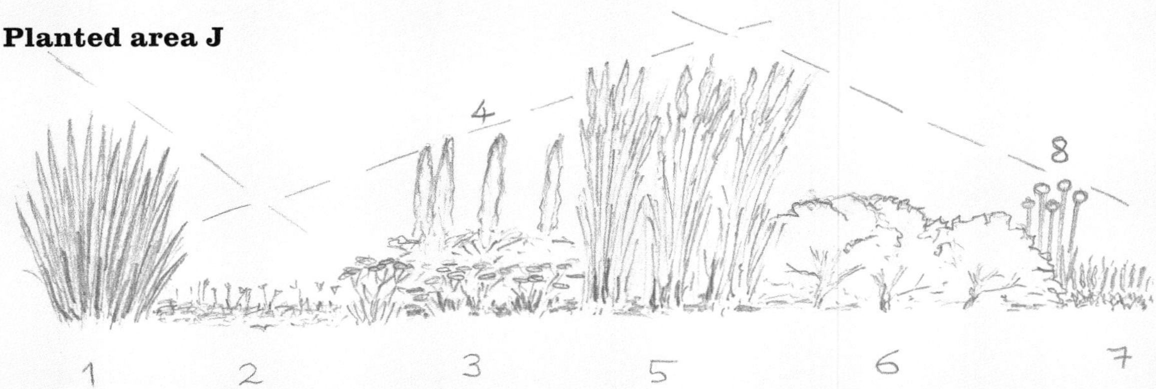

1. *Phormium* 'Bronze Baby' Has very strong architectural merit, making it perfect for marking the end of a planted area, as here, with its bold, sword-like bronze leaves.
1.2m x 80cm

2. *Eschscholzia californica* This plant spreads readily by seed and will fill places that other plants cannot reach. Bright orange flowers, tough as old boots.
25cm x 15cm

3. *Sedum telephium* subsp. *maximum* 'Matrona' The flowers of this particular sedum are secondary to the foliage, which is a glossy, purple bronze, providing a colour link to the *Phormium*.
60cm x 30cm

4. *Eremurus stenophyllus* Adds a vertical element and a sense of drama, with vivid acid-yellow flowers in towering spikes.
1m x 60cm

5. *Calamagrostis* x *acutiflora* 'Karl Foerster' This grass acts like a punctuation mark in the planting, and provides long-term interest from May until December, with its purple flower heads.
1.8m x 60cm

6. *Cistus argenteus* 'Peggy Sammons' Smothered in purple-pink, papery flowers during summer. Contrasts in shape with the *Phormium*.
1m x 1.5m

7. *Salvia* x *sylvestris* 'Tänzerin' Hardy Salvia are some of my favourite plants, and this one is especially effective, with intense purple-blue flowers with a dash of red.
50cm x 30cm

8. *Allium* 'Globemaster' Adds another shape to the planting: the perfectly spherical, metallic purple flower heads, which last in the border as dried flowers for months.
80cm x 20cm

The family garden with a twist

This long, rectangular garden in south-east London is a good example of how a brief can sometimes change mid process. The owners, a young professional couple, discovered that they were going to become parents shortly after commissioning the design. So instead of the ultimate party garden, a garden with wider uses became the desired objective – somewhere to grow a few vegetables, a space for the baby to crawl and eventually toddle, and a place to entertain friends and dine outside too – after all, life doesn't stop when you have children!

The long, rectangular nature of the garden led to an early decision to shift the axis from up and down to across at an angle, immediately creating more of a sense of a journey through the garden and ensuring that elements of the space remained hidden unless explored. The same angle was applied to a number of features: a timber jetty, pergola, flower beds and a raised pool, to help enhance the cohesion in the garden.

A: Hard surface
Paved seating area for seating/outdoor dining.

B: 'Reversed' pergola
This is in effect two single-sided rows of posts that meet in the middle to form a single arch in the centre, with a further arch at the terminal end of the timber jetty on either side of which they are fixed. In a smaller garden where a full-sized pergola might seem oppressively dominant, this kind of structure brings the benefits of framed views into the garden and vertical presence, without overwhelming the space.

C: Timber jetty
Linking the paved dining/seating area with the rest of the garden via a timber jetty gives the initial journey into the garden a real sense of purpose, and also allows the garden to be viewed in a number of ways through various 'windows' formed by the pergola. Keeping the lawn side of the jetty open means that one can step off at any point, which reduces wear and tear that can occur if access is over a small patch of lawn.

D: Seating
Oak bench, made from a single rectangular beam of green (unseasoned) oak.

E: Lawn
A small, but essential lawn space for the children to crawl and toddle around in.

F: Artwork
One of the owners is an artist, so providing an outdoor space for interchangeable artwork was essential, and locating it at the end of the jetty gave it a sense of purpose as a focal point. It can also be glimpsed obliquely from the seating area and a number of other points in the garden.

G: Raised pool
This 3.5m x 1m x 0.45m pool brings the reflective, calming and aural qualities without any of the risk attached to water and small children. The 450mm raised sides are high enough to prevent a toddler from falling or climbing in, and can be augmented with a grille fitted across the surface of the water at a later date.

H: Barbecue and food preparation area

I: Paved area with additional seating

J: Raised beds
These raised beds have been created with the intention of growing a range of vegetables including courgettes, runner beans, spinach and potatoes.

K & M: Planted areas
These areas have been planted with semi-shade-tolerant plants including *Mileum effusum* 'Aureum', *Digitalis purpurea*, *Smyrnium perfoliatum*, *Pulmonaria* 'Opal', *Lilium martagon* var. *album*, *Heuchera* 'Persian Carpet', *Helleborus* x *hybridus*, *Euphorbia jacquemontii* and *Omphalodes cappadocica*.

L: Planted area
(see page 167)
Distinctive waterside planting creates a dramatic display in and around the raised pool (G).

N: Fruit trees
A pair of existing plum trees help to accentuate the angled design of the garden. Originally there were four small trees in the garden, which were thinned out to leave just these two.

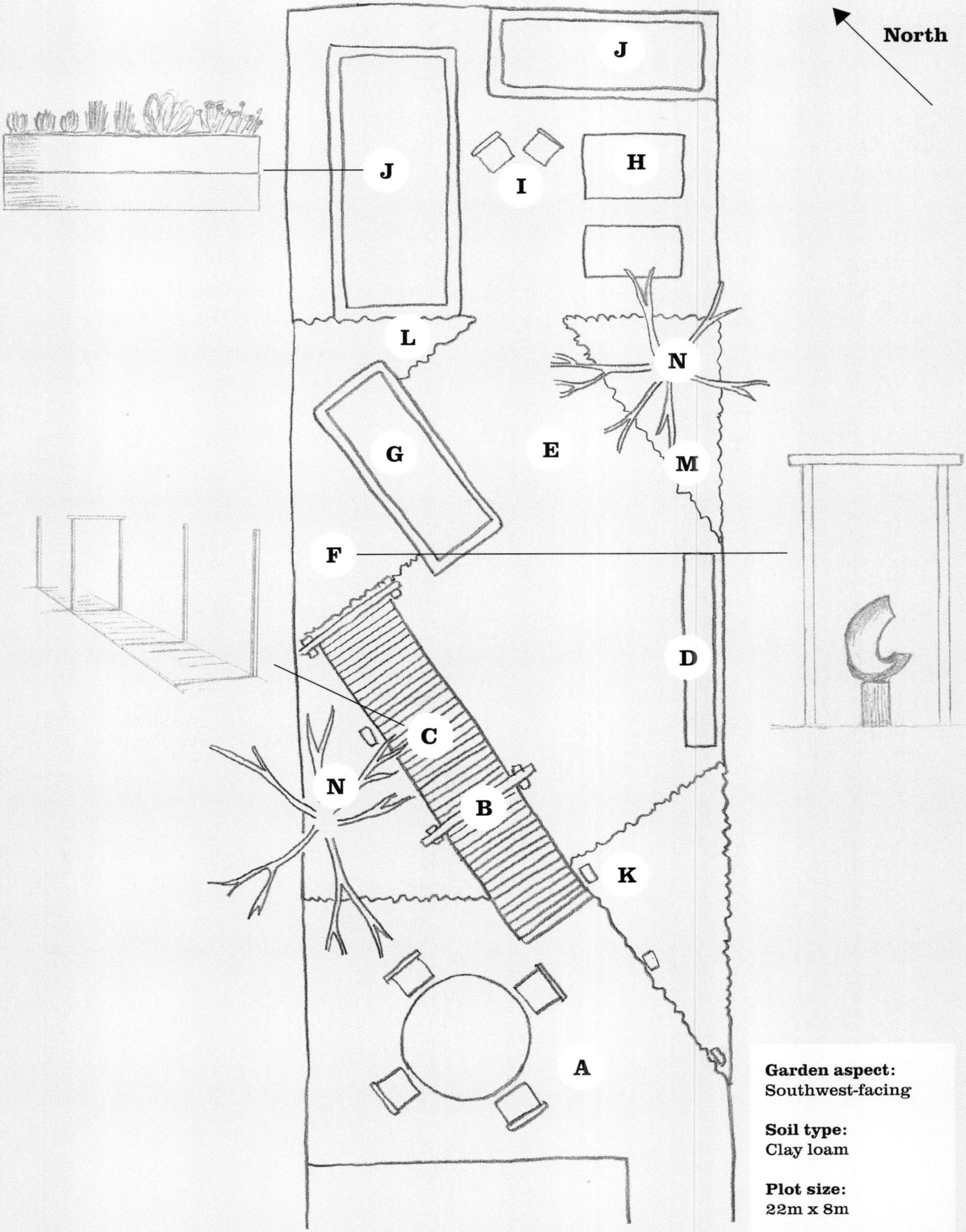

Garden aspect:
Southwest-facing

Soil type:
Clay loam

Plot size:
22m x 8m

Gardening by the seaside

Seaside gardens represent several interesting challenges. This garden on the south coast of England is east-facing, but the proximity of the sea – just one row of houses away – and a lack of other properties within 30m of the boundaries, ensure that it is as sunny as if it were facing due south. The soil in the area is quite poor: sandy, fast-draining and with few nutrients, which when added to the full sun and strong salty winds can lead to plants drying out and wilting rapidly – so appropriate plant selection is vital.

The garden had previously consisted of a large area of lawn with a rock garden and borders to the side, and the owner wanted to reduce the level of maintenance – especially the area of lawn – and increase the number of flowers. She also wanted to create the illusion of privacy by obscuring some of the overlooking windows of properties on the eastern boundary. The plan was very loosely based on the shape of a seahorse, the spine of the animal forming the shape of the path from the house to the garden shed.

A: Doors
Double doors lead from the second bedroom into the garden.

B: Kitchen window

C: Paving
A paved sandstone area features a large selection of containers planted with drought-tolerant plants.

D: Spiral paving detailing
An inset detail of square, polished granite pavers, which also form the edge of the main path is arranged in a spiral. This breaks up what would otherwise be a large expanse of paving and forms the notional 'tail' of the seahorse.

E: Curved pathway
Formed of over-fired, burnt stock pavers – a burnt stock means that the brick has been fired until the sand turns to glass, which as well as giving the brick distinctive black/purple patches also renders it non-slip as it becomes resistant to moss and algae growth – with an edge of polished granite pavers. The brick-red that is the primary colour in the burnt stock is also present in the sandstone paving flags in the main paved area, providing continuity of colour. The curving, looping path makes for a much more interesting journey, even if it is just to a shed!

F: Mini lawns
Although one of the main aims of the redesign was to get rid of the large lawn, two small panels of grass, easily cut with a lightweight rechargeable battery-powered mower, were included in the new design, to help set off the planted borders as well as to provide a space for the grandchildren to play on.

G: 'Amber Beauty' trees
Three *Prunus maackii* 'Amber Beauty' cut across the diagonal axis of the garden. Arranging the trees in this way helped to obscure some of the distant overlooking windows but also brought vertical height and much-needed shade into the heart of the garden, rather than banishing it to the fringes. 'Amber Beauty' is grown mainly for its attractive orange bark, which is especially striking in winter.

H: Planted area (*see page 173*)
The planting scheme here is full of colour and rich with different species.

I: Feature screening
A curved row of round, milled oak posts with clematis (in keeping with the garden's curved theme) partially obscure the shed – a better option than creating a flat-sided fence or trellis in front of the shed, which would only accentuate its presence.

J: Seating area
A bespoke, curving bench located in the dappled shade of one of the three *Prunus* trees.

K: Shed and compost area

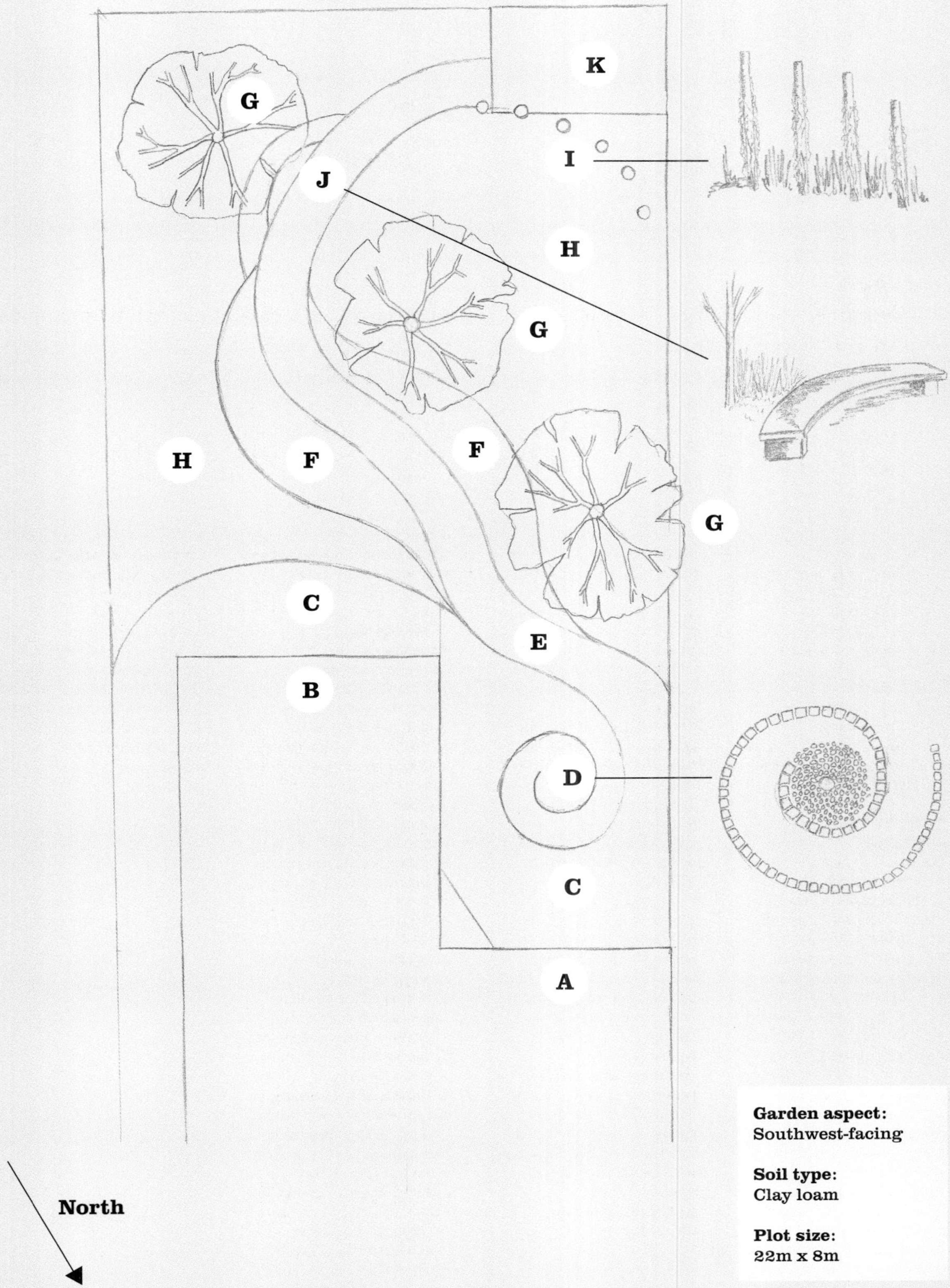

Garden aspect:
Southwest-facing

Soil type:
Clay loam

Plot size:
22m x 8m

The family garden with a twist (p168)

The rectangular raised pool, crafted from planed hardwood railway sleepers (*see area 'G' page 168*) forms the centrepiece of this planting area, counterbalancing the many foliage plants that surround it. The planting is in keeping with the semi-shaded, moist, waterside theme suggested by the pool, though in actuality the soil around it is no moister than anywhere else in the garden as the pool is an entirely self-contained feature raised above ground level, rather than a natural body of water with wet margins. The pool itself is also a planting opportunity, allowing for a range of plants – visually linked to the those that surround it – to be employed that simply cannot be grown elsewhere in the garden. Substantial architectural plants such as *Cyperus papyrus* make a real impact here, adding vertical presence to the planting and forming additional eye-catchers to be glimpsed through the 'windows' of the reversed pergola (*see area 'B' page 168*).

Planted area L

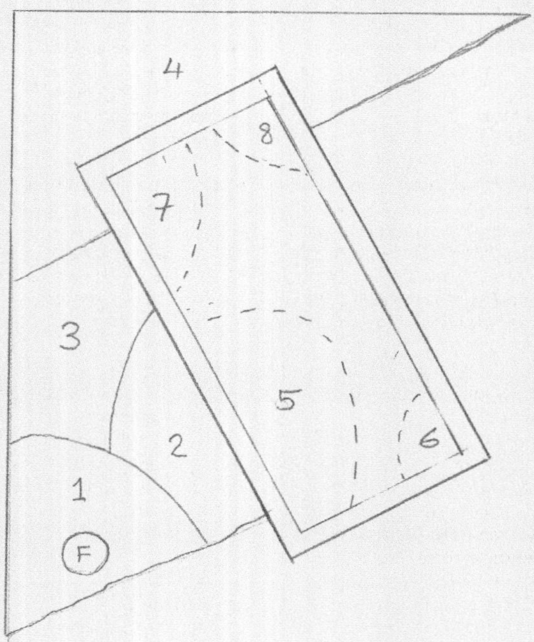

1. ***Deschampsia cespitosa*** **'Goldtau'** This lovely cultivar illuminates semi-shade with golden flowers. 60cm x 60cm

2. ***Hosta*** **'Halcyon'** Attractive, mid-sized blue-grey leaves that are gently tapered to a point. 50cm x 50cm

3. ***Ranunculus aconitifolius*** A very beautiful spring woodland plant, with white flowers and deeply cut leaves. 90cm x 60cm

4. ***Astilbe*** **'Purpurlanze'** With its pretty purple flowers, this is one of the few Astilbe that tolerate drier soils. 1.2m x 70cm

5. ***Nymphaea pygmaea*** **'Alba'** One of the smallest of all the water lilies (up to 25cm wide) and so suitable for a small pond such as this.

6. ***Iris ensata*** This water-loving iris is worth growing just for its sword-like foliage, let alone the purple flowers. 90cm x 30cm

7. ***Cyperus paperifera*** Cyperus papyrus the plant from which early paper was made, this is a great architectural water plant for sheltered, frost-free gardens. 1.8m x 60cm

8. ***Lythrum salicaria*** The lovely native purple loosestrife, a lovely sight when in bloom with its dusky-red-pink flowers. 1.2m x 50cm

Gardening by the seaside (p170)

The owner of the garden wanted lots of flowers in her garden, but the seaside location and relatively poor soil – it's fast draining and doesn't hold on to nutrients very well – meant careful plant selection was essential, good soil preparation vital, and the application of weed-suppressing and moisture-retaining mulch a time- and plant-saving boon. The soil was improved before planting with a 40cm layer of green waste compost from the local council waste-processing facility.

The overall design theme of the garden, a series of curves based around the main sweeping path and bisected by three prunus tree that strike across the garden, led naturally to a series of interlocking drifts of plants that followed the same flow as the path, creating a wave-like effect. Fragrant lavender was planted near the edge of the path and the seating area and a sweeping line of feature *Stipa gigantea* 'Gold Fontaine' planted to form a counter-curve to the main path. The result is a modernist, naturalistic update on the old-fashioned cottage garden.

Planted area H

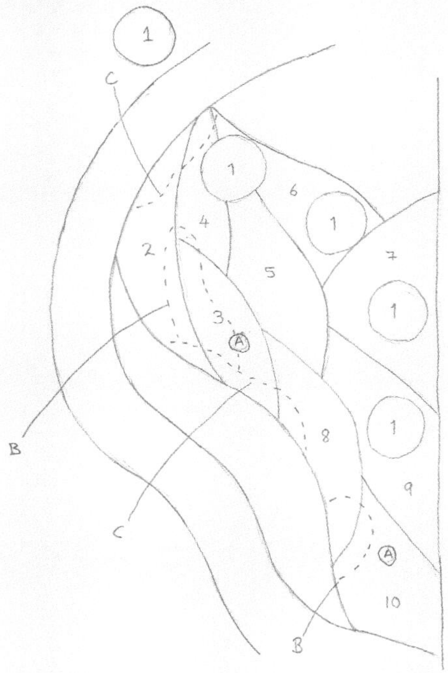

1. *Stipa gigantea* 'Gold Fontaene'
Structural plants, with oat-like flowers on thin stems. Their see-through nature means that, despite their height, they can be planted near to the front of a border.
2m x 1.3m

2. *Lavendula* 'Vera'
Taller than many of the more popular compact lavender, and with an attractive, loose habit that is particularly suitable in a cottage garden themed planting.
60cm x 40cm

3. *Euphorbia amygdaloides* 'Purpurea' A great foliage interest plant, with dark-purple leaves topped by acid-yellow flowers.
50cm x 40cm

4. *Erysimum* 'Bowles' Mauve' This beautiful plant flowers so freely, that, after three or four years, it has flowered itself to death. But it is worth every moment.
75cm x 60cm

5. *Eryngium* x *zabelii* 'Jos Eijking'
A relatively new cultivar with steely silver foliage and intense metallic-blue flowers, this is a perfect seaside plant and incredibly tough and drought-tolerant.
75cm x 50cm

6. *Helenium* 'Waldtraut'
Upright and sturdy, this border stalwart flowers from midsummer into autumn, producing attractive golden-brown flowers that are a superb foil to steely eryngium.
100cm x 80cm

7. *Knautia macedonica*
Tough and beautiful, with deep burgundy-coloured flowers that are much loved by bees and butterflies and, when turned to seed, finches.
1m x 50cm

8. *Salvia* x *sylvestris* 'Blauhügel' A compact hardy salvia that bears plenty of pretty purple-blue flowers over a long period through summer.
45cm x 45cm

9. *Persicaria amplexicaulis* 'Firetail'
Long-flowering, from June until October, with tall spires of bright red flowers over attractive, robust foliage.
80cm x 60cm

10. *Nassella trichotoma*
An extraordinary-looking grass, resembling the fibre-optic table ornaments popular in the 1970s.
50cm x 50cm

A. *Prunus maackii* 'Amber Beauty'
B. *Nigella damascena*
C. *Eschscholzia californica*

Outdoor entertaining

This small south-facing garden in south London benefits from full sun for most of the day, but a slope from the house to the boundary fence make it hard to accommodate the outdoor dining and entertaining spaces that the owners, a young professional couple, want to include. In order to make the garden as useable as possible it has been reconceived as two level areas with a 1m high terrace between them, creating equal-sized zones for the garden's various features – a well-proportioned space for table and chairs, fragrant planting and a large pool. Detailing is vital in a garden of this size, and the pool offers the opportunity to mix high-quality diamond-cut sandstone (with a very smooth finish) with a cedar timber deck.

Whilst the owners very much want to include plants, they have concerns over the amount of time they would need to spend maintaining the garden and worry that an overly complex planting scheme would be beyond their abilities to look after. All planting within this garden has therefore to be as easy to care for as possible.

A: Water feature
In a small space it can be tempting to scale down the size of any planned garden features, but sometimes this can only serve to make the space seem even smaller and fussier. This rectangular pool, surrounded by a cedar timber deck, not only serves to cool down the top terrace of the garden (which, as it is south-facing and sheltered, could otherwise become uncomfortably warm) but also helps to create an illusion of greater space through the reflection of the sun and sky. Beneath the deck is a soakaway system that absorbs and slowly releases all the water run-off from the garden, which helps to reduce the flow of flood water into local water courses, and an underground water tank to take water from the house roof for irrigation.

B: Seating
This seating area for outdoor dining is surfaced with resin-bonded gravel, a stable, sealed aggregate that doesn't move or 'pick up' under foot.

C: Outdoor dining/entertaining
This large circular table is made from the same type of timber – cedar – as the timber deck and bench (I) to give continuity across the various timber elements.

D & K: Planted areas
Simple planting consisting of Lavender (*Lavendula* 'Hidcote') and clipped box pyramids. During summer the fragrance from the lavender pervades the entire top terrace, whilst the box pyramids provide structure, form and year-round presence. In spring the beds are enlivened with large drifts of the white daffodil *Narcissus* 'Thalia'.

E: Terrace wall and picture frame
A 1m high wall constructed from the same stone as the pool edge – diamond-cut sandstone – this time cut as blocks. Although comparatively expensive, using such high-quality materials sparingly ensures their impact is maximised. To the west of the garden a simple 'picture frame' of timber helps to subdivide the dining space and create windows onto the planting around the Magnolia.

F: Steps
These steps are made from the same diamond-cut sandstone as used in the terrace and pool edge.

G: Pleached hornbeam
An aerial hedge of hornbeam that rises above the 1.2m fence on the eastern boundary to a height of 2.3m and provides greater privacy from neighbours without becoming an oppressive nuisance.

H: Lawn
Small lawns can be a nightmare to look after, because of the level of wear and tear over a small area, but the owners were insistent and it provides an additional area for occasional chairs.

I: Seating
This timber bench has been placed next to *Sarcococca confusa*, a highly fragrant, shade-tolerant (this bed is north-facing and in almost complete shade) evergreen shrub that flowers in winter, when it will fill the garden with perfume. In a hot, sunny garden a shaded bench can provide a very welcome break from the heat.

J: Planted area
(*see page 178*)

L: Back door

M: Window

Garden aspect:
South-facing

Soil type:
Heavy clay

Plot size:
10m x 11m

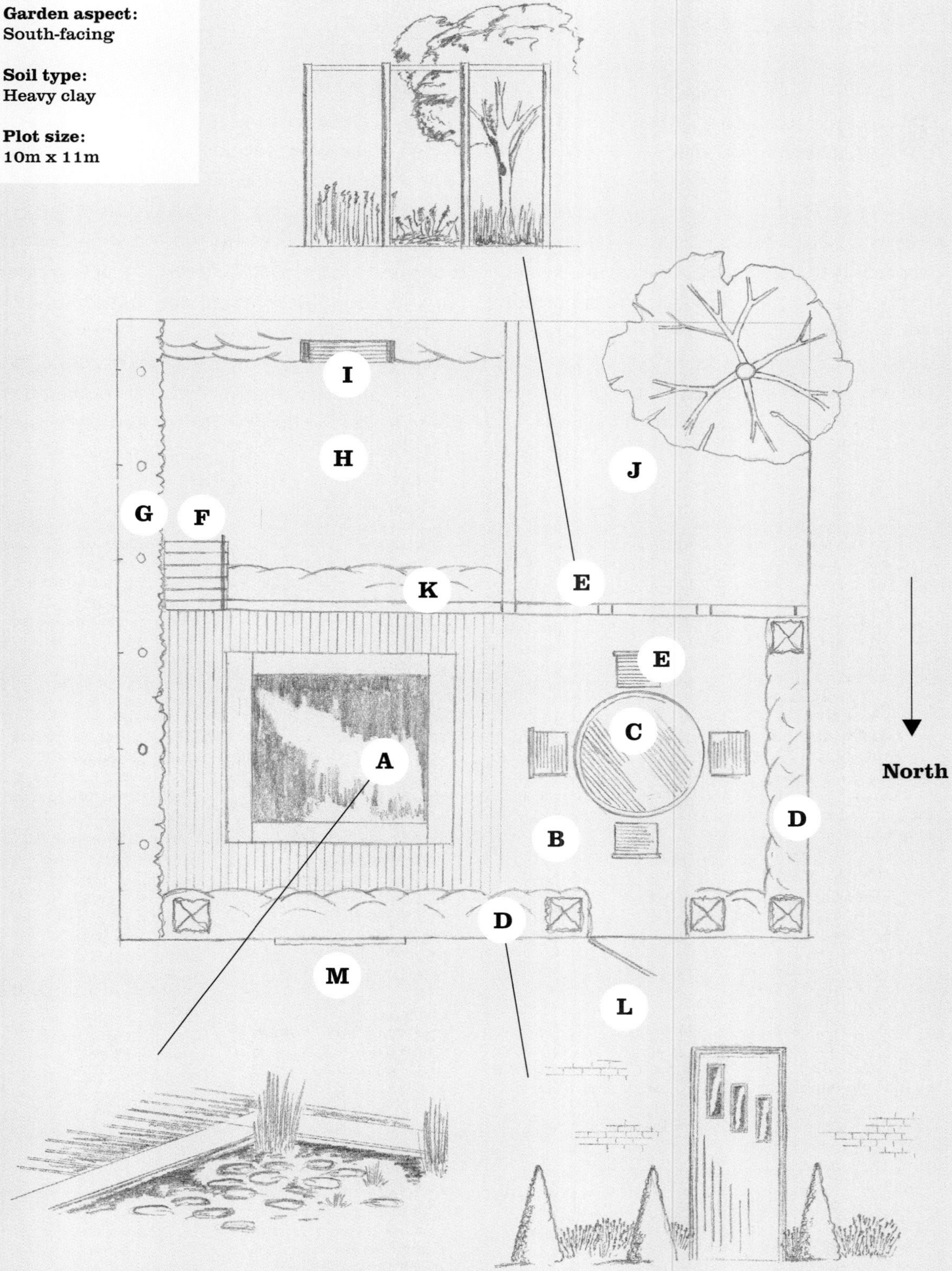

The unconventional front garden

Sometimes the only available, or suitable, space to make a garden is a 'non-traditional' one. This garden, located on a comparatively quite side street in Yorkshire includes many of the features associated with a garden space – seating area, pool, planting – but in this case they are in the front garden, rather than the rear. The reasons for this are twofold; the rear garden of this property is roughly east-facing and rather cool and shaded, plus the owners, a couple with two young children, have dedicated the small rear garden space to their youngsters and creating as many play opportunities as possible.

The front garden, however, offered easier growing conditions – west-facing – and enough space to make a viable ornamental garden that the whole family can enjoy, including an outdoor dining area, herb- and salad-growing space and ornamental pool. The owners have a very limited budget, so every effort has been made to make materials and features look as good as possible for the minimum outlay.

A: Yew trees
A pair of fastigiate yew trees, *Taxus baccata* 'Fastigiata' form offset markers to the garden gate and the terminal point of the first section of garden path. Chopping the path to the front door into two straight sections creates a greater sense of journey without making it impossible to negotiate – paths need to be practical, as well as aesthetically alluring. Repeating two plants with strong structural presence creates surprise and expands the vertical plane in a small space.

B: Planted area
(*see page 179*)
This area of the garden has been developed into a simple potager – a hardworking kitchen garden with beautiful, ornamental displays.

C: Fencing
Fences are rarely things of great beauty, but are often necessary where it is impossible to accommodate a more environmentally and aesthetically pleasing boundary feature such as a native hedge. A 'hit and miss' boundary fence creates a pleasing, permeable barrier and is achieved by removing every other, or every other two, posts, or a combination thereof.

D: Offset pathway
Rather than using expensive paving materials, the paths in this garden have been made using a sub-base of recycled crushed concrete topped with rolled granite chips. To give them a little extra quality, and to give a little extra visual interest, the paths have been edged with a granite sett.

E: Decking
This oak timber decking has been made from new untreated oak railway sleepers cut in half along their length – a cost-effective way of doubling the area of coverage from the same amount of materials. The decking is used around the main seating area but also near the front door to provide continuity of materials.

F & H: Planted areas
Informal planting of cottage garden favourites and other mixed perennials including *Astrantia* 'Hadspen Blood', *Dicentra spectabilis alba*, *Rosa glauca*, *Rosmarinus officianalis*, *Geranium* x *magnificum*, *Crocosmia* 'Lucifer', *Penstemon* 'Andenken an Friedrich Hahn', *Clematis recta*, *Campanula lactiflora* 'Sparkes' Variety', *Veronicastrum* 'Fascination', *Thalictrum* 'Elin' and *Cotinus* 'Grace'.

G: Water features
The two pools of water here appear as though they are a single water feature, crossed with a bridge, while in fact there is a solid path between them. The two pools are linked by an underground pipe so that just one filtration system is required, and a metal grate has been installed 1cm below the water surface to ensure the children are safe from falling in the water.

I: Rowan tree
Pre-existing Rowan tree (*Sorbus* 'Joseph Rock')

J: Yew hedge
Yew (*Taxus baccata*) hedge to 1.6m

K: Outdoor dining/entertaining
Made from the same untreated oak sleepers as the deck, this outdoor dining table has a built-in herb and salad planter in the centre of the table – in effect a large timber box, lined with a waterproof membrane and drilled with drainage holes – that drops into the table. The perfect way to combine outdoor dining and growing, in this case augmented by containers with vegetables growing around the table.

L: Dining room window

M: Hall window

N: Front door

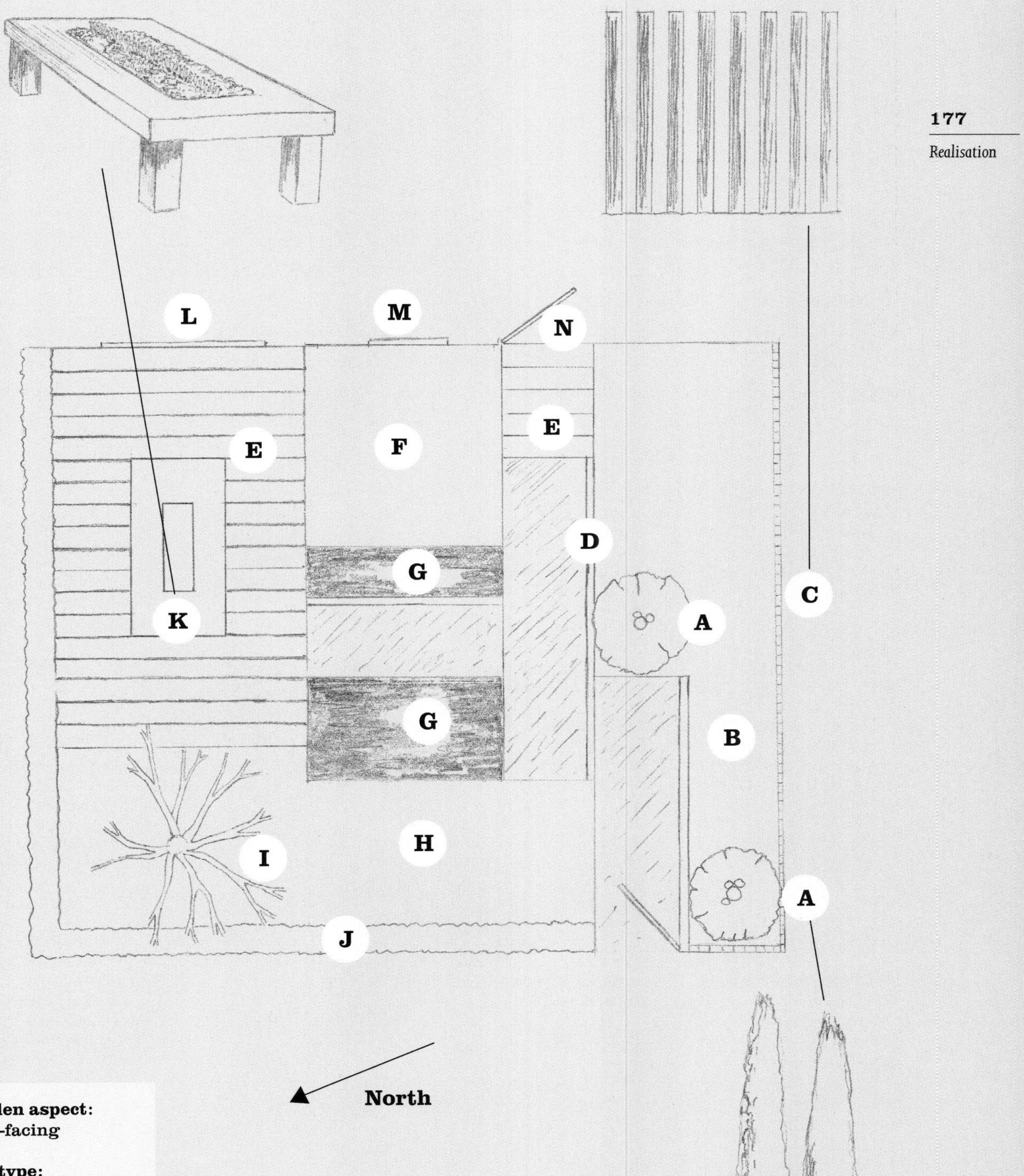

Garden aspect:
West-facing

Soil type:
Alluvial loam

Plot size:
7m x 7m

Outdoor entertaining (p174)

The owners of this garden wanted their limited area for mixed planting to be as texturally interesting, and colourful, as possible. The conditions – part sun, part shade – mean that plant selection is varied across the planted area with appropriate plants to suit. Rather than organising the plants into large drifts, this planting has a slightly looser arrangement and is closer to a matrix planting – where plants drift in and out of each other's space, as if they have self-seeded and mingled together. The effect has echoes of the way plants grow in the wild, whilst still being controlled and enabling plants to be deliberately positioned to maximise their aesthetic impact.

The clay soil and partial shade in this part of the garden was countered by the digging in of 25cm of sharp grit, 20cm of composted bark and 20cm of green waste manure from the local council into the top 40cm of soil. In order not to damage the tree roots, the material was applied to the top of the soil around the tree, in layers and over a period of six months, to allow it to break down into the soil.

Planted area J

1. *Rosa* 'Complicata' A lovely, old-fashioned rose with highly fragrant flowers that fill the garden with scent when in bloom during summer. It's also tough, and quite happy growing in clay-based soils – as are most roses.
1.8m x 1.8m

2. *Miscanthus yakushimensis* 'Little Kitten' More compact than the majority of *miscanthus*, with fine, light-green leaves marked with a creamy white central 'rib'. This ornamental grass bears silvery flowers from late summer.
70cm x 80cm

3. *Gillenia trifoliata* An under-used perennial with pretty, starry white flowers from early summer and great autumn foliage colour.
90cm x 50cm

4. *Stipa tenuissima* The hair grass is an irresistible textural treat, and a great plant for creating a sense of almost watery movement through a planting scheme. Best grown from seed, which it produces prodigiously.
60cm x 30cm

5. *Galega orientalis* Has a lovely, slightly wild look to it and copes with sun and partial shade, producing violet-blue flowers in spring and summer. It can spread a bit if the conditions are right, which may need managing.
1.2m x 60cm

6. *Spiraea japonica* 'Gold Flame' Brings a splash of foliage colour to a semi-shaded spot, producing bronze-red foliage in spring that then turns golden yellow before finally becoming green, at which point it produces dark-pink flowers.
75cm x 75cm

The unconventional front garden (p176)

Along with areas of flowering and foliage plants, the front garden offered the perfect space to grow vegetables and herbs in a 'potager' style – growing small amounts of edible plants mixed with flowering plants. Growing your own food is great fun, and there are environmental benefits in a reduced carbon footprint, along with the health benefits that come from knowing exactly where your food has come from and what has gone into making it grow. And for children there are wonderful educational opportunities – there is nothing quite like seeing a plant grow from seed to maturity, and then eating it!

But in a small space it's vital to grow the crops that benefit the most from being harvested fresh or that have a premium attached to them. So rows of potatoes and cabbages are out, replaced by tasty herbs, hard-to-find vegetables and expensive crops like asparagus. These are mixed with flowering plants that act as companion plants by either drawing pests away from the crop or masking the smell of vegetables so that pests cannot locate them. The alluvial loam in the garden was in good condition but benefitted from the addition of well-rotted manure, dug into the top 50cm of soil

Planted area B

1. *Rosa* 'Gertrude Jekyll' One of the most fragrant of the new English roses. Here it's grown mainly for its scent and beauty although aphids will be attracted to it and can be dealt with using an organic spray or natural predators. 1.3m x 1.3m

2. Purple sprouting broccoli A high-value crop that has a relatively short shelf life, so is well worth growing even if only a few plants. It will crop throughout late winter and early spring, a lean time for home-grown produce.

3. Swiss chard Another winter-cropping plant. The variety 'Bright Lights' is particularly colourful.

4. *Delphinium* 'Black Knight' This imposing border favourite is highly attractive to slugs and snails – helping to draw them away from the veg. 1.6m x 70cm

5. Jerusalem artichoke A tasty, expensive crop and a seriously impressive architectural plant. The flowers are loved by bees.

6. Asparagus Although it takes three seasons from planting before harvesting can commence, this high-value crop is well worth the effort.

7. Garlic Whilst not very expensive, the range of garlic varieties to grow is far wider than can be bought in the shops. The giant elephant garlic is great for roasting.

8. Little Gem lettuce Easy to grow and fast to crop, Little Gems can be grown to maturity in a matter of weeks.

9. Lollo rosso An easy-to-grow salad leaf with a distinctive taste and a great colour contrast to green lettuce.

10. *Allium schoenoprasum* (chives) Lovely to have at hand for garnishing salads, especially potato salad. Its flowers are edible too.

11. *Nasturtium* The peppery leaves are a tasty treat in salads, but this is also a sacrificial plant as aphids are drawn to it in preference to the rest.

12. *Molinia caerulea* subsp. *arundinacea* 'Transparent' A lovely ornamental grass that has height and presence without bulk. Good for cutting but also provides a habitat for a variety of beneficial insects. 1.8m x 60cm

The hybrid garden

With many small domestic gardens the architectural integrity of the house is secondary to creating a garden that works practically and aesthetically. In this case however, the house – a 1960s modernist property designed by the architectural practice SPAN as part of a development in Kent – demands a particular approach, respecting the very strong architectural features of the property, the rear, garden elevation of which is based around a series of cube-like bays. The owners, a professional couple, were very clear about what they wanted – simple structural planting but also lots of flowers and informality – which at times conflicted with an ultra-minimal, modernist garden. The result is something of a hybrid, with areas of wild meadow, planted bays of colour drawn in size and shape directly from the staggered cube bays of the house, and structural hedges to create a series of crisp horizontal lines through the garden.

The rear elevation of this detached house faces almost due north, but the amount of space and absence of neighbouring structures means that there is plenty of sunshine in the main planting areas.

A: Rear elevation of house

B: Decking
Large cedar deck to accommodate dining and sitting out.

C: Hot tub
This hot tub has been built flush with the surrounding deck.

D: Paving
Paved area with four rectangular Portland stone paving flags. These paving flags link aesthetically to a roof terrace on the first floor of the building which is entirely paved with Portland stone. High-quality natural stone can be a great aesthetic asset but is often expensive; however, even a small amount of high-quality stone can make an impact in a garden, if used in a high-profile position and installed with skill.

E: Seating
This Portland stone bench, made from the same stone as above, is inset into a beech (*Fagus sylvatica*) hedge which is in turn backed by a white painted brick wall. The result is a very cool, sleek seating area with views back into the garden and house that also acts as a focal point from the other end of the garden.

F: Planted area
(*see page 184*)

G: Beech hedge
A beech hedge (*Fagus sylvatica*) in three sections of varying lengths echoes the cubed rear section of the house. Each bay that is created by the hedges houses a section of the perennial planting (F). The hedges are also cut at three diminishing heights (2m, 1.7m, 1.4m), to create a strong horizontal architectural form in the garden.

H: Orchard and wildflower meadow
This rectangular section of the garden used to house a lean-to car port and tarmac hard standing for parking. This was ripped up and replaced with a mini orchard and wildflower meadow, which creates a very striking contrast to the sleek, white painted minimalist architecture of the house. The five trees are a mixture of apple, pear and plum and the wildflower meadow was seeded with seed harvested from local meadows. The combination of long grass and flowers beneath old orchard trees harks back to a time when these were common sights in the area of Kent where the house is located.

I: Mown grass pathway
A simple mown grass path meanders through the orchard and wildflower meadow.

J: Seating
This Portland stone bench is identical to the one inset into the beech hedge on the other side of the garden. Using the same feature twice but in a different context creates an interesting dynamic between continuity of materials and contrast of setting.

K: Sculptures
Two sculptures, the work of the same artist and bought as a pair, stand in opposing corners of the orchard meadow. The owners purchased the works from an art college graduation show, an excellent way to buy new art at reasonable prices and support up-and-coming artists.

L: Doors
Double French doors lead out into the garden.

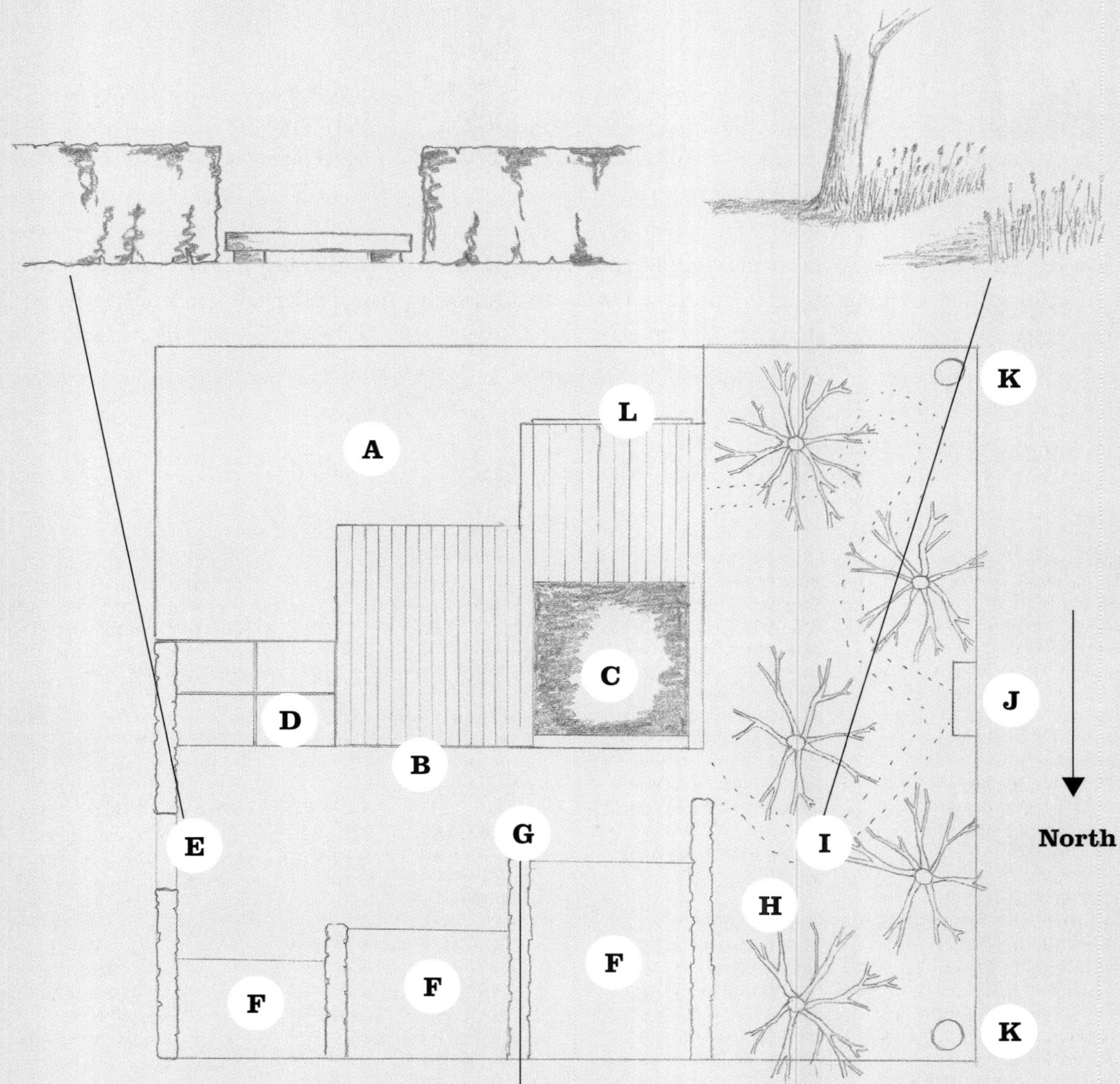

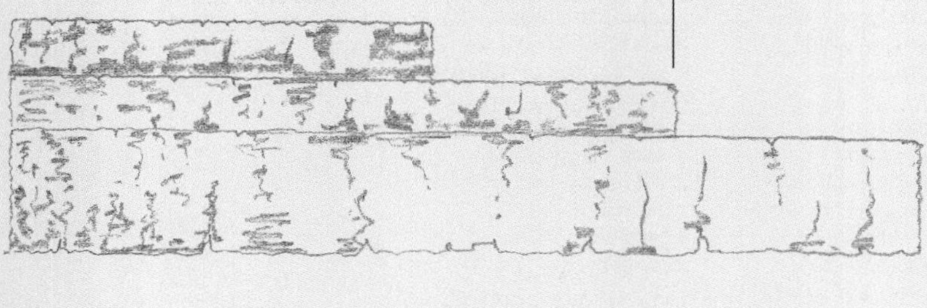

Garden aspect:
North-facing

Soil type:
Sandy loam

Plot size:
30m x 30m

The surprising space

This garden on the Channel Island of Guernsey is a hundred metres or so from the sea and protected from the worst of the weather, to an extent, by 3m high walls that wrap around it. As well as facing the typical seaside challenges of high light levels, salt winds and dry, hot conditions, there's an added complication in that the area which is now garden used to be where the owners' cars were parked, and so consisted of tarmac rather than soil. The garden is also a slightly odd shape, with one boundary wall cutting across the garden at an angle.

To manage the unusual shape and lack of soil, the garden is based around interlocking circles of decked hard standing, edged with low raised walls that enable the imported topsoil (after removing the tarmac there was no soil to replace it with) to be built up by 300mm, giving not just a greater depth of soil to plant into but also a sense, at least when the plants are established, that the seating areas are slightly sunken.

A: Conservatory

B: Water feature
This raised, curved water feature discharges water over a low (450mm) lip into a splash area below, creating a steady, calming and cooling sound – perfect in such a hot, sunny garden.

C: Fire pit
This fire pit turns the sunny seating area into a warm evening retreat, ideal for a drink or meal as the sun sets.

D: Built-in seating

E: Decking
In a very sunny garden a timber deck, such as this oak one, has a slight advantage over stone as it doesn't absorb quite as much heat, so is less likely to become uncomfortably hot to walk on after several hours of exposure to full sun.

F: Planted barrier
Double-skinned hollow wall, 500mm high, built from rendered blocks, that have been filled with topsoil and planted with aromatic drought-tolerant plants including lavender, rosemary, *Artemesia*, *Nepeta* and thyme. A planted hollow wall offers the opportunity to create a barrier with a lower visual impact than a solid wall, and that maximises the space available for plants.

G: Tree grove
Grove of *Elaeagnus* 'Quicksilver', a silver-leaved, highly fragrant drought- and wind-tolerant large shrub/small tree. Planted here they help to break up the outline of the garage (J) and provide a degree of shelter from westerly wind.

H: Trained trees
Two roof-trained *Elaeagnus multiflora* provide form, structure and cast shade in the second seating area, creating a cooler space for sitting in than the sun drenched oak deck.

I: Screening
A row of clipped evergreen holm oak, *Quercus ilex*, provide an additional screen against seas winds, of which this plant is highly tolerant, and also helps to soften the appearance of the 3m tall wall.

J: Garage

K: Planted area
(*see page 185*)

L, M & N: Planted areas
These beds have been planted with sun-loving plants including *Ballota pseudodictamnus*, *Eryngium* x *oliverianum*, *Artemesia arborescens*, *Ozothamnus rosmarinifolius*, *Stipa gigantea*, *Corokia virgata*, *Helianthemum* 'Henfield Brilliant', *Ceanothus* 'Concha', *Sternbergia lutea*, *Helichrysum petiolare*, *Diascia rigescens*, *Verbena rigida*, *Arum creticum*.

O: Box hedging
Clipped box (*Buxus sempervirens*) balls provide structure and repetition around the edge of the main seating area and also, through their shape, reference the interlocking circles that are the theme of the garden.

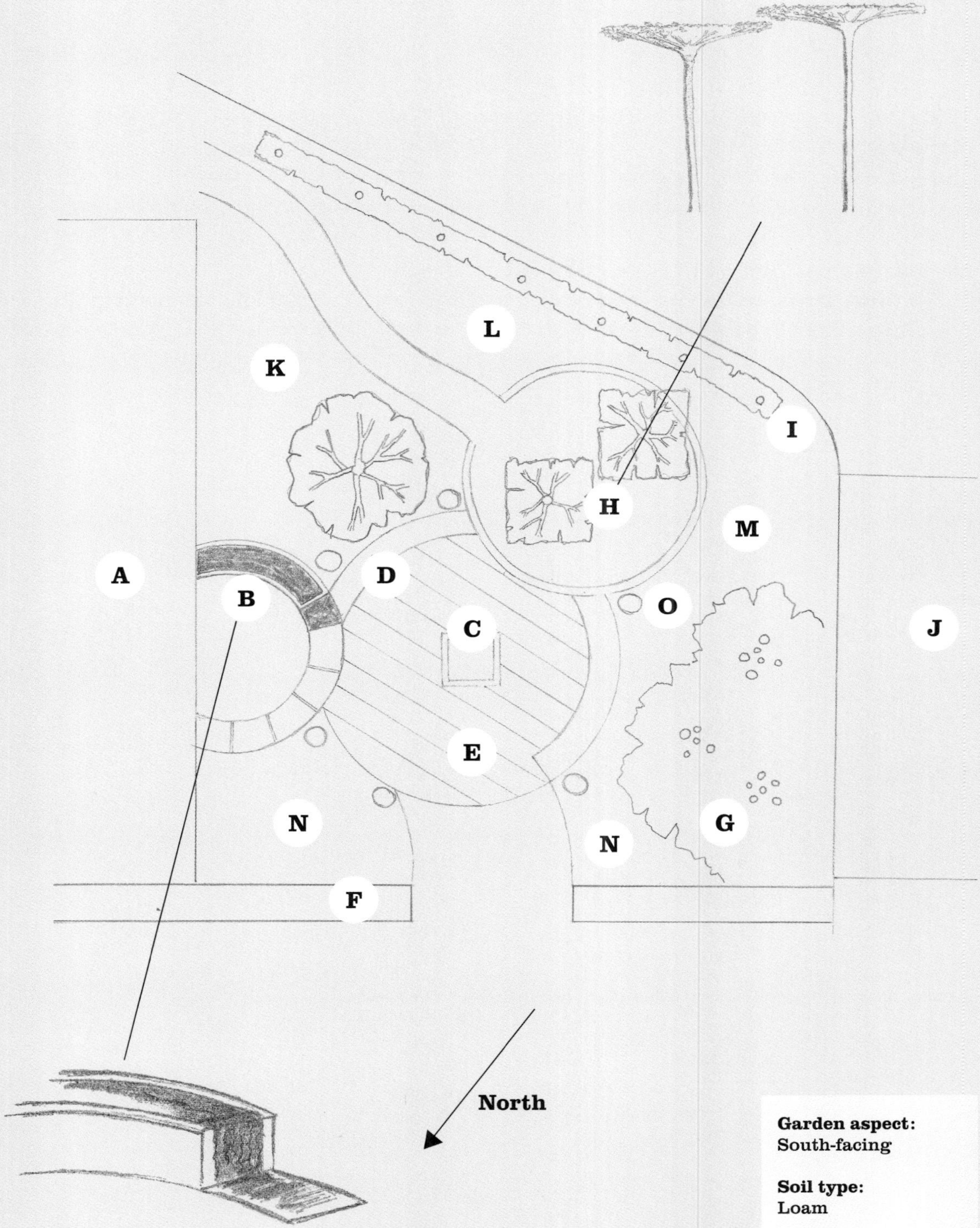

Garden aspect:
South-facing

Soil type:
Loam

Plot size:
18m x 24m

The hybrid garden (p180)

The rectangular planting bays in this garden are hemmed in by hedges and so therefore only readily viewed 'head on'. Consequently the hierarchy of plants, organised by height from front to back as well as considerations of colour, texture and form, is even more important to get right than in a planting where plants can be viewed from several sides. However, structuring everything evenly by height from front to back can appear a little too uniform, and it's good to throw in one or two taller plants nearer the front, providing they aren't too solid-looking but instead have a wafting, translucent feel.

Given that the house is such a strong architectural presence, and the ethos of the garden is one of low-environmental impact, the choice of plants needed to be naturalistic in appearance and suitable for the sandy soil too. It was also important to recognise the aesthetic established by the meadow and orchard, so that when looking across the borders toward the meadow the view was a gradation of planted to natural, rather than a jolting sudden shift. The colour theme in this bay is dusky pinks and reds, with late summer and autumn the principal flowering time.

The sandy soil was improved with well-rotted, chopped farmyard manure dug into the top 30cm, and then mulched with spent mushroom compost.

Planted area F

1. *Eupatorium purpureum*
A statuesque border perennial with a suitably wild look, being a native to damp meadows and woodland edges in North America. It has large, dusky-pink flowers that are adored by bees, butterflies and moths.
2m x 80cm

2. *Miscanthus sinensis* 'Ferner Osten'
Sits perfectly in front of the *Eupatorium* and references its colour with dark-red inflorescences in autumn that fade to a light silver grey by winter. It also has fabulous autumn foliage colour.
1.8m x 70cm

3. *Echinacea purpurea* 'Magnus' Has especially big flowers, with a dark bristly central boss to each one, the rest of the flower dusky purple-pink.
1.2m x 60cm

4. *Geranium psilostemon* A great all-rounder with vivid magenta flowers in summer, each one with a conspicuous dark central 'eye'. During autumn this has some of the best foliage colour of all perennials, turning red, orange and deep burgundy.
1m x 1m

5. *Molinia caerulea* subsp. *caerulea* 'Edith Dudzsus' Dropping this *molinia* into the composition at this point disrupts the continuous flow of diminishing height, making it a lot more interesting as a result. The see-through nature of this grass means that it doesn't obscure the plants behind it.
1.5m x 40cm

6. *Hemerocallis* 'Little Wine Cup' This lovely, low-growing day lily has dark-burgundy flowers in summer and decent foliage that turns butter-yellow in autumn.
30cm x 40cm

The surprising space (p182)

This part of the garden is baked by the sun and is incredibly hot, the heat amplified by the surrounding house and walls which 'bounce' light and warmth back into the garden. Consequently the only plants that can really thrive in such conditions are those on the more extreme edge of sun and drought adaptation. Fortunately the comparative absence of frost and cold winters on the Channel Islands means that exotic plants thrive here.

Stylistically the planting is best described as a xeroscape – a comparatively sparse planting of strongly architectural drought-proof plants along with flowering annuals and bulbs. The trick to making such a planting work aesthetically is to ensure there is space around the plants to allow the architectural presence of agave and echium to come through. Plant too many other plants around them and they lose their impact, becoming cast members rather than the stars of the show.

The soil in the garden needed no improvement to grow these sun-loving specialists, but a mulch comprised of pebbles of varying sizes from 5mm to 30mm, with a few larger paddle stones added in, helps to seal in moisture, keep down weeds and give a suitable rocky look to the garden, whilst also referencing the nearby seaside.

Planted area K

1. *Agave americana* Eventually becomes a beast of a plant, it is monocarpic – meaning it dies after flowering. A hugely impressive structural plant with blue-green, leathery leaves tipped with sharp spikes.
2m x 3m

2. *Echinops bannaticus* 'Taplow Blue' This pollen- and nectar-rich plant is known as the globe thistle, which describes its steely blue flowers well. It's suitable for any sunny spot but incredibly tolerant of really dry conditions, hence its inclusion here.
1m x 60cm

3. *Echium pininana* Another mighty beast that can reach as high as 4m, but more often is rather shorter. Smothered in blue flowers, loved by insects, a complete show off of a plant that impresses on every level and is completely drought-proof.
4m x 90cm

4. *Erodium* x *variabile* 'Ken Aslet' This low-growing sub-alpine plant thrives in hot conditions and well-drained soil. Here it forms a carpet of dusky pink between the major structural plants, reminiscent, if only on a tiny scale, of the incredible flush of wildflowers experienced in areas such as the Veldt in South Africa after the spring rains have fallen.
8cm x 30cm

5. *Leptospermum scoparium* 'Red Damask' The stems and foliage on this shrub are so fine and wiry that when it is in flower it seems to be a mass of dark-red blooms and nothing else.
2m x 2m

6. *Aloe striatula* Has thick, fleshy leaves rather like agave, and although on a smaller scale it is just as architecturally powerful, especially when producing its orange-yellow, waxy, nectar-rich flowers that resemble those of a red hot poker plant.
1m x 1.4m

'The first parts of the garden in Devon developed by Keith and Ros Wiley are now four years old and maturing into their full potential. A garden rarely looks beautiful by accident; behind the beauty is the planning, dreaming and scheming before the eventual act of realisation begins – an act that never stops as long as the garden lives. For the owners of the six gardens featured in *Landscape Man* and for all those who embark on the wonderful, life affirming process that is making a garden, aspiring to beauty will always be the aim, but the enjoyment of the journey is the real pleasure'.

Index

Page numbers in *italic* refer to the illustrations

Index

Acknowledgements

Putting together a book of this nature is very much a team effort, and although my name appears on the front it is the work of many hands. Jane O'Shea, Simon Davis, David Rowley, Samantha Rolfe and Helen Lewis have contributed some of the most important of the many hands at Quadrille Publishing, and those I haven't mentioned I thank too; it has been a joy to work with such a creative team. My book agent Julian Alexander at LAW helped to craft the initial thoughts for the book, held my hand when needed, and introduced me to Quadrille and vice versa. *Landscape Man* is the brainchild of John Silver and Simon Bisset at Red House TV, with whom it has been my great privilege and honour to work for the last 10 months. Simon also took many of the beautiful photographs that grace these pages and has been a travelling companion *par excellence*. The Red House team, who have been like a second family throughout our time together, have made filming the series such a great experience; Nicks Bullard, Matt Cox, Ed St Giles, Lotta Hellzen, Matt Lees, Caroline McCool, Louisa MacInnes and Geri Sweeney. Thanks too to our wonderful contributors for allowing access to their gardens, and their dreams, and the superb camera and sound crews who captured the action, in particular Tony Etwell, Cliff Evans, Daniel Spencer and Mike Stephenson. Finally my warm thanks to Walter Iuzzolino at Channel 4 and my TV agent, Laura Hill at Independent, and to J Barbour and Sons Ltd for being so generous. This book is dedicated to my wonderfully supportive and patient wife, Jane, and our twin babies Amelie and Dylan, who came along just about halfway through the writing of the book and have put everything in perspective.

Picture credits

10-11© Simon Bisset; 15 Liz Eddison/The Garden Collection, design: Chris Beardshaw, RHS Chelsea 2007; 16 Elizabeth Balmforth; 17 Clive Nichols/Clare Matthews; 18 Steven Wooster/The Garden Collection, design: Laara Copley-Smith; 19 above Marcus Harpur, design: Justin Greer, London; 19 below www.modulargarden.com, tel: 020 76190100; 21 www.modulargarden.com, tel: 020 76190100; 22 above Sarah Cuttle; 22 below © Simon Bisset; 23 MMGI/Marianne Majerus, designer: Thomasina Tarling; 24 Andrew Lawson; 25 © MMGI Marianne Majerus, designers: Lynne Marcus and John Hall; 27 © Simon Bisset; 29 above © Simon Bisset; 29 below Thomas Hoblyn Garden Design; 31-35 above © Simon Bisset; 35 below Red House Television Limited; 36-39 © Simon Bisset; 41 John Carey/ Mainstreamimages; 42 © Simon Bisset; 43 above Red House Television Limited; 43 below Ian Kitson Landscape Architecture and Garden Design; 45 © Simon Bisset; 46 Matthew Wilson; 47 Jane Marika Smith; 48 © Simon Bisset; 49 above © Simon Bisset; 49 below Matthew Wilson; 51-53 © Simon Bisset; 54 above ©MMGI/Marianne Majerus, designer: Adam Caplin; 54 below Neil Sutherland/The Garden Collection; 55 Sarah Cuttle; 56 Nicola Stocken Tomkins/The Garden Collection; 57 Sarah Cuttle; 58 © Simon Bisset; 59 above Red House Television Limited; 59 below Jill FenwickInspired Garden Design; 61- 65 above © Simon Bisset; 65 below Matthew Wilson; 66-68 © Simon Bisset; 69 all © Simon Bisset, apart from 69 below right Jane Marika Smith; 70-71© Simon Bisset; 74 © Andrea Jones/private garden, London; 75 Nicola Stocken Tomkins/The Garden Collection; 76 GAP Photos/Suzie Gibbons; 77Nicola Stocken Tomkins/The Garden Collection; 78 above Ray Main/Mainstreamimages/domus furniture.co.uk; 78 below Steven Wooster/The Garden Collection, design: Laara Copley-Smith; 78-79 Clive Nichols/Charlotte Rowe; 81 Marcus Harpur/Saling Hall, Essex; 82 © Simon Bisset; 83 above Photolibrary.com; 83 below Marcus Harpur/Saling Hall, Essex; 84 Marcus Harpur, design: Justin Greer, London; 85 MMGI/Claire Takacs/King Johns Lodge, East Sussex, designers: Jill and Richard Cunningham; 87 above Steven Wooster/The Garden Collection, design: Louise del Balzo; 87 below left A. Green-Armytage/Garden World Images, design: Nava Habet; 87 below right Photolibrary.com; 88-89 Marcus Harpur, design: Judith Sharpe, London; 90 Jerry Harpur/design: Judith Sharpe, London; 91Marie O'Hara/The Garden Collection, The Little Cottage, Lymington; 92 Matthew Wilson; 93 Jerry Harpur/design: Simon Fraser, London; 95 Jane Marika Smith; 97 Andre Wood/Red Cover/Getty Images; 98 above Nicola Stocken Tomkins/The Garden Collection; 98 below Matthew Wilson; 98-99 © MMGI /Marianne Majerus, designer: Declan Buckley; 100 © Andrea Jones/Hermannshof Garden (Schau-und Sichtungsgarten Hermannshof), Germany; 101 Jerry Harpur, design: Lisette Pleasance, London; 103 Jane Marika Smith; 104 Steven Wooster/The Garden Collection, design: Louise del Balzo; 105 above © Andrea Jones/Evening Island. Chicago Botanic Garden.USA/Landscape Architecture and planting by Lisa E. Delplace OVSLA; 105 below left Nicola Stocken Tomkins/The Garden Collection, St Mary's Grove; 105 below right ©Living Etc/Paul Raeside/IPC+ Syndication; 106-107 Jane Marika Smith; 109 above left Jerry Harpur, design: Helen Dillon, Dublin; 109 above right Jane Marika Smith; 109 below left Andrew Lawson/The Dillon Garden, Dublin; 109 below right Liz Eddison/The Garden Collection, design: Cleve West, RHS Chelsea 2008; 110 Andrew Lawson; 111 © MMGI/Marianne Majerus/Hermannshof, Weinheim, Germany, designer: Cassian Schmidt; 112-113 background Jane Marika Smith;112 inset above & below Jane Marika Smith;113 inset Jane Marika Smith; 114 Samantha Rolfe; 115 Greg Ryan/Sally Beyer/Red Cover/Getty Images; 117 Matthew Wilson; 119 above left © Simon Bisset; 119 above right Jane Marika Smith; 119 centre left © Andrea Jones/ The Westland Garden RHS Chelsea Flower Show 2007, design: Diarmuid Gavin & Stephen Reilly; 119 below left Lee Beel Photography; 119 below right Jane Marika Smith; 120-121 background Jane Marika Smith; 120 above inset Jane Marika Smith; 120 centre inset GAP Photos/Marg Cousens; 120 below inset Lee Beel Photography; 121 above & below inset Jane Marika Smith; 122-123 Jane Marika Smith; 124 Jerry Harpur/design: Ursel Gut, Bremen, Germany; 125 above © MMGI/Marianne Majerus, designer: Joe Swift; 125 below Wildscape/Alamy; 126 left © Andrea Jones/ Private coastal garden, Santa Barbara, California, design: Isabelle Greene Associates; 126 right Liz Eddison/The Garden Collection, design: Kate Gould, RHS Chelsea 07; 127 above © Simon Bisset/Thomas Hoblyn Garden;127 below John Carey/Mainstreamimages; 129 Jane Marika Smith; 130 Sarah Cuttle; 131 GAP Photos/Andrea Jones; 132 Liz Eddison/The Garden Collection, design: Tom Stuart-Smith, RHS Chelsea 08; 133 T. sims/Garden World Images; 134 Nicola Stocken Tomkins/The Garden Collection, East Ruston Old Vicarage, Norfolk; 135 above © Simon Bisset/Great Dixter Gardens; 135 below Marcus Harpur/Saling Hall, Essex; 137 Andy Hay (rspb-images.com); 139 Jerry Harpur/design: David Stevens; 140 above Botanica/Photolibrary.com; 140 below © MMGI/Marianne Majerus; 141 Sarah Cuttle; 142-143 Photolibrary.com; 149 Matthew Wilson; 151 Thomas Hoblyn Garden Design; 152 Helen Fickling (Design: Thomas Hoblyn)/RHS Chelsea Flower Show 2009; 153 © Simon Bisset/Thomas Hoblyn Garden; 154-155 Clive Nichols/Charlotte Rowe; 156-157 © MMGI/Marianne Majerus; 158-159 Clive Nichols/Lisette Pleasance; 160-161 Clive Nichols; 162-185 Matthew Wilson; 186-187 © Simon Bisset.

For Jane, Dylan and Amelie, with love.

First published in 2009 by
Quadrille Publishing Limited
Alhambra House
27–31 Charing Cross Road
London WC2H 0LS
www.quadrille.co.uk

This edition produced for The Book People Ltd, Hall Wood Avenue, Haydock, St Helens WA11 9UL

For the book
Editorial Director Jane O'Shea
Art Director Helen Lewis
Project Editor Simon Davis
Designer David Rowley
Picture Research Samantha Rolfe
Special Photography Simon Bisset
Production Director Vincent Smith
Production Controller Ruth Deary

For the television series
Creative Director John Silver
Series Editor Simon Bisset
Directors Matthew Cox and Ed St Giles
Assistant Producers Nicks Bullard and Louisa Macinnes
Production Manager Lotta Hellzen

Cataloguing-in-Publication Data: a catalogue record for this book is available from the British Library.

redhouse

ISBN 978 184400 800 1

Printed and bound in Germany